Nonlinear Deformation Waves

International Union of Theoretical and Applied Mechanics

Nonlinear Deformation Waves

Symposium, Tallinn, Estonian SSR, USSR
August 22–28, 1982

Editors
Uno Nigul and **Jüri Engelbrecht**

With 145 Figures

Springer-Verlag
Berlin Heidelberg New York 1983

Prof. U. Nigul

Dr. J. Engelbrecht

Institute of Cybernetics
Academy of Sciences of the Estonian SSR
200104 Tallinn
Akadeemia tee 21
Estonian SSR, USSR

ISBN-13: 978-3-642-82008-3 e-ISBN-13: 978-3-642-82006-9
DOI: 10.1007/ 978-3-642-82006-9

Library of Congress Cataloging in Publication Data
Main entry under title:
Nonlinear deformation waves.
(IUTAM symposium)
At head of title: International Union of Theoretical and Applied Mechanics.
Proceedings of the IUTAM Symposium on Nonlinear Deformation Waves.
1. Wave-motion, Theory of – Congresses.
2. Nonlinear theories – Congresses.
I. Nigul, U. K. (Uno Karlovich).
II. Engel'brekht, ÎUriĭ K.
III. International Union of Theoretical and Applied Mechanics.
IV. IUTAM Symposium on Nonlinear Deformation Waves (1982 : Tallinn, Estonia)
V. Series: IUTAM-Symposien.
QA927.N659 1983 531.1133 83-523

Softcover reprint of the hardcover 1st edition 1983

2061/3020 5 4 3 2 1 0

Symposium Organizations

Scientific Committee

U.Nigul, USSR, Chairman
J.Achenbach, USA
K.Kawata, Japan
B.Lundberg, Sweden
L.Morland, UK
M.Roseau, France
Z.Wesolowski, Poland

Symposium Sponsors

International Union of Theoretical and Applied Mechanics
Academy of Sciences of the USSR
Academy of Sciences of the Estonian SSR
Institute of Cybernetics of the Acad. Sci. Estonian SSR

Chairmen of the sessions

J.Achenbach, USA
D.Crighton, UK
D.Blackstock, USA
U.Nigul, USSR
J.Engelbrecht, USSR
V.V.Bolotin, USSR
B.Lundberg, Sweden
Z.Wesolowski, Poland
N.Alumäe, USSR

Soviet Organizing Committee

O.M.Belotserkovskii	Member of the Acad.Sci. USSR, Moscow, Co-Chairman
Yu.N.Rabotnov	Member of the Acad.Sci. USSR,Moscow, Co-Chairman
U.K.Nigul	Prof., Tallinn, Vice-Chairman
J.K.Engelbrecht	Prof., Tallinn, Scientific Secretary
H.K.Aben	Corresponding Member of the Acad.Sci. ESSR, Tallinn
N.A.Alumäe	Member of the Acad. Sci. ESSR, Tallinn
V.V.Bolotin	Corresponding Member of the Acad. Sci. USSR, Moscow
S.K.Godunov	Corresponding Member of the Acad. Sci. USSR, Novosibirsk
A.N.Guz	Member of the Acad. Sci. Ukrainian SSR, Kiev
V.K.Dobroselskii	Moscow
A.Yu.Ishlinskii	Member of the Acad. Sci. USSR, Moscow
E.T.Lippmaa	Member of the Acad. Sci. ESSR, Tallinn
L.M.Lyamshev	Prof., Moscow
G.K.Mikhailov	Prof., Moscow
I.F.Obraztsov	Member of the Acad. Sci. USSR, Moscow
R.S.Oraevskii	Moscow
K.K.Rebane	Corresponding Member of the Acad. Sci. USSR, President of the Acad. Sci. ESSR, Tallinn
L.I.Sedov	Member of the Acad. Sci. USSR, Moscow
B.G.Tamm	Member of the Acad. Sci.ESSR, Tallinn
G.K.Tikhonov	Tallinn

Preface

Recent progress in the study of nonlinear wave propagation has been influenced by developments in mechanics, acoustics, hydrodynamics, plasma physics and many other fields of physics.This vast field of research has also given rise to fascinating mathematical ideas: the inverse scattering method and the technique of exterior differential forms being just some to be mentioned. Obviously the theory of nonlinear waves may be interpreted as an interdisciplinary study with the mechanics of continuous media as a theoretical basis.

This was the starting point of the proposal to the General Assembly of the IUTAM to hold an IUTAM Symposium on this topic, made by the USSR National Committee of Theoretical and Applied Mechanics and the Academy of Sciences of the Estonian SSR. Actually the IUTAM Symposium on Nonlinear Deformation Waves was the third meeting of such kind to be held in Tallinn. In 1973, the Academy of Sciences of the Estonian SSR and Gorky State University organized a national Symposium on Nonlinear and Thermal Effects in Transient Wave Propagation. In 1978, the Academy of Sciences of the Estonian SSR organized another national Symposium on Nonlinear Deformation Waves with participants from several other countries. The participants of this Symposium in their final resolution expressed a wish that a similar meeting of definitely international type should take place again in Tallinn in 1982. The IUTAM General Assembly accepted this proposal on its meeting in August, 1980,Toronto and the IUTAM Bureau appointed the Scientific Committee from the representatives of 7 countries (see the Symposium

Organizations). The Scientific Committee suggested preliminarily that the following topics should form the basis of the Scientific Programme:

- nonlinear mathematical models of solids and liquids;
- analytical (asymptotic) and numerical methods in nonlinear wave propagation;
- inverse problems;
- experimental investigations and applications (nondestructive testing, biomechanics, geophysics etc.).

On the Symposium 15 general lectures and 33 short communications were presented. Since the number of applications for short communications was much larger than proposed, the Scientific Committee organized a workshop were 16 communications were discussed in the form of poster reports. In this volume the texts of 14 general lectures and 31 short communications are published.

The Symposium was arranged by the Institute of Cybernetics of the Academy of Sciences of the Estonian SSR. The financial support from the IUTAM, the Academy of Sciences of the USSR, the Academy of Sciences of the Estonian SSR, and the Institute of Cybernetics is gratefully acknowledged. Thanks are also extended to the speakers, to the members of the Scientific Committee, to the Chairmen of the sessions, to the USSR National Committee on Theoretical and Applied Mechanics, and to the members of the Local Organizing Committee.

Our special thanks belong to the Springer-Verlag for the contribution in preparing these Proceedings for the publication.

U.Nigul
J.Engelbrecht

Participants

ABEN,H., Institute of Cybernetics of the Estonian Acad.Sci., Akadeemia St.21, 200104 Tallinn, USSR

ACHENBACH,J.D., Department of Civil Engineering, The Technological Institute, Northwestern University, Evanston, Illinois 60201, USA

AINOLA,L., Department of Mathematics, Tallinn Technical University, Ehitajate St.5, 200026 Tallinn, USSR

AINSAAR,A., Institute of Physics of the Estonian Acad. Sci., Riia St. 142, 202400 Tartu, USSR

ALUMÄE,N., Institute of Cybernetics of the Estonian Acad.Sci., Akadeemia St. 21, 200104 Tallinn, USSR

ANNIN,B.D., Institute of Hydrodynamics, 630090 Novosibirsk, USSR

BAGDOYEV,A.G., Institute of Mechanics of the Armenian Acad. Sci., Barekamutyan St. 24b, 375019 USSR

BAKHVALOV, N.S., Department of Computational Mathematics and Cybernetics, Moscow State University, Lenin Hills, 117234 Moscow, USSR

BALODIS,A., Institute of Polymer Mechanics, Aizkraukles St. 23, 226006 Riga, USSR

BALTOV,A.,Institute of Mechanics and Biomechanics, Bulgarian Academy of Sciences, Sofia 1000, P.O.373, Bulgaria

BARENBLATT,G.I., Institute of Oceanology of the USSR Acad. Sci., Krashkov St.23, 117218 Moscow, USSR

BAZHENOV,V.G., Institute of Mechanics, Gorki State University, Gagarin Ave. 23, Building 6, 603600 Gorki, USSR

BERDICHEVSKII,V.L.,Department of Hydromechanics, Moscow State University, Lenin Hills, 117234 Moscow, USSR

BEREZIN,Yu.A., Institute of Theoretical and Applied Mathematics of the Siberian Branch of the USSR Acad. Sci., 630090 Novosibirsk, USSR

BITTNAR,Z., Technical University in Prague, Building Research Institute,Šolinova 7, 16608 Prague, Czechoslovakia

BJØRNØ,L., Department of Fluid Mechanics, Technical University of Denmark, Building 352, DK-2800, Lyngby, Denmark

BLACK,K., Department of Fluid Mechanics,Technical University of Denmark, Building 352, DK-2800 Lyngby,Denmark

BLACKSTOCK,D.T.,Applied Research Laboratories, The University of Texas at Austin, P.O.Box 8029 Austin,Texas 78712,USA

BLOKHIN,A.M., Institute of Mathematics of the Siberian Branch of the USSR Acad. Sci., 630090 Novosibirsk,USSR

BOLOTIN,V.V., Mechanical Engineering Research Institute, Griboyedov St.4, 101820 Moscow, USSR

BRAGOV,A.M., Institute of Mechanics, Gorki State University, Gagarin Ave. 23, Building 6, 603600 Gorki, USSR

BRAUN, M., University of Duisburg, FB 7, Lotharplatz, 4100 Duisburg, Federal Republic of Germany

BREAZEALE,M.A., Department of Physics & Astronomy, The University of Tennessee, Knoxville, Temnessee 37916,USA

BUCHAR,J., Institute of Metallurgy C.A.S.,Žižkova 22, 616 62 Brno, Czechoslovakia

BRUSHLINSKIĬ,K.V.,Institute of Applied Mechanics, Miusskaya Square 4, 125047 Moscow, USSR

CHESKIDOV,P.A., Institute of Theoretical and Applied Mechanics of the Siberian Branch of the USSR Acad. Sci., 630090 Novosibirsk,USSR

COSTIN,A.,Institutul de Cercetari Stiintifice pentru Protectia Muncii, strada Gen.Budisteanu nr. 15, or.Bucuresti, Romania

CRANDALL,S.H., Department of Mechanical Engineering, Massachusetts Institute of Technology, Cambridge, Massachusetts 02139,USA

CRIGHTON,D.G., The University of Leeds, School of Mathematics, Department of Applied Math.Studies, Leeds LS2 9JT, United Kingdom

DONATO,A,, Istituto di Matematica, Universita di Messina, Via C.Battisti 90, Messina 98100, Italia

ENGELBRECHT,J., Institute of Cybernetics of the Estonian Acad. Sci., Akadeemia St.21, 200104 Tallinn, USSR

EPSTEIN,M.,Department of Mechanical Engineering, University of Calgary, 2500 University Drive, Calgary, Alberta, Canada T2N 1N4

FEDORCHENKO,A.T., Moscow Institute of Technical Physics, Dolgoprudnyi, 141700 Moscow Region, USSR

FELDMANE,E., Institute of Polymer Mechanics, Aizkraukles St. 23, 226006 Riga, USSR

FOMIN,V.M., Institute of Theoretical and Applied Mechanics of the Siberian Branch of the USSR Acad. Sci., 630090 Novosibirsk,USSR

FRIDMAN,V.Ye., Radiophysical Research Institute, Lyadov St.25, 603600 Gorki, USSR

FUSCO,D., Istituto di Matematica, Universita di Messina, Via C.Battisti, 90, Messina 98100, Italia

FRYDRYCHOWICZ,W., Faculty of Mathematics, Warsaw University, Warsaw, Poland

GALIN,G.Ya., Institute of Mechanics of Moscow State University, Michurin Ave. 1, 117192 Moscow, USSR

GODUNOV,S.K., Institute of Mechanics of the Siberian Branch of the USSR Acad.Sci., 630090 Novosibirsk, USSR

GOLSHTEIN,R.V., Institute of Problems of Mechanics, Vernadskii Ave. 101, 117526 Moscow, USSR

GRIMSHAW,R., Department of Mathematics, University of Melbourne, Parkville, Victoria 3052, Australia

GRINFELD,M.A., Institute of Physics of the Earth,B.Gruzinskaya 10, 123810 Moscow, USSR

GURBATOV,S.N., Gorki State University, Gagarin Ave. 23,603600 Gorki, USSR

GUSHCHIN,V.V., Radiophysical Research Institute,Lyadov St.25, 603600 Gorki, USSR

GUZ,A.N., Institute of Mechanics of the Ukrainian Acad. Sci., Nesterov St.3, 252057 Kiev, USSR

GVOZDEV,A.A., Institute of Physics of the Earth,B.Gruzinskaya 10, 123810 Moscow, USSR

GVOZDEVA,L.G., Institute of High Temperatures of the USSR Acad. Sci., Korovinskoye Road,127412 Moscow ,USSR

HARUMI,K., Department of Mathematics and Physics, Faculty of Technology, Gumna University, Kiryu, Gumna, Japan

IORGA,I., Institutul de Cercetari Stiintifice pentru Protectia Muncii, strada Gen.Budisteanu nr.15, or.Bucuresti, Romania

ISHLINSKII,A.Yu.,Institute of Problems of Mechanics, Vernadskii Ave. 101, 117526 Moscow, USSR

IVANOV,Ts., Institute of Mechanics and Biomechanics,Bulgarian Academy of Sciences, Sofia 1000, P.O.B.373, Bulgaria

KAPP,I., Zentralinstitut für Physik der Erde, 15 Potsdam,DDR

KAWATA,K., Institute of Interdisciplinary Research, Faculty of Engineering, University of Tokyo, 4-6-1 Komaba, Meguro-ku, Tokyo, 153, Japan

KAŽIS,R., Laboratory of Ultrasound, Kaunas Technical University, Juro St. 50, 233028 Kaunas, USSR

KEDRINSKII,V.K.,Institute of Hydrodynamics, Lavrentyev Ave.15, 630090 Novosibirsk, USSR

KENK,K., Tallinn Technical University, Ehitajate St.5, 200026 Tallinn, USSR

KONDAUROV,V.I., Institute of Physics of the Earth,B.Gruzinskaya 10, 123810 Moscow, USSR

KORNEV,V.M., Institute of Hydrodynamics of the Siberian Branch of the USSR Acad. Sci., 630090 Novosibirsk, USSR

KOROBEINIKOV,V.P., Institute of Mathematics, Vavilov St. 42, 117966 Moscow, USSR

KOSMODAMIANSKII,A.S.,Donetsk State University, Universitetskaya st. 24, 340055 Donetsk, USSR

KOZHANOV A.I., Institute of Mathematics of the Siberian Branch of the USSR Acad. Sci., 630090 Novosibirsk, USSR

KRASILNIKOV,V.A., Moscow State University, Lenin Hills,117234 Moscow, USSR

KRASILSHCHIKOVA,Ye.A., Institute of Problems of Mechanics of the USSR Acad. Sci., Vernadski Ave. 101, 117526 Moscow, USSR

KUBENKO,V.D., Institute of Mechanics of the Ukrainian Acad. Sci., Nesterov St. 3, 252057 Kiev, USSR

KUKUDZHANOV,V.N., Computing Centre of the USSR Acad. Sci., Vavilov St. 40, 117333 Moscow, USSR

KULIKOVSKII,A.G.,Institute of Mathematics, Vavilov St. 42, 117966 Moscow, USSR

KULBACH,V., Tallinn Technical University, Ehitajate St. 5, 200026 Tallinn, USSR

KUTSENKO,A.N., Department of Electroacoustics and Ultrasound, Odessa Technical University, Shevchenko Ave.1, 270044 Odessa, USSR

KUTSER,M., Institute of Cybernetics of the Estonian Acad.Sci., Akadeemia St. 21, 200104 Tallinn, USSR

KÕIV,M., Institute of Physics of the Estonian Acad. Sci.,Riia St. 142, 202400 Tartu, USSR

KÄERDI,H., Tallinn Technical University, Ehitajate St. 5, 200026 Tallinn, USSR

LEPIK,Ü., Department of Theoretical Mechanics, Tartu State University, Vanemuise St.46, 202400 Tartu, USSR

LOBANOV,Ye.V., Mechanical Engineering Research Institute of the USSR Acad. Sci., Griboyedov St. 4, 101830 Moscow,USSR

LUNDBERG,B., Department of Mechanical Engineering, University of Lulea, 951 87 Lulea, Sweden

MAINARDI,F., Istituto di Fisica "A.Righi", Universita degli Studi di Bologna , 46 Via Irnerio, 40126 Bologna, Italia

MAKOVIČKA,D., Building Research Institute, Šolinova 7. 16608 Prague 6, Czechoslovakia

MALISCHEWSKY,P., Zentralinstitut für Physik der Erde der Akademie der Wissenschaften der DDR, Institutsteil Jena, 69 Jena, Burgweg 11, DDR

MARKOV,V.V., Institute of Mathematics, Vavilov St.42, 117333 Moscow, USSR

MAUGIN,G.A., Laboratoire de Mecanique Theorique, Tour 66, Universite Pierre et Marie Curie, 4 Place Jussieu,75230 Paris Cedex 05, France

METSAVEER,J., Tallinn Technical University, Ehitajate St. 5, 200026 Tallinn, USSR

MIKHAILOV,G.K., The USSR National Committee on Theoretical and Applied Mechanics, Vernadskii Ave. 101, 117526 Moscow, USSR

MOLOTKOV,I.A., Leningrad Branch of the Institute of Mathematics of the USSR Acad. Sci., Fontanka St. 27, 191011 Leningrad, USSR

MOVSISYAN,L.A., Institute of Mechanics of the Armenian Acad. Sci., Barekamutyan St. 24b, 375019 Yerevan, USSR

NASH,W.A., College of Engineering Mechanics, University of Massachusetts, Amherst, Massachusetts 01003, USA

NAUGOLNYKH,K.A., Institute of Acoustics, Shvernik St. 4, 117036 Moscow, USSR

NICOLOVA,I., Institute of Mechanics and Biomechanics, Bulgarian Academy of Sciences, Sofia 1000,P.O.B.373, Bulgaria

NIGUL,U., Institute of Cybernetics of the Estonian Acad. Sci., Akadeemia St. 21, 200104 Tallinn, USSR

NIKITIN,L.V., Institute of Physics of the Earth,B.Gruzinskaya 10, 123810 Moscow, USSR

NILSSON,L., Department of Structural Mechanics, University of Luleå, S-95187 Lulea, Sweden

NOVIKOV,A.A., Institute of Mechanics, Gorki State University, Gagarin Ave. 23, Building 6, 603600 Gorki, USSR

NOWACKI,W., Institute of Fundamental Technological Research, Polish Academy of Sciences, Swietokrzyska 21, 00-049 Warsaw, Poland

OGORODNIKOV,I.A., Institute of Thermal Physics of the Siberian Branch of the USSR Acad. Sci., 630090 Novosibirsk, USSR

OHLSON,N.G., Division of Solid Mechanics, Royal Institute of Technology, S-10044 Stockholm 70, Sweden

OLDENBURG,M., Division of Structural Engineering, University of Lulea,S-95187 Lulea, Sweden

OSTROVSKII,L.A., Institute of Applied Physics of the USSR Acad. Sci., Ulyanov St.46, 603600 Gorki, USSR

OSOKIN,A.A., Moscow Institute of Technical Physics, Dolgoprudnyi, 141700 Moscow Region, USSR

ÕPIK, I., Academy of Sciences of the Estonian SSR, Kohtu St.6, 200103 Tallinn, USSR

PARKER,D.F., Department of Theoretical Mechanics, University of Nottingham, University Park, Nottingham NG7 2RD, United Kingdom

PAZELSKII,V.V., Presidium of the Siberian Branch of the USSR Academy of Sciences, 630090 Novosibirsk, USSR

PELINOVSKII, Ye.N., Institute of Applied Physics of the USSR Acad. Sci., Ulyanov St.46, 603600 Gorki,USSR

PIRNER,M., Techniky a Zkusebni, Ustav Stavebni, Vyšehradska 3, Praha 2, Czechoslovakia

PLATOVA,T.M., Department of Theory of Elasticity, Tomsk University, Lenin Ave. 36, 634010 Tomsk, USSR

PLACHENOV,A.B., Leningrad Branch of the Institute of Mathematics of the USSR Acad. Sci., Fontanka St.27, 191011 Leningrad, USSR

POLYAKOVA,A.L., Institute of Acoustics, Shvernik St.4,117036 Moscow, USSR

POVERUS,L., Tallinn Technical University, Ehitajate St.5, 200026 Tallinn, USSR

PURTOV,Yu.V., Institute of Mechanics, Gorki State University, Gagarin Ave. 23, Building 6, 603600 Gorki, USSR

RAKHMATULIN, Kh.A., Moscow State University, Lenin Hills, 117234 Moscow, USSR

RAVASCO,A., Institute of Cybernetics of the Estonian Acad.Sci., Akadeemia St.21, 200104 Tallinn, USSR

REBANE,K., Presidium of the Estonian Academy of Sciences, Kohtu St.6, 200103 Tallinn, USSR

RIMSKII,V.K., Institute of Mathematics of the Moldavian Acad. Sci., Kishinev, USSR

ROSEAU,M., Universite Pierre et Marie Curie, Mecanique Theorique, Tour 66, 4 Place Jussieu, 75230 Paris Cedex 05, France

RUDENKO,O.V., Faculty of Physics, Moscow State University, Lenin Hills, 117234 Moscow, USSR

RYBAK,S.A., Institute of Acoustics, Shvernik St.4, 117036 Moscow, USSR

RÄÄMET,R., Tallinn Technical University, Ehitajate St.5, 200026 Tallinn, USSR

SABODASH,P.F., Department of Theoretical Mechanics, The Odessa Technological Institute of Food Industry, Odessa, USSR

SAMSONOV,A.M., Department of Mathematical Physics, Institute of Technical Physics, Politekhnicheskaya St.26, 194021 Leningrad, USSR

SEAMAN,L., Stanford Research Institute, Menlo Park,California 94025, USA

SEDOV,L.I., Moscow State University, Lenin Hills, 117234 Moscow, USSR

SELEZOV,I.T., Institute of Hydromechanics of the Ukrainian Acad.Sci., Zhelyabov St.8/4, 252057 Kiev, USSR

SIBGATULLIN,N.R., Moscow State University, Lenin Hills,117234 Moscow, USSR

SINGH,M., Faculty of Engineering, Department of Mechanical Engineering, The University of Calgary, 2500 University Drive N.W., Calgary, Alberta, Canada T2N 1N4

STEPANENKO,M.B., Mining Institute of the Siberian Branch of the USSR Acad. Sci., 630090 Novosibirsk, USSR

STEPANYANTS,Yu.A., Institute of Applied Physics of the USSR Acad. Sci., Ulyanov St.46, 603600 Gorki, USSR

STOROZHEV,V.I., Donetsk State University, Universitetskaya 24, 340055 Donetsk, USSR

STULOV,A.S., Institute of Cybernetics of the Estonian Acad. Sci., Akadeemia St.21, 200104 Tallinn, USSR

SUGIMOTO,N., Department of Mechanical Engineering, Faculty of Engineering Science, Osaka University, Toyonaka, Osaka 560, Japan

SUTYRIN,V.G., Department of Hydromechanics, Moscow State University, Lenin Hills, 117234 Moscow, USSR

SUVOROVA,Yu.V.,Mechanical Engineering Research Institute, Lyapunov St.3, 117312 Moscow, USSR

SVESHNIKOVA,Ye.I., Department of Hydromechanics, Moscow State University, Lenin Hills, 117234 Moscow, USSR

SZEKERES,A., Technical University Budapest, Budapest,Hungary

ŠTARAS,A., Faculty of Mathematics, Vilnius State University, Partizanu St.24, Vilnius, USSR

TANIUTI,T., Department of Physics, Nagoya University,Chikusa-ku, Nagoya, Japan

VAINIKKO,I., Department of Theoretical Mechanics, Tartu State University, Vanemuise St. 46, 202400 Tartu, USSR

VAKULENKO,S.A., Leningrad Branch of the Institute of Mathematics of the USSR Acad. Sci., Fontanka St. 27, 191011 Leningrad, USSR

VATER,B., Institut für Mechanik, Berlin-Adlershof, Rundower Chaussee 5, DDR

VEKSLER,N.D., Institute of Cybernetics of the Estonian Acad. Sci., Akadeemia St.21, 200104 Tallinn, USSR

VIKTOROV,V.V. Institute of Problems of Mechanics of the USSR Acad.Sc., Vernadskii Ave. 101, 117526 Moscow, USSR

VIRKUS,H., Building Research Institute, Estonia Ave.7, Tallinn, USSR

WESOLOWSKI,Z., Institute of Fundamental Technological Research, Polish Academy of Sciences, Swietokrzyska 21, 00-049 Warsaw, Poland

WOLOSZYNSKA,K., Institute of Fundamental Technological Research, Polish Academy of Sciences, Swietokrzyska 21, 00-049 Warsaw, Poland

ZABOLOTSKAYA,Ye.A., Department of Wave Processes, Moscow State University, Lenin Hills, 117234 Moscow, USSR

ZHUK,A.P., Institute of Mechanics of the Ukrainian Acad.Sci., Nesterov St.3, 252057 Kiev, USSR

ZVOLINSKII,N.V., Institute of Problems of Mechanics of the USSR Acad. Sci., Vernadskii Ave. 101, 117526 Moscow, USSR

Final Address of the Chairman of the Scientific Committee, U. Nigul

It is my feeling that I may take the responsibility to declare that the participants of this meeting are thankful to the IUTAM for the decision of carrying out this Symposium in Tallinn, as well as for the financial aid. I am also pleased to say that the activities of the Scientific Committee and of the Local Organizing Committee have been warmly welcomed and supported by the USSR National Committee on Theoretical and Applied Mechanics,by the Academy of Sciences of the USSR,and by the Academy of Sciences of the Estonian SSR. On behalf of the participants of the Symposium I am very pleased, indeed, to thank these institutions. In spite of our best intentions and efforts we should have been quite helpless, if we had not had this hall and several other rooms in this building at our disposal. Let me thank "Eesti Projekt" and, personally Mr.H.Rohtmaa, Director, for providing excellent working conditions for our Symposium.
According to our records during the period of preparing this Symposium I and Jüri Engelbrecht have approached altogether 47 persons at the Institute of Cybernetics of the Academy of Sciences of the Estonian SSR and in some other institutions and asked: "Come along, please, and help!". Everyone of them has answered: "OK, I shall!". They did and they did their best.
Let us applaud for these 47 persons!
Dear Jüri (Engelbrecht), let me ask you to shake hands with me and to state that, independently of it whatever the documents read on our position in this or other Committee, according to my sincere opinion we have shared the duties and responsibility as the absolutely equal acting Co-Chairmen during this meeting.
Good luck to everybody! Thank you!

Epilogue

At the successful completion of the IUTAM Symposium on Nonlinear Deformation Waves, the members of the Scientific Committee would like to pay tribute to their energetic chairman Uno Nigul for his excellent work on putting together the scientific program, as well as for his invaluable and persistent efforts to assist the participants from abroad. They would also like to express their appreciation to the members of the Soviet Organizing Committee, in particular to the Scientific Secretary, Jüri Engelbrecht, and to the Local Organizing Committee at the Institute of Cybernetics of the Academy of Sciences of the Estonian SSR for the excellent local arrangements. Finally, they would like to express gratitude to the Institute of Cybernetics for financial support of the Symposium.

The choice of Tallinn for this Symposium was indeed an auspicious one. The activities concerning nonlinear waves in this beautiful town are well known in other parts of the world.

The Symposium brought together more than a hundred participants from the USSR and some forty participants from other countries. It provided an only too rare opportunity for personal exchange of scientific information and general scientific discussion.

Many papers at this Symposium were concerned with the development of governing equations which provide appropriate descriptions of wave propagation phenomena. Other papers presented advanced mathematical methods to obtain solutions of specific problems. Still other papers presented advanced experimental methods and results of experimental investigations. Waves in solids,liquids and gases were discussed. Altogether the program provided an excellent overview of current research activities in the area of nonlinear deformation waves.

The study of wave propagation phenomena is enjoying an ever increasing attention. A wide spectrum of applications of wave phenomena in the earth, in the oceans, in the atmosphere, in machinery, in biological systems, in acoustical and electrical devices, and in many other systems has stimulated this increasing interest in the study of wave motion.

This Symposium has continued a sequence of IUTAM Symposia specifically devoted to wave propagation phenomena. The most recent of these was in 1977. Let us hope that the frequency of these Symposia will increase nonlinearly so that the next IUTAM Symposium on Mechanical Wave Phenomena will take place sooner than five years from now.

J.Achenbach, K.Kawata, B.Lundberg,
M.Roseau, Z.Wesolowski

Contents

PART 1 ANALYTICAL METHODS AND EVOLUTION EQUATIONS

PART 2 WAVES IN HEREDITARY AND PLASTIC MEDIA

PART 3 INVERSE PROBLEMS

PART 4 NONLINEAR ACOUSTICS

PART 5 COMBINED FIELDS AND APPLICATIONS

PART 1
ANALYTICAL METHODS AND EVOLUTION EQUATIONS

Recent Developments of the Reductive Perturbation Method

Tosiya TANIUTI

Department of Physics
Nagoya University
Nagoya, Japan

Summary

The reductive perturbation method,that is an asymptotic method of reduction of general nonlinear evolution equations to some integrable nonlinear evolution equations,is outlined and its recent developments are summarized briefly with emphasis on applications to the quasi one-dimensional nonlinear wave-propagation.

1. Introduction

Even since Gardner, Greene, Kruskal and Miura [1] discovered the complete integrability of the Korteweg-de Vries (KdV) equation, studies of the completely integrable nonlinear evolution equations have been done extensively. However, the basic physical systems are not completely integrable. Hence we have developed an asymptotic method which enables us to reduce the general systems to some completely integrable nonlinear equations, which we call the reductive perturbation method [2-4]. In the present lecture the recent developments of the reductive perturbation methods are summarized briefly.

2. The Reductive Perturbation Method

2.a. The Weakly Dispersive System

Let U be a column vector with n real elements $u_1, u_2, \cdots, u_n$ and consider the following nonlinear equation for U

$$U_{,t} + A\,U_{,x} + K\,U_{,xxx} = 0 \quad . \tag{1}$$

Here, the subscripts t and x denote the partial differentiation, A and K are n×n matrices, which are real vector valued functions

Nonlinear Deformation Waves
IUTAM Symposium Tallinn 1982
Editors: U. Nigul, J. Engelbrecht

of U and in a certain domain of the U space (Ω) all the eigenvalues of A$\{\lambda_j\}$ are real and distinct, necessary analyticities of A, $\{\lambda_j\}$ and K with respect to U are assumed. When eq.(1) is linearized about a constant vector $U^{(0)}$ in Ω by

$$U = U^{(0)} + \delta U\cdot \exp i(kx-\omega t) \quad ,$$

ω is given by means of the linear dispersion relation as a function of (real) k. If ω can be expanded for small $|k|$ as

$$\omega = \sum_{m=0}^{\infty} \mu^{(m)} k^{2m+1} \tag{2}$$

eq.(1) may be called the weakly dispersive nonlinear equation (in the neighbourhood of $U^{(0)}$); here $\mu^{(0)}$ is one of the eigenvalues of $A_0(\equiv A(U^{(0)}))$, say $\lambda_{j0}(\equiv\lambda_j(U^{(0)})$. If eq.(2) is valid, there exist n real distinct ω corresponding to $\{\lambda_{j0}\}$: ω and δU are obtained by solving the linear equation

$$(-\omega I+kA_0-k^3K_0)\delta U = 0 \quad , \tag{3}$$

where $K_0\equiv K(U^{(0)})$; if δU is assumed as

$$U = \sum_{m=0} \delta U^{(m)} k^{2m} \quad , \tag{4}$$

for O(k) eq.(3) gives

$$(A_0-\mu^{(0)})\delta U^{(0)} = 0$$

$$\text{e.g.}\quad \mu^{(0)} = \lambda_{j0} \quad \text{and} \quad \delta U^{(0)} = R_j \quad ,$$

where R_j is the eigenvector of A_0 for λ_{j0}, being normalized with the adjoint (left) eigenvector L_j as

$$(L_i\cdot R_j) = \delta_{ij} \quad , \tag{5}$$

and for $O(k^3)$ it yields

$$(A_0-\lambda_{j0})\delta U_1 - (\mu^{(1)}I+K_0)R_j = 0 \quad ,$$

which is multiplied by L_j from the left to give

$$\mu^{(1)} = -(L_j K_0 R_j) \tag{6}$$

and can be solved for $\delta U^{(1)}$ as

$$\delta U^{(1)} = R_j + \sum_{\ell=1}^{n}{}' (L_\ell K_0 R_j)(\lambda_{\ell 0}-\lambda_{j0})^{-1} R_\ell \quad , \tag{7}$$

and all the higher-order terms are determined successively. Let the gradient in the U space be denoted by ∇_u, then $\lambda'_{j0} \equiv \nabla_u \lambda_{j0} \cdot R$ becomes the derivative of λ_j at $U^{(0)}$ in the R-direction. If both λ'_{j0} and $\mu^{(1)}$ do not vanish, eq.(1) can be reduced to the KdV equation, in the asymptotic sense: U approaches asymptotically $U^{(0)}$ as $|x| \to \infty$ and a perturbation of small but finite amplitude be considered in correspondence to the linear approximation for small k [2-5]. For the perturbation U is expanded about $U^{(0)}$ as

$$U = U^{(0)} + \varepsilon U^{(1)} + \varepsilon^2 U^{(2)} + \cdots \quad , \tag{8}$$

where ε is a positive small parameter. In comparison with eq.(4), eq.(8) implies that the wavelength $\sim\varepsilon^{-1/2}$, and we introduce the transformation

$$\xi = \varepsilon^{\frac{1}{2}}(x-\lambda_{j0}t) \quad , \tag{9.a}$$

$$\tau = \varepsilon^{\frac{3}{2}}\, t \quad . \tag{9.b}$$

Hereafter the subscript j will be omitted. Introducing (8) and (9) into eq.(1), in a similar way used to solve (3) we have, for $O(\varepsilon^{3/2})$,

$$U^{(1)} = u^{(1)} R \quad , \tag{10}$$

where $u^{(1)}$ is a scalar function of ξ and τ and $u^{(1)} \to 0$ as $|\xi| \to \infty$, and in the next order $O(\varepsilon^{5/2})$ we get

$$u^{(1)}_{,\tau} + \lambda'_0 u^{(1)} u^{(1)}_{,\xi} - \mu^{(1)} u^{(1)}_{,\xi\xi\xi} = 0 \quad , \tag{11}$$

in which we have used the relation

$$\lambda_0' = L(R\cdot\nabla_u A_0)R \quad . \tag{12}$$

It is to be noticed that $\lambda_0' u^{(1)}$ is the deviation of the characteristic speed λ_j (from λ_{j0}) of the hyperbolic system for K=0. The mode with finite λ_{j0}' will be called genuinely nonlinear. In general, solutions of higher order equation to give $U^{(n)}$ $(n\geq 2)$ involve secular terms. It was shown by Kodama and Taniuti [6-8] that the secular terms in the multi-soliton solutions can be removed by modifying the KdV equation (11) to

$$u^{(1)}_{,\tau} + \lambda_0' u^{(1)} u^{(1)}_{,\xi} - \mu^{(1)} u^{(1)}_{,\xi\xi\xi} + \sum_{m=2}^{\infty} a^{(m)} \sigma_{m+1,\xi}$$

$$= 0 \quad . \tag{13}$$

Here σ_m $(m=3,4,\cdots)$ are the higher order conserved densities of the KdV equation (11) and $a^{(m)}$ $(m>2)$ are constants and can be determined by the dispersion relation (2) so that $a^{(m)} \propto \varepsilon^{m-1} \cdot \mu^{(m)}$. Equation (13) is completely integrable and may be written in the canonical form by means of the conserved quantities $I_m = \int \sigma_m d\xi$:

$$u^{(1)}_{,\tau} + \sum_{m=3} b^{(m)} \left(\frac{\delta I_m}{\delta u^{(1)}}\right)_{,\xi} = 0 \tag{13'}$$

where $b^{(m)}$ are constants.

The extension to the case that waves with different modes coexist was done by Oikawa and Yajima [9,10]: let eqs.(9) and (10) be extended to

$$U^{(1)} = \sum_j u_j^{(1)} R_j \quad , \tag{14}$$

$$\xi_j = \varepsilon^{1/2}(x - \lambda_{j0} t - \varepsilon^{1/2}\zeta_j) \quad , \tag{15.a}$$

$$\tau = \varepsilon^{3/2}\, t \quad , \tag{15.b}$$

where ζ_j is a function of x and t to be determined later, then the equation in the lowest order is satisfied by $u_j^{(1)}$'s which are functions of ξ_j and τ, respectively, and in the next order assuming $U^{(2)}=\sum_{j=1}^{n} u_j^{(2)} R_j$ and requirring that $U^{(2)}$ be non-secular, we find that each $u_j^{(1)}$ is governed by the respective KdV equation

$$u_{j,\tau}^{(1)} + \lambda'_{j,j0} u_j^{(1)} u_{j,\xi_j}^{(1)} - \mu_j^{(1)} u_{j,\xi_j\xi_j\xi_j}^{(1)} = 0 \tag{16}$$

provided that ζ_j is given by

$$\sum_{\ell \neq j}^{n} \{\Lambda_{j\ell} \frac{\partial \zeta_j}{\partial \xi_\ell} + \lambda'_{j,\ell 0} u_j^{(1)}\} u_{j,\xi_j}^{(1)} = 0 \quad , \tag{17}$$

in which $\Lambda_{j\ell} \equiv \lambda_{\ell 0} - \lambda_{j0}$ and $\lambda'_{j,\ell 0}$ is the directional derivative of λ_j in the R_ℓ-direction, $(\nabla_u \lambda_j)_0 R_\ell$, consequently ζ_j eliminates the deviation of the j-th characteristic in the direction normal to R_j.

The method applies also to the quasi one-dimensional propagation. This may be shown by means of the following equation for a scalar potential ϕ,

$$\frac{\partial}{\partial t}(\Delta\phi - \phi) + \beta \frac{\partial \phi}{\partial x} - (\frac{\partial \phi}{\partial y} \frac{\partial}{\partial x} - \frac{\partial \phi}{\partial x} \frac{\partial}{\partial y}) \Delta\phi = 0 \quad , \tag{18}$$

where β is a constant responsible for an inhomogeneity in the y-direction and $\Delta \equiv (\partial^2/\partial x^2) + (\partial^2/\partial y^2)$. The equation is used as a model equation for the Rossby wave in geophysics and the drift wave in plasma physics. In what follows, β is equated to zero, instead the unperturbed state is assumed non-uniform. Noticing that functions of x only or y only are trivial solutions of this equation, we consider a disturbance about $\phi^{(0)}(y)$,

$$\phi = \phi^{(0)} + \delta\phi \quad ,$$

and impose, on $\delta\phi$, the periodic boundary condition at y=0 and $y=\bar{\ell}$. If ϕ is a stream function, $\phi^{(0)}(y)$ gives a steady flow in the x-direction and the disturbance may be a wave propagating along the x-axis through a channel of width $\bar{\ell}$. In this case we attempt to reduce eq.(18) to the KdV equation. First, we rewrite eq.(18) as follows,

$$S\phi_{,t} + A\phi_{,x} + K\phi_{,xxx} + I\phi_{,xxt} + J\phi_{,xx} = 0 \quad . \tag{18'}$$

Here S, A, I, J and K are operators defined as

$$S = S(y,y') = \delta''(y-y') - \delta(y-y') \quad , \quad I = \delta(y-y') \quad ,$$

$$A = A(y,y') = \int \delta'(y-y'')\phi(|y'')dy''\delta''(y-y')$$

$$- \int \delta'''(y-y'')\phi(|y'')dy''\delta(y-y') \quad ,$$

$$K = K(y,y') = \int \delta'(y-y'')\phi(|y'')dy''\delta(y-y') \quad , \quad J = -\phi_{,x}\delta(y-y') ,$$

where $\phi(|y) \equiv \phi(x,t,y)$ and

$$S\cdot\phi = \int S(y,y')\phi(|y')dy' \quad \text{etc.}$$

Then the linear dispersion relation for

$$\delta\phi = Y(y)\exp[i(k_x x-\omega t)]$$

follows immediately:

$$(-S\omega+A_0 k_x - K_0 k_x^3 + I k_x^2 \omega)Y = 0 \tag{19}$$

where $A_0 \equiv A(\phi^{(0)})$, $K_0 \equiv K(\phi^{(0)})$.

Solving equation (19) goes in parallel with solving (3), that is, $\mu^{(0)}$ is given by one of the eigenvalues λ_{j0} of the equation

$$(S\lambda_{j0}-A_0)R_j = 0 \quad , \tag{20}$$

and

$$\mu^{(1)} = -L_j(K_0+I\lambda_{j0})R_j = -\int_0^{\ell} R_j^2\,dy \quad , \tag{21}$$

in which $L_j \propto R_j/(\lambda_{j0}-v^{(0)})$ is the adjoint eigenfunction and the orthonormal conditions

$$L_i\cdot S\cdot R_j = \int_0^{\ell} L_i R_j (v^{(0)} - v^{(0)''})/(\lambda_{j0}-v^{(0)})dy = \delta_{ij} \tag{22}$$

have been used. Here $v^{(0)} \equiv -\phi^{(0)'}$ and $v^{(0)}$ is assumed well-behaved so that all the eigenvalues $\{\lambda_{j0}\}j=0, 1, 2, \cdots$ are real and distinct, $\mu^{(m)}$ $(m=0,1,2,\cdots)$ are bounded and R_j's are normalizable by eq.(22) and constitute a complete set of function. Also from eq.(12) it follows

$$\lambda_{j0}' = L_j(R_i \cdot \frac{\delta A}{\delta \phi})_0 \; R_j \quad ,$$

where

$$R_j \cdot (\frac{\delta A}{\delta \phi})_0 \equiv \int \delta'(y-y")R_j(y")dy"\delta"(y-y')$$
$$- \int \delta'''(y-y")R_j(y")dy"\delta(y-y') \quad ,$$

hence

$$L_j R_j \cdot (\frac{\delta A}{\delta \phi})_0 \; R_j$$
$$= \int_0^{\ell} R_j^3[(v^{(0)"}-\lambda_{j0})/(v^{(0)}-\lambda_{j0})]'/(v^{(0)}-\lambda_{j0})dy \; . \quad (23)$$

For the weakly dispersive and genuinely nonlinear mode (with finite value of λ_{j0}') eq.(18) can be reduced to the KdV equation by means of the expansion (8) for ϕ and the transformations (9), that is, assuming $\phi^{(1)}=u^{(1)}(\xi,\tau)R_j$ we obtain the KdV equation (11). An explicit form of $\phi^{(0)}$ which satisfies all the conditions so far assumed may be given by [11]

$$v^{(0)} = c_0 + f\,V(y) \quad , \qquad (24)$$

where c_0 is constant, f is a parameter which is small compared to c_0 but much greater than ε and

$$V(y) = \sum_{n=1} a_n \sin k_n y + \sum_{n=1} b_n \cos k_n y$$

$$k_n = 2n\pi/\bar{\ell} \quad .$$

In this case λ_{j0} and R_j can be obtained in powers of f:

$$\lambda_{j0} = \tilde{\lambda}_{j0} + f\tilde{\lambda}_{j1} + \cdots \quad ,$$

$$R_j = \tilde{R}_{j0} + f\tilde{R}_{j1} + \cdots ,$$

in which $\tilde{\lambda}_{j0}$ and $\tilde{R}_{j0}$ are given in terms of the wavenumber in the y direction $k_j=2j\pi/\bar{\ell}$ $(j=0,1,2,\cdots)$, as follows,

$$\tilde{\lambda}_{j0} = c_0 k_j^2/(1+k_j^2) ,$$

$$\tilde{R}_{j0} \propto \begin{cases} \sin k_j y \\ \\ \cos k_j y \end{cases} ,$$

and corresponding to each of the two independent solutions of $\tilde{R}_{j0}$ we have

$$\tilde{\lambda}_{j1} = (1+k_j^2)^{-1} \langle \tilde{R}_{j0} | V''+k_j^2 V | \tilde{R}_{j0} \rangle ,$$

$$\tilde{R}_{j1} = (c_0-\tilde{\lambda}_{j0})^{-1} \sum_n{}' \langle \tilde{R}_{n0} | V''+k_j^2 V | \tilde{R}_{j0} \rangle \tilde{R}_{n0} ,$$

where the summation is carried out over all the zeroth-order eigenfunctions $\tilde{R}_{n0}$ except the j-th eigenfunction under consideration. Consequently we get

$$\mu_j^{(1)} = c_0/(1+k_j^2)^2 + O(f) ,$$

and corresponding to $\tilde{R}_{j0} \propto \begin{cases} \sin k_j y \\ \cos k_j y \end{cases}$,

$$\lambda'_{j0} \propto -f[1/(\sqrt{8/\ell}\cdot c_0)] k_{3j} (k_j^2-k_{3j}^2) \begin{Bmatrix} b_{3j} \\ a_{3j} \end{Bmatrix} + O(f^2) .$$

In the case that waves with different modes coexist, it is readily shown that the expansion (14) for $\phi^{(1)}$ and the transformation (15) lead to eqs. (16) and (17), in which in the same way as eq. (23) was derived, $\lambda'_{j,\ell 0}$ is obtained by means of the relation

$$\lambda'_{j,\ell 0} = L_j \left(R_\ell \cdot \frac{\delta A}{\delta \phi}\right)_0 R_j .$$

As an example, for $\phi^{(0)}$ given by eq. (24), the persistence of two solitons with different (k_j) modes was demonstrated by Watanabe [12].

2.b. The Weakly Dissipative System

The reduction of the weakly dissipative system proceeds in parallel with that of the weakly dispersive system: the equation corresponding to eq.(1) may be given by

$$U_{,t} + A\,U_{,x} + K\,U_{,xx} = 0 \quad . \tag{25}$$

If the linear dispersion relation gives the expansion of ω,

$$\omega = \sum_{m=0} i^m \mu^{(m)} k^{m+1}$$

with real valued $\mu^{(m)}$, then $\mu^{(0)}$ is equal to one of the eigenvalues λ_{j0} and the dispersion equation

$$(-\omega I + kA_0 + ik^2 K_0)\,\delta U = 0$$

can be solved by assuming $\delta U = \sum_{m=0} \delta U^{(m)} k^m$, so that

$$\mu_j^{(1)} = L_j K_0 R_j \quad .$$

If all the $\mu_j^{(1)}$ are negative, the system may be called weakly dissipative. In this case, by means of the expansion (8) and the transformation

$$\xi = \varepsilon(x - \lambda_{j0} t) \quad , \tag{26.a}$$

$$\tau = \varepsilon^2 t \quad , \tag{26.b}$$

eq.(25) can be reduced to the Burgers equation

$$u^{(1)}_{,\tau} + \lambda_0' u^{(1)} u^{(1)}_{,\xi} + \mu^{(1)} u^{(1)}_{,\xi\xi} = 0 \quad .$$

In the higher order, the secular terms appear, which can be eliminated by modifying the Burgers equation to the following integrable equation [13],

$$u^{(1)}_{,\tau} + \lambda_0' u^{(1)} u^{(1)}_{,\xi} + \mu^{(1)} u^{(1)}_{,\xi\xi} - (2\mu^{(1)}/\lambda_0') \sum_{m>3} a_{m-1} \frac{\partial}{\partial\xi} C_m$$

$$= 0 \quad ,$$

in which

$$a_m = \varepsilon^{m-1}\mu^{(m)} \quad ,$$

and

$$C_m = -\left(\frac{\partial^m\psi}{\partial\xi^m}\right)/\psi \quad ,$$

where ψ is given by the equation

$$\psi,_\tau = -\sum_{m>2} a_{m-1}\frac{\partial^m\psi}{\partial\xi^m}$$

and is linked with $u^{(1)}$ through

$$u^{(1)} = (2\mu^{(1)}/\lambda_0')(\psi,_\xi/\psi) \quad .$$

The reduction of the case that waves with different modes co-exist is also closely similar to that of the dispersive system [9,10].

2.c. The Strongly Dispersive System

For the strongly dispersive system Eq.(1) may be extended to

$$U,_t + A\,U,_x + B + K\,U,_{xxx} = 0 \quad , \tag{27}$$

here B is a column vector with n elements which are function of U. Then the constant state $U^{(0)}$ is given by $B(U_0)=0$, and the dispersion equation becomes

$$W_1\delta U = 0 \tag{28}$$

where

$$W_\ell \equiv -i\ell\omega I + i\ell k A_0 + \nabla_u B_0 - i(\ell k)^3 K_0 \quad ,$$

and ω is determined by det $W_1=0$; it may be assumed that in a certain range of k, there exist n real distinct roots $\omega_j(k)$ and corresponding n linearly independent eigenvectors R_j, which will be normalized by the same equation as (5). It is readily seen that the slow modulation of the plane wave with the wavenumber k, $U=R(k)e^{ikx-\omega(k)t}\ \delta u(x,t)+c.c.$, is governed by the Schrödinger equation,

$$i\delta u,_{\tau} + (\omega''/2)\,\delta u,_{\xi\xi} = 0 \quad ,$$

where the prime denotes the differentiation with respect to k, ξ and τ are introduced by $\xi=\varepsilon(x-\omega' t)$, $\tau=\varepsilon^2 t$ and ε expresses the order of the slow variation [3,4]. If det $W_\ell \neq 0$ for $|\ell|\geqq 2$, the system (27) is called strongly dispersive. In this case, using the same expansion as (8) and assuming

$$U^{(1)} = u_1^{(1)}(\xi,\tau)R\,e^{i\theta} + c.c. \quad ,$$

$$U^{(2)} = \sum_{\ell=-2}^{2} U_\ell^{(2)}(\xi,\tau)e^{i\ell\theta} + c.c. \quad , \quad (\theta\equiv kx-\omega t) \quad ,$$

we obtain [3-5]

$$iu_{1,\tau}^{(1)} + \frac{\omega''}{2}u_{1,\xi\xi}^{(1)} + Q|u_1^{(1)}|^2\,u_1^{(1)} = 0 \quad , \qquad (29)$$

provided that det $W_0 \neq 0$. Here Q is given by

$$Q = iL[ik\{2(\nabla_u\bar{A}_0\cdot R^*)R_2^{(2)} - (\nabla_u\bar{A}_0\cdot R_2^{(2)})R^* + (\nabla_u\bar{A}_0\cdot R_0^{(2)})R$$
$$+ (\nabla_u\nabla_u\bar{A}_0:RR^*)R - \frac{1}{2}(\nabla_u\nabla_u A_0:RR)R^*\}$$
$$+ \nabla_u\nabla_u B_0:(RR_0^{(2)}+R^*R_2^{(2)}) + \frac{1}{2}\nabla_u\nabla_u\nabla_u B_0 \vdots RR^*R] \quad ,$$

in which $R_2^{(2)}$ and $R_0^{(2)}$ are responsible for the contributions from the second and zeroth harmonic, respectively, namely

$$R_2^{(2)} = -W_2^{-1}[ik(\nabla_u\bar{A}_0\cdot R)R + \frac{1}{2}\nabla_u\nabla_u B_0:RR]$$

$$R_0^{(2)} = -W_0^{-1}[ik\{(\nabla_u\bar{A}_0\cdot R^*)R - c.c.\} + \frac{1}{2}(\nabla_u\nabla_u B_0:R^*R+c.c.)]$$

and $\bar{A}_0\equiv A_0-k^2K_0$. If Q is real, eq.(29) becomes the nonlinear Schrödinger equation, which is completely integrable. If det $W_0=0$, the slow mode, $U_0^{(2)}$, is not determined in the second order, but in many cases, it can be determined in the third order.

In this case, however, the group velocity ω' often coincides with the phase velocity of the slow mode, then the resonance-coupling between the fast and slow mode takes place and if $\omega''Q>0$, the slow mode is strongly excited by the enhanced modulational instability, as the result the ordering $U_1^{(2)} \sim U_0^{(2)}$ ceases to be valid [14]. Here, this is shown by means of the quasi one-dimensional propagation governed by eq.(18) [15]. Assuming the fixed boundary condition at $y=0$ and $y=\bar{\ell}$ and $v^{(0)} = c_0$ yields (for $\beta=0$)

$$W_\ell = -i\ell(\omega-k_x c_0)[\delta'' - \{\ell^2 k_x^2 + (\omega/(\omega-c_0 k_x))\}] \quad ,$$

consequently

$$\omega_j = k_x\{c_0(k_j^2+k_x^2)/(1+k_x^2+k_j^2)\} \quad ,$$

where $k_j = j\pi/\bar{\ell}$, and $R_j \propto \sin k_j y$. Hence, if k_x is not small the system is strongly dispersive, and also $W_0 \equiv 0$.
Let $\phi(x,t,y)$ be expanded as

$$\phi = \phi^{(0)}(y) + \sum_{m=1} \varepsilon^m \phi^{(m)} \quad ,$$

$$\phi^{(m)} = \sum_\ell \phi_\ell^{(m)}(\xi,\tau|y)e^{i\ell\theta} \quad ,$$

then $\phi_1^{(1)}$ takes the form, $\phi_1^{(1)} = u_1^{(1)}(\xi,\tau)Y_1^{(1)}(y)$, where for the j-th mode (k_j), $Y_1^{(1)} = R_j$. It is then shown that the second order equations yield $\phi_1^{(2)} \propto R_j$ and $\phi_2^{(2)} = 0$, while $\phi_0^{(2)}$ is determined in the third order by the equation

$$[-(\lambda_g - c_0)\frac{d^2}{dy^2} + \lambda_g]\phi_{0,\xi}^{(2)} + 2k_x^2 |u_1^{(1)}|^2_{,\xi} (R_j^2)' = 0 \quad ,$$

which is integrated to give

$$\phi_0^{(2)} \propto |u_1^{(1)}|^2[-\{4k_x^2 k_j/(1+k_{2j}^2)\}](\lambda_g-\lambda_0)^{-1} \sin k_{2j}y \, , \quad (30)$$

where $\lambda_g = d\omega/dk_x$ and $\lambda_0 = (k_{2j}^2 c_0)/(1+k_{2j}^2)$ that is the phase velocity of the 2j-th slow mode (with $k_x=0$). The equation (30) demonstrates that the amplitude $\phi_0^{(2)}$, consequently Q in eq.(29), diverges for the resonance $\lambda_g = \lambda_0$. The explicit form of Q is quite complicated, but it is noticed that $(d^2\omega/dk_x^2)Q$ is positive for the following two domains of k_x:

$$0 < k_x^2 < k_r^2 \quad , \qquad 3k_j^2 < k_x^2 < 3(k_j^2+1) \quad ,$$

where

$$k_r^2 = \frac{3}{2}(2k_j^2+1)[\{1 + 4k_j^2(k_j^2+1)/3(2k_j^2+1)^2\}^{\frac{1}{2}} - 1] \quad ,$$

and $\lambda_g \to \lambda_0$ as $k_x \to k_r$, hence the modulational instability is enhanced at the resonance. As k_x approaches k_r, the ordering assumed so far ceases to be valid. Nozaki et al. [15] examined the two cases, the weak and strong resonances. In the weak resonance the slow mode ϕ_0 $(\ell=0)$ is larger than $O(\varepsilon^2)$ but less than $O(\varepsilon)$~the fast mode ϕ_1 $(\ell=1)$, for example, $\lambda-\lambda_0 \sim O(\varepsilon^{2/3})$ and $\phi_0 \sim \phi_1^2/(\lambda_g-\lambda_0) \sim \varepsilon^{4/3}$. In this case it is shown that $u_1^{(1)}$ and $u_0^{(4/3)}$ are given by the coupled equations of the form similar to the Zakharov equation [16],

$$iu_{1,\tau} + \frac{1}{2}\left(\frac{d^2\omega}{dk_x^2}\right)_{k_x=k_r} u_{1,\xi\xi} + \tilde{Q}u_0u_1 = 0 \quad , \qquad (31.a)$$

$$u_{0,\tau} + \delta\lambda u_{0,\xi} = R|u_1|^2{}_{,\xi} \quad , \qquad (31.b)$$

where $\tilde{Q}$, R are constants, $\delta\lambda=\varepsilon^{-2/3}(\lambda_0-\lambda_g)$, $\xi=\varepsilon^{2/3}(x-\lambda_g t)$, $\tau=\varepsilon^{4/3}t$ and the superscripts of u_0 and u_1 are omitted. The system of equations (31) is integrable. For the strong resonance $\lambda_g-\lambda_0 \sim O(\varepsilon)$, ϕ_0 and ϕ_1 become the same order, $O(\varepsilon)$, and it is shown [15] that for $\phi^{(0)}(y)$ given by eq.(24) the system is reduced to the coupled equations of the KdV equation and the nonlinear Schrödinger equation which appear also in the interaction of the Langmuir wave and the ion-acoustic wave in plasma [17].

The reductive perturbation method for the quasi one-dimensional propagation of the strongly dispersive wave was first given by Taniuti [18] in a quite general form and it was applied by VanDam [19] to the electrostatic plasma-wave and the sound wave, also the application to the nonlinear Klein-Gordon equation in multi-dimensions was done by Watanabe and Taniuti [20].

A modification of the nonlinear Schrödinger equation to eliminate the secular terms in the higher order approximations was given by Kodama [21] using the sine-Gordon equation as an example. Also in the case that different modes coexist, the reduction to the nonlinear Schrödinger equation of each mode is possible in a similar way to the weakly dispersive case [9,10]. A review of the reductive perturbation method was published as Supplement of the Progress of Theoretical Physics, No.55 (1974).

References

1. Gardner, C.S.; Greene, J.M.; Kruskal, M.D.; Miura, R.M.: Method for solving the KdV equation. Phys. Rev. Lett. 19 (1967) 1095-1097.

2. Taniuti, T.; Wei, C.C.: Reductive perturbation method in nonlinear wave propagation. J. Phys. Soc. Jpn. 24 (1968) 941-946.

3. Taniuti, T.: Reductive perturbation method and far fields of wave equations. Suppl. Prog. Theor. Phys. No.55 (1974) 1-35.

4. Taniuti, T.; Nishihara, K.: Nonlinear Waves. Tokyo, Iwanami (1977)(in Japanese); translated and edited by Taniuti, T.; Jeffrey, A., London, Pitman (to be published).

5. Taniuti, T.; Yajima, N.: Perturbation method for a nonlinear wave modulation. J. Math. Phys. 10 (1969) 1369-1372.

6. Kodama, Y.; Taniuti, T.: Higher order approximation in the reductive perturbation method I. J. Phys. Soc. Jpn. 45 (1978) 298-310.

7. Kodama, Y.; Taniuti, T.: Higher order approximation in the reductive perturbation method I (errata). 45 (1978) 1765 -1766.

8. Kodama, Y.; Taniuti, T.: Higher order corrections to the soliton-velocity and the linear dispersion relation. Physica Scripta 19 (1979) 486-489.

9. Oikawa, M.; Yajima, N.: Interactions of solitary waves — A perturbation approach to nonlinear systems. J. Phys. Soc. Jpn. 34 (1973) 1093-1099.

10. Oikawa, M.; Yajima, N.: Generalization of the reductive perturbation method to multi-wave systems. Suppl. Prog. Theor. Phys. No.55 (1974) 36-51.

11. Nozaki, K.; Taniuti, T.; Watanabe, K.: Solitons in a convective motion of a low-β plasma. J. Phys. Soc. Jpn. 46 (1979) 983-990.

12. Watanabe, K.: Two dimensional behavior of solitons of drift

waves or Rossby waves. J. Phys. Soc. Jpn. 50 (1981) 15-16.

13. Kodama, Y.; Taniuti, T.: Higher order approximation in the reductive perturbation method III. J. Phys. Soc. Jpn. 47 (1979) 1706-1716.

14. Nishihara, K.: Nonlinear interaction between strongly dispersive wave and weakly dispersive one. Technology Report of Osaka Univ. 28 (1978) 25-34.

15. Nozaki, K.; Taniuti, T.; Watanabe, K.: Solitons in a convective motion of a low-β plasma II. J. Phys. Soc. Jpn. 46 (1979) 991-1001.

16. Zakharov, V.E.: Collapse of Langmuir waves. Sov. Phys. - JETP 35 (1972) 908-914.

17. Nishikawa, K.; Hojo, H.; Mima, K.; Ikezi, H.: Coupled nonlinear electron-plasma and ion-acoustic waves. Phys. Rev. Lett. 33 (1974) 148-151.

18. Taniuti, T.: Reductive perturbation method in nonlinear wave propagation IV. Report of Department of Phys., Nagoya Univ., DPNU-29, Nov., 1971 (unpublished).

19. VanDam, J.M.: Nonlinear wave modulation in a plane bounded plasma. J. Phys. Soc. Jpn. 34 (1973) 1633-1640.

20. Watanabe, K.; Taniuti, T.: Reductive perturbation method for wave modulation in multi-dimensional space. J. Phys. Soc. Jpn. 42 (1977) 1396-1403.

21. Kodama, Y.: Higher order approximation in the reductive perturbation method II. J. Phys. Soc. Jpn. 45 (1978) 311 -314.

Energy Integrals in the Theory of Shock Wave Stability

S.K.GODUNOV, A.M.BLOKHIN

Institute of Mathematics,
Siberian Branch of the USSR
Academy of Sciences, Novosibirsk

Summary

The present paper is concerned with problems arising in the process of investigating energy integrals dissipating on shock waves, for linearized equations of gas dynamics. Here, preliminarily studied is the question of the correctness of the mixed problem for the wave equation for which dissipating integrals are effectively constructed.
Considered in the paper is also the problem of how the results of the investigation of the linearized problem can be transferred onto the general quasi-linear problem in a domain by a moving shock wave. Here considerable attention is paid by obtaining a priori estimates of solutions in the quasi-linear case without the assumption of weak distortedness of a shock wave and of the smallness of perturbations.

The problem of shock wave stability has been thouroughly investigated in the 50-60-s by many authors due to the appearance of computers and to the possibility of numerical calculation of complex gas dynamic flows containing shock waves. The physical set-up of the problem is as follows: let a supersonic stationary gas flow be separated from a subsonic stationary gas flow by a strong break surface - a shock wave with the equation x=0. Let, at the initial moment, a perturbation be imposed on the shock wave front. There is a question: how does this perturbation behave in a course of time?
Let us understand the problem of the shock wave stability as a mixed problem for equations of gas dynamics with boundary conditions on a shock wave. We shall investigate the correctness of such a mixed problem by means of methods of dissipative energy integrals. Note that in papers [1-3] this problem in the linear set-up has been investigated with the help of

Nonlinear Deformation Waves
IUTAM Symposium Tallinn 1982
Editors: U Nigul, J. Engelbrecht

Fourier method and Laplace transformation.
In the linear set-up the problem is formulated as follows. Considered is a mixed problem for a system of acoustic equations in a domain $t > 0$, $x > 0$, $|y| < \infty$:

$$A\,U_t + B\,U_x + C\,U_y = 0, \tag{1}$$

where

$$A = \begin{vmatrix} M^2 & 0_2 & 0 & 0 \\ 0 & M^2 & 0 & 0 \\ 0 & 0 & 1 & 0 \\ 0 & 0 & 0 & 1 \end{vmatrix}, \quad B = \begin{vmatrix} M^2 & 0_2 & 1 & 0 \\ 0 & M^2 & 0 & 0 \\ 1 & 0 & 1 & 0 \\ 0 & 0 & 0 & 1 \end{vmatrix}, \quad C = \begin{vmatrix} 0 & 0 & 0 & 0 \\ 0 & 0 & 1 & 0 \\ 0 & 1 & 0 & 0 \\ 0 & 0 & 0 & 0 \end{vmatrix}, \quad U = \begin{vmatrix} u \\ v \\ p \\ s \end{vmatrix},$$

with boundary conditions for $x = 0$

$$u + d\cdot p = 0, \quad s + p = 0, \quad v_t - \lambda\cdot p_y = 0, \tag{2}$$

and with the initial data for $t = 0$

$$U(0,x,y) = U_0(x,y). \tag{3}$$

System (1) was obtained by linearizing gas dynamics equations relative to the constant basic solution after the direct shock wave and by a certain way of dimensionlessness of the linearized system, and boundary conditions (2) were obtained as a result of linearizing Rankine-Hugoniot conditions on a shock wave; u, v, p, s are small perturbations of components of the speed, pressure and enthropy ; $M(0 < M < 1)$ is some constant.
Let d, λ be arbitrary real coefficients. We demand that for mixed problem (1), (2), (3), Lopatinsky condition should be held meaning that this problem does not admit construction of incorrectness examples, such as Hadamard example [4]. Let us reduce the mixed problem (1), (2), (3) to a simpler one.
For mixed problem (1), (2) two facts are valid.

1. Multiply in a scalar way system (1) by the vector

$$\begin{vmatrix} -\xi \\ -\eta \\ M^2(\tau+\xi) \\ 0 \end{vmatrix}, \text{ where } \tau = \frac{\partial}{\partial t}, \quad \xi = \frac{\partial}{\partial x}, \eta = \frac{\partial}{\partial y}.$$

After some simple calculations we obtain the fact that the function p satisfies the wave equation

$$\left\{M^2(\tau+\xi)^2 - \xi^2 - \eta^2\right\} p = 0. \qquad (4)$$

2. Multiply in a scalar way system (1) by the vector

$$\begin{vmatrix} -\tau \\ 0 \\ M^2\tau \\ 0 \end{vmatrix}.$$

Consider the obtained expression at x = 0

$$\left\{M^2(1+d)\,\tau^2 - \beta^2\tau\xi \quad + M^2\lambda\,\eta^2\right\} p=0, \quad \beta^2=1-M^2. \qquad (5)$$

Consider mixed problem (4), (5) with the initial data for t=0

$$p(0,x,y)=p_0(x,y), \quad \frac{\partial p}{\partial t}(0,x,y) = -\frac{\partial p_0}{\partial x}(x,y) - \frac{\partial u_0}{\partial x}(x,y) - \frac{\partial v_0}{\partial y}(x,y). \qquad (6)$$

We find out at what values of the parameters d, λ mixed problem (4), (5), (6) (and the initial mixed problem (1), (2), (3)) is set up incorrectly, i.e. it has solutions proportional to $\exp(\hat{\tau}t + \hat{\xi}x + \hat{\eta}y)$ with $\mathrm{Re}\,\hat{\tau} > 0$, $\mathrm{Re}\,\hat{\xi} \leq 0$, $\mathrm{Re}\,\hat{\eta} = 0$. From [1, 5] it follows that the plane (d, λ) can be divided into the following three domains:

1) Domain I is the incorrectness domain of mixed problem (4), (5), (6) (see Fig.).

2) Domain II is a domain whose exponential solutions fade in a course of time ($\mathrm{Re}\,\hat{\tau} < 0$, $\mathrm{Re}\,\hat{\xi} \leq 0$, $\mathrm{Re}\,\hat{\eta} = 0$) (see Fig.):

$$\lambda < 0, \quad \beta^2 d + \lambda M^2 > 0. \qquad (7)$$

3) Domain III is a domain whose exponential solutions do not decrease in a course of time ($Re\,\hat{\tau} = Re\,\hat{\xi} = Re\,\hat{\eta} = 0$) (see Fig.).

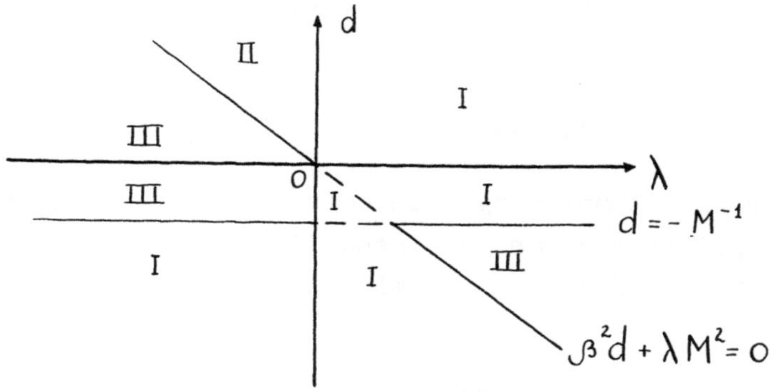

Fig. The plane (d, λ) - plane of coefficients in boundary conditions (2).

Domain III in paper [1] is called the domain of spontaneous radiation of sound by the break. It can be shown that in domain III mixed problem (4), (5), (6) admits propagation of perturbations near the boundary with the rate exceeding the sound speed. Note that for polytropic gas with adiabatic curve index γ inequalities (7) are held [6].

Domain II is the domain of the correct set-up of mixed problem (4), (5), (6) (and with it - of the initial mixed problem (1), (2), (3)). To prove this statement, a dissipative energy integral was constructed in papers [5, 6] from which there follows the belonging for each t of the solution of mixed problem (1), (2), (3) to the space W_2^2, if the initial data belong to the same space.

Consider the way the dissipative energy integral for mixed problem (4), (5), (6) is constructed. To obtain estimates of the solution and derivatives of the solution, constructed is an extended system. To do this, let us introduce into our consideration differential operators $L_1 = \tau/\beta^2$, $L_2 = \xi - M^2\tau/\beta^2$, $L_3 = \eta$, by means of them mixed problem (4), (5) is rewritten as follows:

$$\{\tilde{L}_1^2 - L_2^2 - \tilde{L}_3^2\}\, p = 0,$$

$$\{\tilde{L}_1 - a \pm L_2\}\, L_{\pm} = 0 \text{ or } \{mL_1^2 + nL_2^2 - \frac{\beta}{M^2} L_1 L_2\}\, p = 0 \text{ for } x=0,$$

where

$$\tilde{L}_1=ML_1,\tilde{L}_3=L_3/\beta,\ L_{\pm}=\{a_1L_1+a_2^{\pm}L_2\}p,\quad a_1=\frac{m}{M},\ a_2^{\pm}=-\frac{n}{a_{\pm}},\ n=-\lambda/\beta,$$

$$m=\beta d+\lambda M^2/\beta,\quad a_{\pm}=\frac{M}{2m}(\gamma_1\pm\sqrt{\gamma_1^2-4mn}),\quad \gamma_1=\beta/M^2.$$

Consider, for simplicity, the case $\gamma_1^2-4mn>0$.(Note that for the polytropic gas the inequality holds). Below we use the fact that if a function p satisfies equation (4), then the vector

$$Y=\begin{vmatrix}\tilde{L}_1\\ L_2\\ \tilde{L}_3\end{vmatrix}\ P$$

satisfies the following system [5]:

$$\{D\cdot\tau+Q\cdot\xi+R\cdot\eta\}\ Y=0, \tag{8}$$

where $D=(E-M^2Q)/\beta^2$,

$$E=M\cdot\begin{vmatrix}1 & -m_1 & -n_1\\ -m_1 & 1 & 0\\ -n_1 & 0 & 1\end{vmatrix},\quad Q=\begin{vmatrix}m_1 & -1 & 0\\ -1 & m_1 & n_1\\ 0 & n_1 & -m_1\end{vmatrix},\quad R=\begin{vmatrix}n_1 & 0 & -1\\ 0 & -n_1 & m_1\\ -1 & m_1 & n_1\end{vmatrix};$$

m_1, n_1 are some constants. Matrix $D>0$ if $m_1^2+n_1^2<1$. In this case system (8) is symmetric t-hyperbolic (due to Friedrichs).

Of two systems of (8) kind we make up the following extended system:

$$\left\{\begin{vmatrix}D & 0\\ 0 & D\end{vmatrix}\tau+\begin{vmatrix}Q & 0\\ 0 & Q\end{vmatrix}\xi+\begin{vmatrix}R & 0\\ 0 & R\end{vmatrix}\eta\right\}Y_p=0, \tag{9}$$

where

$$Y_p=\begin{vmatrix}\tilde{L}_1L_+\\ L_2L_+\\ \tilde{L}_3L_+\\ \tilde{L}_1L_-\\ L_2L_-\\ \tilde{L}_3L_-\end{vmatrix},$$

putting $n_1=0$, $m_1=a_+/(1+a_+^2)$ for the first system of (8) kind, and putting $n_1=0$, $m_1=a_-/(1+a_-^2)$ for the second one. By means of system (9) we can estimate all the second derivatives of p. Indeed, system (9) is symmetric t-hyperbolic due to Friedrichs since $D>0$, and the boundary conditions for x=0 are dissipative for this system, because

$$-\left(\begin{vmatrix} Q & 0 \\ 0 & Q \end{vmatrix} Y_p, Y_p\right)\Big|_{x=0} = a_+(a_1L_1L_2p+a_2^+L_2^2p)^2+\frac{a_+}{1+a_+^2}\cdot$$

$$\cdot(a_1L_1\widetilde{L}_3p+a_2^+L_2\widetilde{L}_3p)^2+a_-(a_1L_1L_2p+a_2^-L_2^2p)^2+\frac{a_-}{1+a_-^2}\cdot$$

$$\cdot(a_1L_1\widetilde{L}_3p+a_2^-L_2\widetilde{L}_3p)^2>0.$$

Note that we have constructed the dissipative energy integral for mixed problem (4), (5), (6) not on the whole domain II, but in some part of it: $m,n>0$, $-4mn+\beta^2/M^4>0$. However, it is easy to construct the dissipative energy integral in the rest of domain II [5]. It is possible to construct the dissipative energy integral in domain III. It turns out that if the point d, λ belongs to domain III, then, to construct the dissipative energy integral for mixed problem (4), (5), (6) in the space W_2^2, it is necessary to require greater smoothness of the initial data. A priori estimate for the initial mixed problem (1), (2), (3) is obtained in a similar way. Its deduction is described in detail in papers [5, 6].

The above mixed problem (4), (5), (6) is a special case of the problem about dissipative energy integrals for the vector wave equation

$$u_{tt}-u_{xx}-u_{yy}=0,\quad u=\begin{pmatrix} u_1 \\ \vdots \\ u_n \end{pmatrix},\quad t>0,\ x>0,$$

with the boundary condition $A\cdot u_t+B\cdot u_x+C\cdot u_y=0$ for x=0 (A,B,C are quadratic matrices). This problem was investigated in detail in paper [7]. The scalar case was considered before in [8]. Demand that the matrices A,B,C do not admit the possibility to construct incorrectness examples of the Hadamard type,

and that such incorrectness should not appear for arbitrarily small perturbation of matrix coefficients A,B,C. This condition is used to be called the Lopatinsky uniform condition.
It turns out that for the Lopatinsky uniform condition to be held it is necessary and sufficient that the matrix A+C is non-degenerate and that $2n \times 2n$ matrix Λ

$$\Lambda = \begin{pmatrix} A+C & 0 \\ 0 & A+C \end{pmatrix}^{-1} \begin{pmatrix} 0 & A+C \\ -A+C & 2B \end{pmatrix}$$

has all the characteristic roots lying strictly in the left halfplane.
Construction of integrals is based on the following operator identity. Having assigned arbitrary $n \times n$ hermitian matrices K,L,M,N, we make up the following hermitian matrices:

$$P = \begin{vmatrix} K & L & M \\ L & K & iN \\ M & -iN & K \end{vmatrix}, \qquad Q = \begin{vmatrix} L & K & iN \\ K & L & M \\ -iN & M & -L \end{vmatrix}, \qquad R = \begin{vmatrix} M & -iN & K \\ iN & -M & L \\ K & L & M \end{vmatrix}$$

and see that

$$(\tau P - \xi Q - \eta R) \begin{vmatrix} \tau I \\ \xi I \\ \eta I \end{vmatrix} = (\tau^2 - \xi^2 - \eta^2) \begin{vmatrix} K \\ L \\ M \end{vmatrix}$$

(I is a unit $n \times n$ matrix).
Having substituted τ , ξ , η by differentiation operators $\frac{\partial}{\partial t}$, $\frac{\partial}{\partial x}$, $\frac{\partial}{\partial y}$, and having applied the obtained identity to the vector u which satisfies the wave equation, we see that the derivatives u_t, u_x, u_y of this vector are connected by the symmetric system

$$(P \frac{\partial}{\partial t} - Q \frac{\partial}{\partial x} - R \frac{\partial}{\partial y}) \begin{vmatrix} u_t \\ u_x \\ u_y \end{vmatrix} = 0,$$

and the vector $U = \begin{vmatrix} u_t \\ u_x \\ u_y \end{vmatrix}$, made up of them, enters quadratic forms of the following energy integral written in the differential form:

$$\frac{\partial}{\partial t}(PU,U)-\frac{\partial}{\partial x}(QU,U)-\frac{\partial}{\partial y}(RU,U) = 0.$$

It is necessary to select K,L,M,N in such a way that $(PU,U)>0$ and so that on all $U=\begin{Vmatrix}u_1\\u_2\\u_3\end{Vmatrix}$, connected by the condition $Au_1+Bu_2+Cu_3=0$, inequality $(QU,U)>0$ was held as well. For such a choice of arbitrary matrix coefficients the written out integral can really be considered as the energy integral for the vector wave equation, dissipative on the assigned on the boundary condition. It turns out that if the boundary condition satisfies the Lopatinsky uniform condition, then the construction of such dissipative energy interval is always possible and is realized by means of the following procedure proposed and grounded in [7].

Having assigned the strictly positive hermitian matrix G, we solve the Ljapunov equation

$$\Lambda^* H + H\Lambda = -G.$$

Since Λ has all the roots strictly in the left halfplane, the solution H of this equation does exist uniquely and is a hermitian and strictly positive definite matrix. Having written H in the form

$$H = \begin{vmatrix} K-M & -L-iN \\ -L+iN & K+M \end{vmatrix},$$

we define hermitian matrices K,L,M,N which take part in constructing the energy integral we are interested in. The proof of the described construction is given in [7].

Consider the mixed problem for a quasi-linear system of equations of gas dynamics with boundary conditions on shock wave. For the given problem one can construct a dissipative energy integral, the technique of the construction developed for the linear mixed problem being transferred for the case of a quasi-linear system.

Equations of gas dynamics are convenient to be considered in a special curvilinear system of coordinates $\xi = (\xi_1, \xi_2)$ in which the shock wave coincides with the coordinate line

$\xi_2=0$. For $\xi_2>0$ on the shock wave a uniform gas flow accumulates whose speed vector is parallel to the x axis. Let for t=0 ξ be an orthogonal curvilinear system of coordinates, the coordinate lines ξ_1=const being straight lines. Further we assume that for $t>0$ the coordinate lines ξ_1=const remain fixed. The system of gas dynamics equations can be written in the following form [9]:

$$A\,U_t + \sum_{i=1}^{2} B_i U_{\xi_i} + H = 0. \qquad (10)$$

Here,

$$A=A(U)= \begin{vmatrix} \rho & 0 & 0 & 0 \\ 0 & \rho & 0 & 0 \\ 0 & 0 & 1/a^2 & 0 \\ 0 & 0 & 0 & 1 \end{vmatrix}, \quad U = \begin{vmatrix} u_1 \\ u_2 \\ u_3 \\ u_4 \end{vmatrix}; \; B_1=B_1(U,F,\xi),$$

$$B_2=B_2(U,F,F_t,F_{\xi_1},\xi)$$

are symmetric matrices; $H=H(U,F,\xi)$; $u_3=p$, $u_4=s$; ρ, $p = \pi(\rho,s)$, s are density, pressure, entropy; $a^2 \left(= \frac{\partial \pi}{\partial \rho} > 0\right)$ is the squared sound speed; u_1, u_2 is the projection of the speed vector upon the straight line ξ_1=const and upon the direction orthogonal to the straight line ξ_1=const; $F = \int_0^t \mu_2^*(t,\xi_1)dt$; $\mu_1^*(\equiv 0)$, $\mu_2^*(t,\xi_1)$ are contravariant components of the motion speed of the movable coordinate system. Rankine-Hugoniot conditions on shock wave are convenient to be written as follows

$$u_1 = -u_\infty \sin\varphi_0 + \frac{\Delta G(p)}{e^f}, \quad u_2 = u_\infty \cos\varphi_0 - \frac{F_{\xi 1}\,G(p)}{e^f},$$

$p=p(V)$ is Hugoniot adiabatic curve,

$$F_t + \frac{u_\infty \cos\varphi_0}{\Delta} F_{\xi 1} - e^f G_1(p)/\Delta + u_\infty \sin\varphi_0 = 0, \qquad (11)$$

where $G(p)=\sqrt{(p-p_\infty)(V_\infty - V)}$, $G_1(p)=G(p)V_\infty/(V_\infty - V)$, $V=1/\rho$,

$$\Delta = \Delta(\xi_1, F) = e^{f_0(\xi_1)} - \varphi_0'(\xi_1)\cdot F; \quad f_0(\xi_1), \varphi_0(\xi_1)$$

are some assigned functions; $e^f = \sqrt{(F_{\xi_1})^2 + \Delta^2}$; u_∞, ρ_∞, $V_\infty = 1/\rho_\infty$, s_∞, $p_\infty = \pi(\rho_\infty, s_\infty)$, $a_\infty^2 (= \frac{\partial \pi}{\partial \rho}(\rho_\infty, s_\infty))$ are the parameters of the accumulating flow.
For t=0 the initial data are assigned

$$U(0,\xi) = U_0(\xi)(\xi_2 < 0), \quad F(0,\xi_1) = 0. \tag{12}$$

The domain

$$\rho > 0, \qquad \frac{\partial \pi}{\partial \rho} > 0$$

is the natural domain of change of thermodynamic magnitudes ρ, p, s. System (10) in such a case is symmetric t-hyperbolic (due to Friedrichs).
Let Ω be a domain in the space t, ξ ($\xi_2 < 0$), bounded by a side surface $S_{6\alpha}$, by pieces of planes t=0(w(0)), t=T(w(T)), x=0(S_b). Let w(t) be the cut of Ω by the plane t-const, $0 < t < T$. Introduce the following integral into our consideration:

$$J_1(t) = J_1(U,F) = \sum_{k=0}^{4} \sum_{i_1=0}^{3} \sum_{\substack{i+j+l \\ = i_1}} \iint_{w(t)} \left(\frac{\partial^{i_1} u_k}{\partial t^i \, \partial \xi_1^j \, \partial \xi_2^l}\right)^2 d\xi_1 d\xi_2 +$$

$$+ \sum_{k=0}^{3} \sum_{i+j=k} \int_{\tau(t)} \left(\frac{\partial^k F}{\partial t^i \partial \xi_1^j}\right)^2 d\xi_1,$$

where $\tau(t)$ is the cut of a piece of the plane x=0(S_b) by the plane t=const, $0 < t < T$.
In paper [9], for mixed problem (10), (11), (12) the following statement was proved:
1) Let the mixed problem have the solution $U(t,\xi) \in C^1(\bar{\Omega})$, $F(t,\xi_1) \in C^2(\bar{S}_b)$, besides,

$$\|U\|_{C^1(\bar{\Omega})} < N, \qquad \|F\|_{C^2(\bar{S}_b)} < N.$$

2) Let the initial data $U_o(\xi) \in W_2^3(w(0))$, besides,

$$J_1(0) < \infty .$$

Then there takes place the following estimate

$$J_1(t) < N_1(J_1(0), N, T) < \infty, \quad 0 \leq t \leq T \tag{13}$$

From (13) it follows that for any $t(0 < t \leq T)$, $U(t,\xi) \in W_2^3(w(t))$, $F(t,\xi_1) \in W_2^3(\mathcal{S}(t))$. Besides, integrating inequality (13) from 0 to T, we obtain

$$\max_{0 \leq t \leq T} J_1(t) + \|U\|^2_{W_2^3(\Omega)} + \|F\|^2_{W_2^3(S_b)} < K(N, J_1(0), T) < \infty. \tag{13'}$$

Indeed, one can show that there takes place a finer estimate:

$$\max_{0 \leq t \leq T} J_1(t) + \|U\|^2_{W_2^3(\Omega)} + \|F\|^2_{W_2^4(S_b)} + \tag{13''}$$

$$+ \iint \sum_{l=1}^{4} \sum_{i+j+k=3} \left(\frac{\partial^3 u_e}{\partial t^i \, \partial \xi_1^j \, \partial \xi_2^k} \right)^2 d\xi_1 dt < K(N, J_1(0), T) < \infty$$

besides, to prove the uniqueness theorem of the classical solution of the mixed problem (10), (11), (12), it is sufficient to have estimate of (13') kind, and to prove the theorem of the classical solution exsitence for small t, one should use a finer estimate (13'').

In paper [10] proved is the theory, of the uniqueness of the classical solution of mixed problem (10), (11), (12). The theorem about the existence of the classical solution for small t is proved with the help of estimate (13''), according to paper [11].

References

1. Djakov, S.P. On shock wave stability. Ž. experimental'noi i teoreticheskoi fiziki, 1954, t.27, No 3/9/, 288-296.
2. Iordansky, S.V. On stability of a plane stationary shock

wave. Applied mathematics and mechanics, 1957, t.21, vyp. 4, 465-472. (In Russian).

3. Zaidel', R.M. Development of perturbations in plane shock waves. Applied mechanics and theoretical physics, 1967, No 4, 30-39 (In Russian).
4. Godunov, S.K. Equations of mathematical physics. Moscow, "Nauka", 1979, 392 p. (In Russian).
5. Blokhin, A.M. On correctness of mixed problem for acoustics system of equations with boundary conditions on shock wave. Novosibirsk, 1979 (Preprint Computer Centre, Siberian Division of the USSR Academy of Sciences, 141, in Russian).
6. Blokhin, A.M. Mixed problem for acoustics system of equations on shock wave. Izvestija SO AN SSSR, technical sciences, 1979, No 13, vyp.3, 25-33 (In Russian).
7. Gordienko, V.M. Un problème mixte pour l'equation vectorielle des ondes: Cas de dissipation de l'energie;Cas mal poses. - C.R.Acad.Sc.Paris, 1979, t.283, Serie A, p.547-550.
8. Godunov, S.K.; Gordienko, V.M. Mixed problem for wave equation. Trudy Seminara Soboleva S.L., 1977, No 2, p.5-32 (in Russian).
9. Blokhin, A.M. Estimate of energy integral of mixed problem for gas dynamics equations with boundary conditions on shock wave. Siberian Math.Journal, 1981, t.22, No 4, 23-51 (In Russian).
10. Blokhin, A.M. Energy integrals in mixed problem for gas dynamics equations with boundary conditions on shock wave. Collection of scientific papers "Investigations of hydrodynamic stability by means of computers", (ed. K.I.Babenko, Institute of Applied Mathematics after M.V.Keldysh, AN SSSR, 1981, 198-228, in Russian).
11. J.Leray. Hyperbolic differential equation, part two. Princeton, 1953.

Solitons in Active Media

L.A. OSTROVSKY
Institute of Applied Physics
of the Academy of Sciences of the USSR
Gorky, USSR

Summary
The paper gives a brief survey of recent investigations of solitons in nonlinear media with energy pump. The processes are described within the framework of various generalizations of the Korteweg-de-Vries equation with additional terms responsible for the amplification and dissipation effects. It is shown that the soliton stage proves to be a typical one in a rather wide class of slowly amplified nonlinear waves. In particular, an explosive soliton instability and stationary "autosolitons" are possible. Specific examples refer to internal gravitational waves in a stratified shear flow, ion-sound waves in plasma and to the "parametric" interaction of electromagnetic solitons with a traveling "pump" wave in nonlinear lines.

Solitons become so popular now that "solitonics" is already divided into two branches: exact methods (the inverse scattering method et al.) and "physical solitonics". One of the most important topics of the latter is the behavior of solitons in real weakly nonconservative media. Damping of solitons was thoroughly considered in literature. Less attention was paid to the soliton behavior in amplificating media, able to pump the energy into the wave. Such problems are rather typical for hydrodynamics, plasma physics, laser physics, radiophysics, etc. The present paper deals with some aspects of this problem in connection with "ordinary" solitons in nonlinear dispersive media with small nonconservativity*

1. A SIMPLE MODEL. DAMPING AND AMPLIFICATION OF SOLITONS.

We proceed from the generalized form of the Korteweg-de Vries

* We do not consider here the localized waves existing due to active nonlinearity like pulses in nerve fibre or in neuristor lines because they actually are not solitons in the sense accepted now.

Nonlinear Deformation Waves
IUTAM Symposium Tallinn 1982
Editors: U. Nigul, J. Engelbrecht

equation (KdV)

$$\varphi_t + \alpha\varphi\varphi_x + \beta\varphi_{xxx} = \varepsilon\hat{L}(\varphi). \tag{1}$$

where ε is the small parameter and $\hat{L}$ is the operator describing the influence of nonconservative factors.

Let the initial condition of the problem be close to the solution of equation (1) with the zero right-hand part. Then it can be expected that at least at some finite time interval the solution will remain close to the soliton, but will possess varying parameters, i.e.

$$\varphi(x,t) = A\,\mathrm{sech}^2\left[\Delta^{-1}\left(x - \int V_s\,dt\right)\right] \tag{2}$$

where $V_s = \alpha A/3$, $\Delta = (12\beta/\alpha A)^{1/2}$, and $A = A(\tau = \varepsilon t)$.

The methods of an asymptotic analysis of equation (1) with respect to solitons are well developed now (see, for example, [1,2]). In the first nontrivial approximation the parameters of solution (2) can be determined from the equation following from (1) after its multiplication by φ and integration in the interval greatly exceeding the soliton length

$$\frac{4}{3}\sqrt{\frac{12\beta}{\alpha}}\;\frac{d(A^{3/2})}{d\tau} = \langle\varphi\hat{L}(\varphi)\rangle \tag{3}$$

where $\langle\ \rangle = \int_{-\infty}^{\infty} dx$. Note that applicability of equation (3) is limited by the condition that the right part of (1) should be small compared with every term of the left part.

The second important note is associated with substitution $t \to -t$, $x \to -x$ corresponding to the backward motion of the wave with respect to the initial one, or in a sense "wave reversal to past". It is easy to see that the left part of (1) changes its sign at such a substitution, and if $\hat{L}(\varphi)$ is conserved (that is typical of nonconservative terms) this substitution means a passage from dissipation to amplification and vice versa.

Specific physical examples will be given below. Now we consider a rather general model equation for which

$$\varepsilon\hat{L}(\varphi) = g\varphi - \nu\varphi_{xx} \tag{4}$$

Equation (3) for the soliton amplitude is

$$\frac{dA}{dt} = \frac{4}{3} g A + \frac{4}{45} \frac{\alpha \nu}{\beta} A^2, \tag{5a}$$

whence

$$A = \frac{A_0 a}{(a + b A_0) e^{-at} - b A_0} . \tag{5b}$$

where $A_0 = A(t=0)$, $a = \frac{4}{3} g$, $b = \frac{4 \alpha \nu}{45 \beta}$. The case of soliton damping within the framework of equations of the type (5a) were considered earlier [3,4]. In the case when there are only "low-frequency" losses ($b=0, a<0$), the soliton amplitude decreases exponentially: $A = A_0 \exp(-\frac{4}{3}|g| t)$, with a decrement 4/3 times as much as that of the long linear wave. In the case of "high-frequency" or "viscous" losses ($a=0, b<0$) (1) has the form of the Korteweg-de Vries-Burgers equation (KdVB) and the soliton decays according to the law

$$A = A_0 (1 - b A_0 t)^{-1} \tag{6}$$

It follows from here that for $t \to \infty$, $A \approx |bt|^{-1}$, i.e. the soliton "forgets" its initial amplitude. A similar effect is known for weak shock waves in nonlinear acoustics [5]. The applicability conditions of the soliton solution (5b) for cases of low-frequency and high-frequency damping have the forms correspondingly

$$\sqrt{A} \gg g \sqrt{12\beta/\alpha^3} \quad a) \qquad \sqrt{A} \gg \nu / \sqrt{12 \alpha \beta} \quad b) \tag{7}$$

Therefore, in both cases solutions in the form of weakly damping solitons exist only for a limited period of time after which the type of solution must change. The next stage was studied by Gorshkov and Papko [6] and it proved to be qualitatively different for low-frequency and high-frequency losses. In the first case nonlinearity becomes negligible and the solution tends to the asymptotics of linearized KdV equation and may be approximately described by the Airy function. Thus, the soliton transforms into an oscillating train damping due to both dissipation and dispersion spreading. In the second case nonlinearity remains essential at all stages, but the dispersion term $\beta\varphi_{xxx}$ can be neglected and this stage is described by the known Burgers equation. As a result the soliton transforms into a spreading triangular pulse.

Now we return to the problem of soliton amplification when a and b in (5b) **are** positive. The rule of time reversal mentioned above is valid here and the asymptotic behavior of the solution is quite different. Even for $b=0$, when the soliton amplitude increases exponentially, in the course of time the relative value of the dissipative term decreases and the applicability interval of the solution is not limited now until the soliton amplitude grows to such a value that applicability of equation (4) itself which is usually deduced for weakly-nonlinear motions will be violated.

For high-frequency losses the situation is more "dramatic": it is clear **from formula (6) at** $b>0$ that at a finite time value $t=t_e=(bA_0)^{-1/2}$ an "explosion" occurs - the soliton amplitude becomes infinite. An explosion behavior is typical also of complete solution (5b) for $a,b>0$. Here, if the initial amplitude is rather small ($bA_0 \ll a$), the first stage will be an exponential growth, which is changed by an explosion increase.

At last, of interest are the cases when parameters a and b have different signs. If $a<0$, $b>0$,there is such a critical value of the initial amplitude $A_{0cr}=b/|a|$, that for $A_0<A_{0cr}$ the soliton amplitude decreases while at $A_0>A_{0cr}$ it increases explosively. In other words, a rigid amplification regime takes place. But if $a>0, b<0$, then $A_{cr}=|b|/a$ is a stationary equilibrium state, so that solitons with a small initial amplitude increase and with a large one - decrease tending to the equilibrium state. The latter can be called a stationary autosoliton in accordance **with the concept of** "autowave processes" widely discussed now. The main types of processes associated with the damping and increase of solitons are shown in fig.1. An asymptotic analysis of solitons can, in principle, be easily generalized to wider class of periodic "cnoidal" waves. For a small amplitude such a wave is close to a sinusoid, and with an amplitude increase it approximates the soliton sequence. Thus a series of growing solitons proves to be an "intermedi-

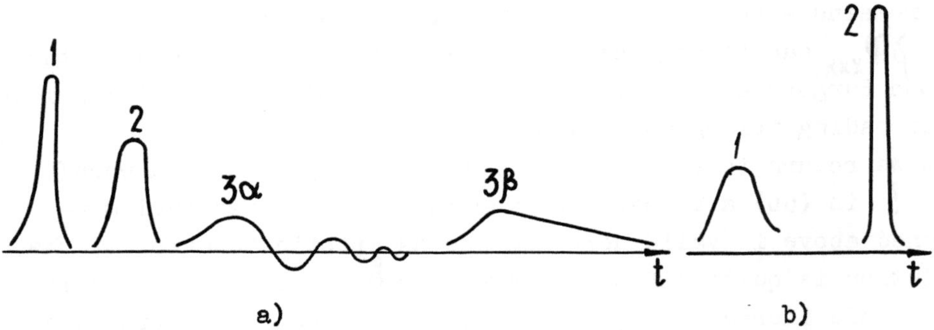

Fig.1. A scheme of soliton evolution: a)in the dissipative medium (1,2,3α - the case of low-frequency losses; 1,2,3β - the case of high-frequency losses), b)in amplifying medium

ate asymptotics" for a more general class of periodic waves. Note that the considered type of equation describes a number of real physical situations. At the same time the operator $L(\varphi)$ turns to be often more complicated. We shall discuss below two examples of "flow" instabilities referring to hydrodynamics and plasma physics.

2. Shear Instability in a Stratified Liquid and Generation of Internal Waves.

The problems of hydrodynamic instability in stratified liquids are discussed in a large body of literature. The simplest models here are associated with liquid consisting of the homogeneous layers separated by a tangential discontinuity and a density jump. As it is known, a Kelvin-Helmholtz instability is possible in such a model; this instability is especially strongly revealed for short-wave perturbations and leads to destruction of a sharp boundary because of liquid turbulization in its vicinity. But long waves can remain stable due to stabilizing effect of the density stratification. However, in the long-wave region there may exist waves with negative energy (NEW) well known in electronics and plasma physics and intensively studied recently for hydrodynamical problems (note, though, that the concept of such waves in hydrodynamics was in fact introduced by Benjamin as early as in 1963 [9]). Energy dissipation of any type leads to an increase of such waves rather than to their damping.

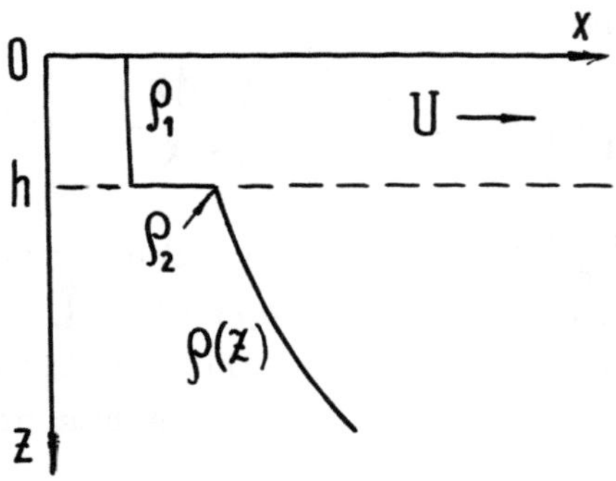

Fig.2

Consider here the following simple model (fig.2). A homogeneous liquid layer with density ρ_1 and thickness h slides with velocity U above a fixed liquid half-space so that at the interface ($z=h$) there is a density jump from ρ_1 to $\rho_2>\rho_1$, besides the lower liquid's density grows exponentially with depth : $\rho(z>h)=\rho_2 \exp[\sigma(z-h)]$ i.e. the lower liquid is characterized by the constant Brunt-Väisäla frequency $N=|g\rho_z/\rho|^{1/2}=(g\sigma)^{1/2}$, where g is the gravity acceleration. Due to this stratification internal waves can be radiated down from the discontinuity level interface and energy consumption on radiation is sufficient for increase of NEW propagating along the interface.

A detailed consideration of such a "radiative" instability is given in the paper by the author, Stepanyants and Tsimring [10], and here we mention only few results referring to the topic of the present paper.

The solution of the linearized equations of hydrodynamics presented in the form $f(z)\exp i(\kappa x+mz-\omega t)$ together with boundary conditions at $z=0,h$, and $z\to\infty$ leads to the dispersion equation relating ω and κ. Generally speaking ω may be complex for real κ, that means here the instability existence. For the condition $U>U_c=[(\rho_2-\rho_1)gh/\rho_1]^{1/2}$ the instability takes place, in particular, for long waves with $\kappa h\ll 1$. As a mechanism limiting the instability we take into account turbulization of the upper layer leading to NEW damping. As a result for long waves we obtain the following dispersion equation

$$\omega=(U-U_c)\kappa+\beta\kappa^3+i(\varepsilon-\mu\kappa+\delta\kappa^2). \tag{8}$$

where β is the dispersion parameter (of the order of $U_c h^2$), ε and β are parameters of turbulent losses, and μ is the radiative instability parameter (of the order of Nh).

Then, following the well-known methods it is easy to obtain the evolution equation in the form

$$\eta_t + \alpha\eta\eta_s + \beta\eta_{sss} = -\varepsilon\eta + \delta\eta_{ss} + \frac{\mu}{\pi}\int_{-\infty}^{\infty}\frac{\eta_x\,dx}{s-x} \qquad (9)$$

Here $s = x-(U-U_c)t$, $\alpha = \frac{3}{2}U_c/h$, η is the interface deviation from the equilibrium state. The operator $\hat{L}$ in the right part (9) differs from the case (4) by the additional integral term. Equation (3) for the soliton amplitude has now the form

$$\frac{dA}{dt} = -\frac{4}{3}\varepsilon A - \frac{4}{45}\frac{\alpha\delta}{\beta}A^2 + \frac{2{,}92}{\pi}\mu\left(\frac{\alpha}{12\beta}\right)^{1/2}A^{3/2} \qquad (10)$$

A simple analysis of this equation shows that for a sufficiently large μ it has three equilibrium states $A_{1,2,3}$, from which $A_1=0$ and $A_3 > A_2$, A_2 being unstable and A_1, A_2 - stable. Therefore, amplification proves to be rigid and a stationary autosoliton exists. If ε and δ are small enough, $A_3 \sim \mu/\delta$, $A_2 \sim \varepsilon/\mu$ and $A_2 \ll A_3$. Then in the interval $A_2 \ll A \ll A_3$ the solution has an "explosive" form:

$$A = A_0(1-t/t_e)^{-2} \qquad (11)$$

where $t_e = \frac{2\pi}{1{,}46\mu}\left(\frac{3\beta}{\alpha A_0}\right)^{1/2}$. However later, the soliton again approaches the stationary level A_3.

This simple theory, in spite of serious limitations (on the loss parameters, flow velocity, etc.) still gives, evidently, a rather rare possibility of an analytic description of the nonlinear stage of the hydrodynamic instability for essentially nonharmonic pulselike processes. Estimates show that such a mechanism can play a definite role in generation of oceanic internal waves.

3. Amplification of Ion-acoustic Solitons in Plasma by a Charged Particle Beam

The following example refers to the interaction between an ion-acoustic soliton and an ion beam in nonisothermal plasma. Note that in the plasma-beam system (considered in the hydrodynamic approximation) stationary solitons are possible propagating without amplification or damping. However, a nonstationary interaction of a soliton with a beam results in soliton amplification. As it is shown in our paper [11], this process can be described by equation of the form (1) for an electric potential

$$\varphi_\tau + \frac{1}{2}\varphi\varphi_\xi + \frac{1}{2}\varphi_{\xi\xi\xi} = \gamma\frac{\partial \rho_b}{\partial x} \tag{12}$$

where φ is the dimensionless potential, $\xi = \tau - x'$, τ, x' are dimensionless time and coordinate, V is the ion sound velocity in plasma, γ is the beam-plasma coupling parameter (it is assumed here that the particle density in the beam is much less than in plasma); ρ_b is the particle density perturbation in the bean.

According to (12), the soliton amplitude variation is described by the equation

$$\frac{dA}{d\tau} = -\frac{0{,}3\gamma}{\sqrt{A}}\int_{-\infty}^{\infty}\varphi\frac{\partial \rho_b}{\partial \xi}\,d\xi \tag{13}$$

However, now we encounter the problem of determining the relation between ρ_b and φ. This problem was first solved in the quasi-hydrodynamic approximation [11]: writing down the hydrodynamic equations for the beam and linearizing them (due to small beam perturbations), we obtain the integral relation between ρ_b and φ and as a result (13) acquires the form

$$\frac{dA}{d\tau} = -\frac{\gamma}{\sqrt{A}}\int_{-\infty}^{\infty}d\xi\,\varphi(\xi,\tau)\int_0^{\tau}dt_1\int_0^{t_1}dt_2\,\varphi_{\xi\xi\xi}(\xi,t_2) \tag{14}$$

where φ has the form of a soliton. The main peculiarity of this equation is that its right part depends explicitly on the slow variable τ and, consequently, a first-order equation for A as above cannot be obtained. In the general case the solution of (14) needs a use of a computer. The solution shows that the soliton undergoes a moderate amplification (up to 2 - 2.5 times), and after that its amplitude remains constant. Such amplification can be called a transient one: a plasma soliton perturbs the beam receiving its energy. This perturbation is divided into a free part (beam wave) and forced wave (traveling with a soliton velocity). After their separation a stationary soliton with a constant amplitude remains in the beam-plasma system.

However the quasi-hydrodynamic approximation is valid only for a sufficiently large difference between velocities of the beam (V) and the ion-sound wave (C_s). If these values are close enough, the resonance effects associated with reflection and trapping of particles should be taken into account. We have performed such an analysis [12]. To describe the beam we used the path integral method writing down equations of motion of individual beam particles and restored the distribution function over their coordinates and velocities by using of which then calculated the density variations. The equations obtained were solved numerically. Examples of calculation results are given in fig.3. Probably, the main conclusion is the fact that for $V > C_s$ (i.e. when with the soliton amplitude increase its velocity reaches the beam velocity) the soliton amplitude always approaches the asymptotic value $A_{st} \sim 3\delta$, when velocities of the soliton and the beam become equal. If the initial difference of these velocities is sufficiently large (δ is large or A_0 is small), then, as it is seen from fig.3, a distinct quasi-hydrodynamic stage can be singled out: a small initial transient amplification and a quasi-equilibrium plateau. However, later the kinetic effects manifest themselves in this case as well, and the amplitude increases much stronger, up to the value A_{st} . In this case calculations show the presence of reflected particles in the beam.

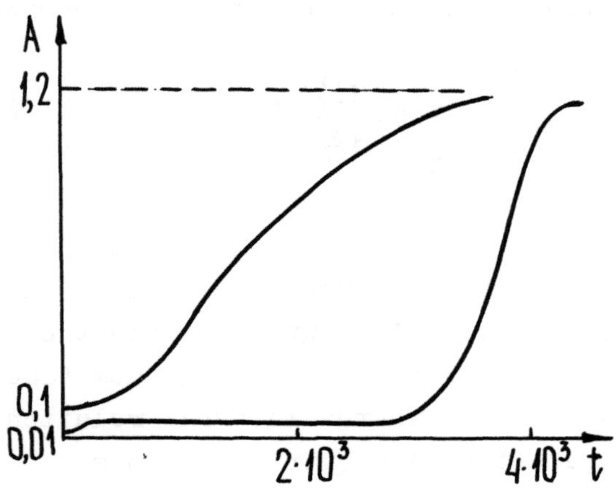

Fig.3
Soliton amplitude increase in plasma for various initial amplitudes and

Note, that though in the given example we cannot do without a computer, the use of approximate equations of the type (13) makes the calculations considerably easier and the result more evident.

It is also noteworthy, that the results obtained are at least in a qualitative agreement with the experimental data of Okutzu et al. [13] on amplification of ion-sound pulses by a beam. For one of the cases described by them we even succeed to perform a quantative comparison showing a good agreement of theory and experimental data.

4. Parametric Amplification of Solitons

In the context of this paper of interest is a different mechanism of soliton amplification associated with their interaction with a long "pump" wave moving synchronously with the soliton (fig.4a). Depending on the soliton position (phase) with respect to the pump the latter performs a positive or negative work; in its turn, the phase varies due to the soliton velocity dependence on its amplitude and on the pump field value ("pedestal") in the place of its location. The analysis of such processes may be reduced to the solution of the type (1) equation with varying coefficients corresponding to the traveling wave, for example

$$\varphi_t + \alpha(\xi)\varphi\varphi_s + \beta(\xi)\varphi_{sss} + p(\xi)\varphi = 0 \qquad (15)$$

where $s = x - ct,\ \xi = s - \delta t,\ c$ is the linear long wave velocity, δ is the small difference between velocities of the pump and the

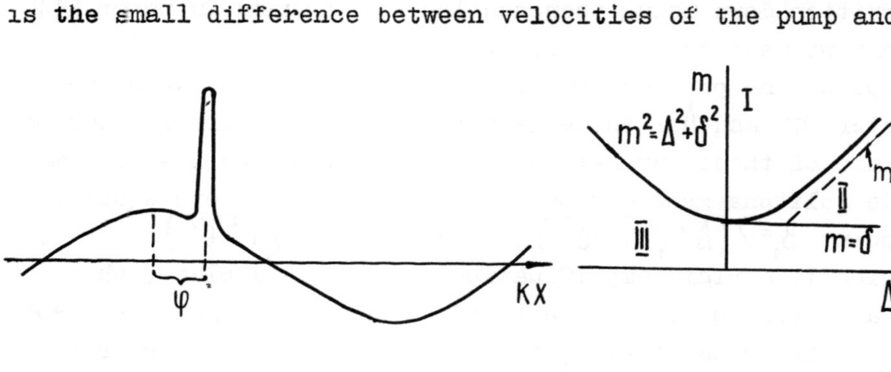

a) b)

Fig.4.

long linear wave in this system. The asymptotic procedure gives an equation for the soliton amplitude differing from the considered above by the dependence on x, t of coefficients α, β and p, whose values should be taken at the soliton location point. This means that the equation for A must be solved together with the soliton trajectory equation $dx/dt = V_s$, where V_s is the soliton velocity. Assume that α and β undergo small sinusoidal variations (for example, $\beta = \beta_0 + \beta_1 \cos K_0 \xi$, where $\beta_1 \ll \beta_0$) and $p = p_0 + p_1 \cos K_0 \xi$, where p_0, p_1 are small values of the same order (usually p_0 describes low-frequency losses and p_1 appears due to variable pump). In this case the soliton evolution is described by the following equations [14]

$$\frac{dB}{d\zeta} = -\frac{2}{3} B (m \sin\varphi + \delta), \tag{16a}$$

$$\frac{d\varphi}{d\zeta} = B - m\cos\varphi - S, \tag{16b}$$

where $\varphi = K_0 \xi$ is the soliton phase with respect to pump, $\zeta = cKt$, $B = 2\alpha_0 A/3\beta_0$, $m \sim p_1$ is the pump parameter, $\delta \sim p_0$ is the wave damping rate in the linear approximation. These equations describe the soliton interaction as a localized "particle" with distributed pump (they are really similar to the equations for particles in the cyclic accelerator). They were written down using phenomenological grounds in paper [14] and here we mention only a few results.

The type of the process depends on the relation between the parameters m and Δ : three main regions can be singled out on the plane of these parameters (fig.4b). In region I small-amplitude solitons grow approaching the stable nonzero equilibrium point $B_1 = 2(\Delta + \sqrt{m^2-\delta^2})$, $\varphi_1 = \arccos(-\sqrt{m^2-\delta^2})$. In region II a rigid amplification regime takes place, when small-amplitude solitons are damping, while stronger ones tend to the equilibrium level B_1 (in this case there also exists an unstable equilibrium position $B_2 = 2(\Delta - \sqrt{m^2-\delta^2})$). At last, in region III all perturbations damp. Note, that the processes of increasing and damping of solitons may have an oscillatory character. In limited systems - resonators - such processes lead to the generation of soliton sequences - both regular and stochastic ones. These processes - "parametric generation of pulses" were studied experimentally for electromagnetic waves in lines with semiconductor diodes. The results of the experiments are given in papers [14,15].

Notes on Envelope Solitons

The class of problems briefly reviewed here does not cover the whole subject of "autosolitonics". Other interesting problems were also considered recently, which we can only mention because of lack of place.

Note, in particular, papers on amplification of envelope solitons in the form of localized wave packets. Fabrikant [16] investigating excitation of water waves by a weak wind, derived a nonlinear Schrödinger equation with a conservative right part describing the energy pump from wind to wave. A specific solution of this equation is a slowly varying envelope soliton reaching the stable equilibrium state. Earlier Shapiro and Shevchenko [17] considered the problem of amplification of Langmuir solitons in plasma by an electron beam. Another interesting case was considered recently by Kolchugina et al. [18] for waves in magnetoactive plasma. Solitons in the form of electron "envelope waves" coupled with acoustic-type soliton may exist in this case. Here two types of solitons are possible corresponding roughly to an ion and magnetic sound which have an essentially different width. At the presence of a microwave pump supplying the soliton with energy, there are possible soliton self-oscillations in the form of periodic "jumps" of the soliton parameters between these two states.

At last, similar effects for "Sine-Gordon" solitons in Josephson diodes, laser systems, etc. deserve special consideration. Such solitons in themselves are in fact autowaves, since they exist under the conditions of energy pump from a non-equilibrium medium. Deceleration of such solitons due to dissipation was repeatedly studied [19]. Negative losses can be also included in the Sine-Gordon equation, which instead of deceleration must accelerate the soliton to a free wave velocity; this process can be also described in the framework of an asymptotic procedure. (As for "principally active" solitary waves, mentioned in the beginning, note that sometimes the form of such waves turns to be soliton-like [20]. However, such solution does not possess a free parameter and its slow evolution in a homogeneous system is impossible).

One may think that in the nearest future problems of "autosolitonics" will be studied more intensively.

References

1. Gorschkov K.A., Ostrovsky L.A., Pelinovsky E.N.: Some problems of asymptotic theory of nonlinear waves. Proc. IEEE 62 (1974) 1511-1517.

2. Gorschkov K.A., Ostrovsky L.A.: Interactions of solitons in nonintegrable systems: Direct perturbation method and applications. Physica 3D (1981) 428-438.

3. Pelinovsky E.N.: Nonlinear ion-sound waves in inhomogeneous dispersive media. Izv. VUZov - Radiofizika 14 (1971) 1281-1283.

4. Ostrovsky L.A.: Short-wave asymptotics for weak shock waves and solitons in mechanics. Intern. Journ. of Nonlinear Mech. 11 (1976) 401-406.

5. Zarembo L.K., Krasil'nikov V.A.: Introduction to nonlinear acoustics. Moscow: Nauka 1966.

6. Gorschkov K.A., Papko V.V.: Damping of solitons and intermediate asymptotics. Izv VUZov - Radiofizika 21 (1976) 432-439.

7. Vasil'ev V.A., Romanovsky Yu.M., Yakhno V.: Wave-particle interaction in nonequilibrium media. UFN 128 (1979) 625-666.

8. Turner D.T.: Buoyancy effects in fluids. Cambridge Univ. Press 1973.

9. Benjamin T.B.: The threefold classification of unstable disturbance in flexible surface bounding inviscid flow. J.Fluid Mech. 16 (1963) 434-450.

10. Ostrovsky L.A., Stepanyants Yu.A., Tsimring L.Sh.: Radiation instability in stratified shear flow. Intern.Journ. of Nonlinear Mech. (to be published).

11. Ostrovsky L.A., Petrukhina V.I., Fainshtein S.M.: On amplification of ion-sound by a charged-particle beam. ZhETF 69 (1975) 2051-2055.

12. Gil'man O.L., Ostrovsky L.A., Fainshtein S.M.: Resonance amplification of ion-acoustic solitons in a plasma-beam

system. IEEE Trans. on Plasma Science. PS-9 (1981) 286-289.

13. Okutsu E., Nakamura M., Nakamura Y., Iton F.: Amplification of ion-acoustic solitons by an ion beam. Plasma Phys. 20 (1978) 561-568.

14. Gorschkov K.A., Ostrovsky L.A., Papko V.V.: Parametric amplification and pulse generation in nonlinear distributed systems. Izv. VUZov - Radiofizika 16 (1973) 1195-1204.

15. Gorschkov K.A., Ostrovsky L.A., Papko V.V.: Solitons in radiophysics. Physica Scripta 20 (1979) 357-363.

16. Fabrikant A.L.: On nonlinear water waves under a light wind and Landau-type equations near the stability threshold. Wave motion 2 (1980) 355-360.

17. Shapiro V.D., Shevchenko V.N. Izv. VUZov - Radiofizika 19 (1976) 767-791.

18. Kol'chugina I.A., Litvak A.G., Sergeev A.M.: Langmuir turbulence in magnetoactive plasma and self-oscillation soliton. Pis'ma v ZhETF 35 (1982) 510-514.

19. ParmentierR.: Fluxons in distributed Josephson contacts. In: Solitons in action, ed. by Lonngren K., Scott A. Academic Press 1978.

20. Kyashko S.V., Rabinovich M.I., Reutov V.P.: Explosive instability and generation of solitons in an active medium. ZhTF 12 (1972) 2458-2465.

The Evolution Equations in the Theory of Nonlinear Waves

J. ENGELBRECHT

The Institute of Cybernetics, Estonian Acad. Sci.,
Tallinn, Estonia, USSR

Summary

The evolution equations obtained as a result of certain asymptotic analyses play a significant role in wave mechanics while beside the distortion of single pulses the evolution equations and the corresponding asymptotic formulae describe also the coupling and interaction effects. In this report the study of nonlinear waves in a continuous medium is extended to investigate: (i) the correspondence between the dispersion relations of the full system and of the derived evolution equations, (ii) the structure of the evolution equations in the case of an active medium, (iii) the two-dimensional evolution equations.

1. Introduction

The evolution equations are a powerful tool in wave mechanics almost in any branch of physics dealing with waves [1-4]. Obtained as a result of a certain asymptotic analysis, the evolution equations describe explicitly the behaviour of single waves and many elegant and coherent methods have been elaborated during the past decade to attack the problem of solving them. The most intriguing problems are those connected with the nonlinear wave propagation because of complicated nonlinear effects, not speaking about the elegance of the nonlinear equations and their solutions. The following discussion is centered around the general theory of evolution equations. In Section 2 the scheme of constructing the evolution equations is presented. Attention is drawn to the fact that the successive analysis must consider the possibilities to take the interaction and coupling effects into account. Section 3 is devoted to the dispersion problems. As far as the evolution equations give the certain asymptotic description of a wave process the correspondence between the

Nonlinear Deformation Waves
IUTAM Symposium Tallinn 1982
Editors: U. Nigul, J. Engelbrecht

waves modelled by the full system and the waves modelled by the evolution equations must be clearly stated. On the basis of simple examples it is shown how the frequencies and the wavenumbers are related in both approaches. Such a successive analysis explains the existence of several anomalies following from the dispersion relations of the evolution equations.

The next part, Section 4 considers the fundamental problems of the evolution equations in an active media. As a rule, up to now the large class of wave motions in active media is modelled by parabolic equations that permit special wave-like solutions. The possibilities to replace the parabolic models by hyperbolic ones are discussed. The dispersion relations of both models as well as the steady-wave solutions are analysed. The pulse transmission in a nerve fibre is considered in more detail in order to demonstrate such an approach. The hyperbolic model gives the possibility to avoid the degeneracy introduced by the truncation of the higher terms in the transmission line equations. The two-dimensional evolution equations are discussed briefly in Section 5. A numerical algorithm suitable for solving the two-dimensional evolution equation subjected to arbitrary initial conditions is given.

2. The Evolution Equations

There are several asymptotic methods developed for the analysis of wave propagation. Among the others, the methods based on the analysis of wave front kinematics correspond precisely to the nature of the wave propagation. The ray method has played an important role in linear acoustics and optics, for example. After the fundamental papers of Taniuti and Wei [5] and Miura [6], the ideas of ray method have been widely used also in nonlinear theory [4]. The nonlinearity and the other second-order effects involved lead to the evolution equations in the form of partial differential equations. The main principle for finding the asymptotic solution of the full (basic) system can be formulated as follows: the evolution equations of the full system in terms of the amplitude factors along the correspondent rays are constructed on the basis of wave front kinematics of the associated system. Actually, beside the ray method, the method of matched separation of initial equations together with the method of stretched coordinates is used. The importance of the successive analysis must be pointed out because it also permits to solve the interaction problem between single waves and to take the possible coupling into account. Coupled waves are described by the coupled system of evolution equations. These problems have been discussed in more detail by the author earlier [7].

The general form of a two-dimensional evolution equation describing the behaviour of a single longitudinal wave is

$$\frac{\partial}{\partial\xi} R(v) = b \left(\frac{\partial^2 v}{\partial\tau_2^2} + \frac{n}{\tau_2}\frac{\partial v}{\partial\tau_2}\right) \quad (2.1)$$

where $v = v(\xi,\tau_1,\tau_2)$ is the amplitude parameter, $\tau_1 = \tau_1(X_1)$, $\tau_2 = \tau_2(X_2)$, $\xi = c_0 t - X_1$ - are the ray coordinates, $R(v)$ is an operator that describes the evolution along the axis τ_1, $b =$ $=$ const., $n = 0,1$ for plane and cylindrical wave-beams, respectively, and $c_0 = \lambda_1$ is the sound velocity. For one-dimensional processes

$$R(v) = 0 \quad (2.2)$$

holds. The operator $R(v)$ may be of various type being dependent on the properties of the medium (of Burgers, of Korteweg-de Vries, of integro-differential or of some other type). A usual form of this operator is

$$R(v) = \frac{\partial v}{\partial\tau_1} + a_{01} v \frac{\partial v}{\partial\xi} + a_0 v + a_1 \frac{\partial v}{\partial\xi} + a_2 \frac{\partial^2 v}{\partial\xi^2} + \ldots \quad (2.3)$$

where a_{01}, a_0, ... are coefficients depending on material properties and may be the functions of the coordinate X_1.

The nonlinear interaction can be taken into account introducing the ray coordinates with the corresponding phase functions ϕ_i [8,4]

$$\xi_i = c_i t - X_i - \varepsilon\phi_i(X_i,t). \quad (2.4)$$

For two one-dimensional longitudinal waves ($i = 1,2$) [4]

$$\phi_i = (-1)^{i+1}\ \frac{3}{4}\ k_1 c_0^{-1} \int_0^{\xi j} v_j d\xi;\ i,j = 1,2;\ i \neq j \tag{2.5}$$

where k_1 is the parameter of the nonlinearity.

The main rules for head-on collision and for interaction between the incident and reflected pulses in the course of reflection from an interface (boundary) are given in [7].

The coupled system of two waves is governed by a system (one-dimensional case) [4]

$$R_1(v_1, v_2) = 0, \tag{2.6a}$$

$$R_2(v_2, v_1) = 0, \tag{2.6b}$$

$$v_1 = v_1(\xi_1, \tau_1),\ v_2 = v_2(\xi_2, \tau_1),$$

$$\xi_1 = c_1 t - X_1,\ \xi_2 = c_2 t - X_1,\ \tau_1 = \tau_1(X_1)$$

where c_1 and c_2 are the corresponding velocities from the linear associated system.

3. Dispersion and the Evolution Equations

The existence of harmonic wavetrains in a dispersive medium and the correspondence between the wavenumber(s) and the frequency lead to several physical consequencies. The phase and group velocities introduced for describing the wave process give an excellent possibility to analyse the behaviour of single wavetrains. Usually the full system of equations and the evolution equations are analysed separately and often the correspondence between the various mathematical models is not clearly stated.

There are two possibilities to obtain the evolution equations for dispersive systems. First, for the weakly dispersive systems the ray coordinates are introduced using the velocities as eigenvalues of the corresponding matrix of the linear associated system [4,5]. The result is then an evolution equation with higher derivatives (KdV equation, for example) or an evolution equation of an integro-differential type [1,4]. Second, for the strongly dispersive systems the ray coordinates do not involve the usual sound velocities but the group velocities determined from the corresponding linear dispersion relation [8]. Beside it, the frequency shift and the orbit modifications are taken into account that may rise due to interaction of different modes and their higher harmonics. The best example of the corresponding evolution equation is of Schrödinger type.

Nevertheless, the evolution equations are now so widely used in many branches of physics that they form a separate class of equations investigated thoroughly by many researchers and the correspondence between the initial full system and the evolution equations is often not so clearly stated. The existence of standing waves or cutoff frequencies, for example, may be easily explained taking into account the moving frame in the nondependent variables of evolution equations. Let us start from the trivial example of the one-dimensional nondispersive wave propagation.

The usual linear wave equation

$$\frac{\partial^2 U}{\partial t^2} - c_0^2 \frac{\partial^2 U}{\partial X^2} = 0, \qquad c_0 = \text{const} \tag{3.1}$$

gives for the harmonic wavetrain

$$U = a_0 \exp(ikX - i\omega t), \; a_0 = \text{const} \tag{3.2}$$

where k is the wavenumber and ω is the frequency, the simple dispersion relation

$$\omega = \pm c_0 k. \tag{3.3}$$

The corresponding evolution equation in ray coordinates $\tau = \varepsilon X$, $\xi = c_0 t - X$ is

$$\frac{\partial v}{\partial \tau} = 0 \tag{3.4}$$

where $v = \partial U_0/\partial t$ is the amplitude parameter, ε is a certain small parameter [4] and the subscript denotes the approximation number. The phase, written in new coordinates is

$$i(kX - \omega t) = i(-k_* \xi + \omega_* \tau) \tag{3.5}$$

with

$$k_* = \omega c_0^{-1}, \tag{3.6}$$

$$\omega_* = k\varepsilon^{-1} - \omega\varepsilon^{-1}c_0^{-1}. \tag{3.7}$$

Assuming now

$$v = v_0 \exp(-ik_*\xi + i\omega_*\tau) \tag{3.8}$$

the dispersion relation yields

$$\omega_* = 0. \tag{3.9}$$

This result giving zero phase and group velocities is certainly not trivial but involves according to (3.7) $\omega = c_0k$, i.e. the simplest evolution equation (3.4) describes the connected wave. Usually such a problem is considered for a wavetrain with respect to a Galilean coordinate system $X' = X + Vt$ moving with the velocity V [9]. Then

$$\omega_* = \omega - kV \tag{3.10}$$

giving the magnitude of the Doppler frequency shift. If $V = c_g$ then this frequency is termed as the slip frequency [9]. For the trivial case considered here the Galilean transformation and the ray coordinate transformation, in principle, coincide but for the more complicated cases (i) the asymptotical character of the evolution equation and (ii) the character of the transformation (either space-space or space-time kind [4]) must be taken into account. In any case, the observer constrained to a particular wavefront could travel with this velocity and maintain his position when (3.6) and (3.7) hold.

If the dispersive medium is considered then the equation of motion

$$\frac{\partial^2 U}{\partial t^2} - c_0^2 \frac{\partial^2 U}{\partial X^2} - c_0^2\ell_0^2 \frac{\partial^4 U}{\partial X^4} = 0, \quad \ell_0^2 = \text{const} \tag{3.11}$$

gives

$$\omega = \pm c_0k(1 - \ell_0^2k^2)^{1/2}. \tag{3.12}$$

The corresponding evolution equation is the linearized KdV

equation

$$\frac{\partial v}{\partial \tau} - \frac{\ell_0^2}{2\varepsilon} \frac{\partial^3 v}{\partial \xi^3} = 0. \tag{3.13}$$

Assuming now the solution in the form (3.8) the dispersion equation

$$\omega_* = \frac{1}{2} k_*^3 \ell_0^2 \varepsilon^{-1} \tag{3.14}$$

holds. Bearing in mind the asymptotic procedure for constructing the evolution equation and the formulae (3.6), (3.7) the dispersion relation (3.12) yields easily from (3.14) when ℓ_0^2 is considered as a small parameter. If suppose (3.12) and (3.14) being not related the group velocities found separately

$$c_g^* = \frac{\partial \omega_*}{\partial k_*} = \frac{3}{2} k_*^2 \ell_0^2 \varepsilon^{-1}, \tag{3.15}$$

$$c_g = \frac{\partial \omega}{\partial k} = c_0(1 - 2 \ell_0^2 k^2)(1 - \ell_0^2 k^2)^{-1/2} \tag{3.16}$$

are completely different.

The interesting situation arises for two-dimensional processes described by the evolution equation (2.1). The dispersion relations for two-dimensional KdV equation were obtained and analysed by Engelbrecht [10] in terms of ray coordinates. Here some results of this research will be modified to analyse the physical side of the process. Let the equation (2.1) be of the form

$$\frac{\partial}{\partial \xi}\left(\frac{\partial v}{\partial \tau_1} + \frac{k_1}{2\varepsilon c_0} v \frac{\partial v}{\partial \xi} + \frac{m\ell_0^2}{2\varepsilon} \frac{\partial^3 v}{\partial \xi^3}\right) =$$

$$= n \frac{c_2^2}{c_0^2}\left(\frac{\partial^2 v}{\partial \tau_2^2} + \frac{1}{\tau_2} \frac{\partial v}{\partial \tau_2}\right) \tag{3.17}$$

where k_1 is the parameter of nonlinearity, c_2 is the velocity of transverse waves and $m = \pm 1$ indicates the character of the dispersion (positive or negative). The coefficient $n = 1$ gives the Zabolotskaya-Khokhlov equation for wave-beams and $n = -1$ - the Kadomtsev-Petviashvili equation for shallow water. Assuming now the solution

$$v = \varepsilon v_0(\Theta)Y_0(\tau_2) + \varepsilon^2 v_1(\Theta)Y_1(\tau_2) + \ldots,$$

$$\Theta = -ik_*\xi + i\omega_*\tau_1 \tag{3.18}$$

the dispersion relation for the linear problem is

$$\omega_* = a^2nc_2^2c_0^{-2}k_*^{-1} - \frac{1}{2}\,m\ell_0^2\varepsilon^{-1}k_*^3 \tag{3.19}$$

where a is the parameter of the transversal distribution given by

$$Y_0 = I_0(a\tau_2). \tag{3.20}$$

Here I_0 is the Bessel function of the zeroth order. The next approximation for $v_1(\Theta)$ is easily obtained by the perturbation scheme [10].

In the terms of the initial coordinates the relation (3.19) yields

$$k = \frac{\omega}{c_0} - \frac{\varepsilon a^2nc_2^2}{c_0\omega} - \frac{m\ell_0^2\omega^3}{2c_0^3}\,. \tag{3.21}$$

For small deviations from the one-dimensional nondispersive case, it corresponds approximately to

$$\omega = c_0k\,\left(1 - \frac{1}{2}\,m\ell_0^2k^2 - \varepsilon na^2c_2^2c_0^{-2}k^{-1}\right)^{-1}. \tag{3.22}$$

The influence of the transverse distribution is essential for small k's, i.e. for long waves. The cases $\omega_* = 0$, $\partial\omega_*/\partial k_* = 0$ follow from (3.19) for certain values of n and m [10]. For example, if

$$k_* = \left(-\,\frac{2\varepsilon a^2nc_2^2}{mc_0^2\ell_0^2}\right)^{1/4}, \tag{3.23}$$

then $\omega_* = 0$. Substituting $\omega_* = 0$ into the relation (3.7) we get $\omega = c_0k \neq 0$. Consequently, taking into account (3.6), for this particular case $k = k_*$ holds. In other words, the slip frequency ω_* is zero and the waves described by the dispersion relation (3.21) are nonslipping.

The discussion about the dispersion in evolution equations was started by Jeffrey and Kakutani [11] who have made use of phase velocities determined from the linear dispersion relations.

4. Active Medium and Evolution Equations

In recent years much attention has been paid to wave propagation in media where the energy is restored by other physical processes. These are, for example, the processes in nerve fibres, in chemical systems and in heat transfer systems [12]. The main mathematical model discussed thoroughly is based on the nonlinear diffusion equation proposed by Kolmogorov, Petrovsky and Piscunov [13] and travelling wave solutions are sought. These waves are usually termed as self-similar [12].

Despite of the tremendous success in the theory of self-similar waves there are certain shortcomings in it due to the properties of the widely used mathematical models. The best example is connected with the celebrated Hodgkin-Huxley model for the pulse transmission in a nerve fibre. The convergent solution of this parabolic model can be obtained for one, critical, propagation rate [14] determined with the correctness of 5-6 digits [15]. This physically uncorrect situation can be avoided using a more perfect model. In this case the hyperbolic transmission line model introduced by Lieberstein [16] and by Engelbrecht [17] gives the possibility to describe the wave propagation without such a contradictive property caused by the infinite velocity of the parabolic model. Another example of an infinite velocity is in the heat conduction. However, the more complete model involving the memory effects removes the paradox and there are certain consequences in wave propagation including changes of phase velocities and attenuation coefficients [18]. Both the examples can be described in mathematical terms as follows: "When truncation introduces degeneracies it is only prudent to try putting the higher terms back" (Poston and Stewart [19]).

In connection with the evolution equations the most important question is the following: as far as the evolution equations describe the single wave distortion best what is the role of evolution equations for processes in active media? To answer this question we shall start from the simple considerations.

The basic one-dimensional model for a two-component active medium with energy restoring is

$$\varepsilon_1 \frac{\partial U}{\partial t} = D_1 \frac{\partial^2 U}{\partial X^2} + F(U,V), \qquad (4.1a)$$

$$\frac{\partial V}{\partial t} = \varphi(U,V) \qquad (4.1b)$$

where ε_1 = const, D_1 = const. The active properties of the medium are described by the nonlinear function $F(U,V)$. Usually $\varepsilon_1 << 1$ [12]. If $F(U,V) = F(U)$, i.e. an one-component medium is considered, the model (4.1a) is able to describe only the monotonously changing wave profile. The main part of the equation (4.1a) is, however, of parabolic character and the question of its dispersion properties arises. If the hyperbolicity (the final velocity of the wave process under consideration) is taken into account then the equation (4.1a) may be rewritten in the form

$$\frac{\partial^2 U}{\partial t^2} + \varepsilon_1 \frac{\partial U}{\partial t} = D_1 \frac{\partial^2 U}{\partial X^2} + F(U,V). \qquad (4.2)$$

Here $D_1^{1/2} = c_0$ has the dimension of the velocity. The wave-like behaviour of the equation (4.2) is clear but due to the second term in the left side the attenuation is obvious. If $F(U,V) = 0$, then the only possible wave-like solution described by (4.1a) is given in Fig. 1a while the wave profile described by (4.2) is arbitrary (see Fig. 1b) dictated first by arbitrary boundary

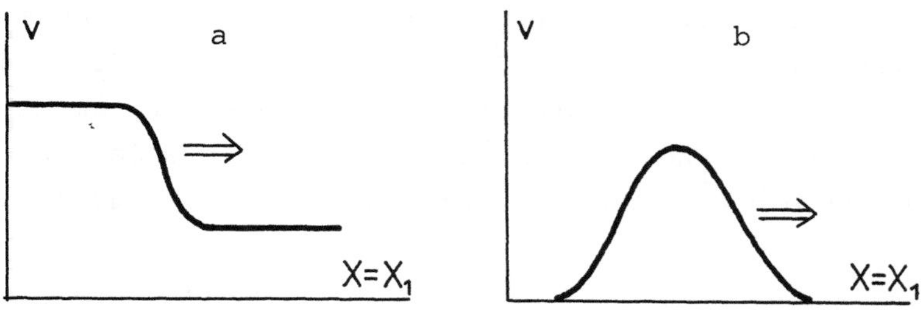

Fig. 1

conditions and then distorted due to the properties of the medium (if $c_0 \neq$ const, for example). The dispersion properties of such models are completely different. Let us consider (i) the equation (4.1a), (ii) the equation (4.2), and (iii) the equation (4.2) with $\varepsilon_1 = 0$ that corresponds to the purely hyperbolic case. If, for example, $F(U,V) = F(U) = bU$, b = const then the dispersion relations are

$$\omega = -\varepsilon_1^{-1} i (D_1 k^2 - b), \tag{4.3a}$$

$$\omega = -\frac{1}{2} i \pm \left[-\frac{1}{4} + \varepsilon_1^{-1} (D_1 k^2 - b) \right]^{1/2}, \tag{4.3b}$$

$$\omega = \pm \left[\varepsilon_1^{-1} (D_1 k^2 - b) \right]^{1/2}, \tag{4.3c}$$

respectively. The case (i) leads to the imaginary frequency, the case (ii) to the complex frequency with the constant (nondependent of k) imaginary part and the case (iii) to the real frequency if $D_1 k^2 > b$ holds. Only for very small k the case (iii) gives the complex frequency. If $F(U)$ is a nonlinear function the situation is much more complicated. The steady waves $U = f(ct - X) = f(\xi)$ for the same cases are described by the equations

$$D_1 \frac{d^2 f}{d\xi^2} - \varepsilon_1 c \frac{df}{d\xi} + F(f) = 0, \tag{4.4a}$$

$$(D_1 - c^2) \frac{d^2 f}{d\xi^2} - \varepsilon_1 c \frac{df}{d\xi} + F(f) = 0, \tag{4.4b}$$

$$(D_1 - c^2) \frac{d^2 f}{d\xi^2} + F(f) = 0, \tag{4.4c}$$

respectively. The standard model for self-similar waves [12] leads to the oscillation equation while the other models contain more possibilities. If $D_1 \neq c^2$ then the equation (4.4b) describes the similar oscillatory movement but with another, smaller "damping" coefficient. The equation (4.4c) is the standard oscillation equation. If $D = c^2$ then the equations (4.4b) and (4.4c) are simplified

$$-\varepsilon_1 c \frac{df}{d\xi} + F(f) = 0, \tag{4.5a}$$

$$F(f) = 0, \tag{4.5b}$$

respectively. Consequently, the standard hyperbolic model permits then only a special constant solution while the case (ii) gives the solution

$$\xi = \xi_0 + \varepsilon_1 c \int \frac{df}{F(f)} + C_1, \quad C_1 = \text{const} \tag{4.6}$$

in the implicit form.

The first step to introduce the scheme elaborated for the quasi--linear systems is to choose the ray coordinates according to the velocities of the linear associated system.

It must be pointed out that there is no possibility to determine any such a velocity from the equation (4.1a) (case (i)). Nevertheless, let us introduce the ray coordinates

$$\xi = c_0 t - X, \quad \tau = \varepsilon X, \quad \varepsilon \sim \varepsilon_1, \quad c_0 = D_1^{1/2} . \tag{4.7}$$

Let $F(U,V) = 0$. Then the equation (4.1a) yields

$$\frac{\partial v}{\partial \xi} = 0 \tag{4.8}$$

and the equation (4.2)

$$\frac{\partial v}{\partial \tau} + \frac{1}{2} c_0^{-1} v = 0. \tag{4.9}$$

Obviously the evolution equation (4.8) describes the steady profile (see Fig. 1a) but the equation (4.9) does not. In the latter case the profile is influenced by the attenuation but the wave profile is dictated by the initial conditions. This is the dynamical sidé of the process and the energy restoring is governed by another equation for the dependent parameter V (see, for example, the equation (4.1b)). Consequently, the initial model (4.1a) allows only one possible profile while another model (4.2) describes a large class of wave profiles. The obvious question arises now - is not the model (4.1a) rather stiff describing only one possible profile? If yes then it is natural to use the model (4.2) giving more freedom to the dynamical side of the process while the active medium equation governs this or another way the energy restoring. Anyhow, although the answer is not clear yet, the serious truncation errors [14] will be avoided under such a scheme. Let now $F(U,V) \neq 0$. It follows from (4.2) and (4.1b)

$$\frac{\partial}{\partial \xi}\left(2c_0 \frac{\partial v}{\partial \tau} + v\right) = \frac{1}{\varepsilon c_0} F(v,V), \tag{4.10a}$$

$$c_0 \frac{\partial V}{\partial \xi} = \varphi(v,V), \tag{4.10b}$$

respectively, giving the description of the wave process in terms of evolution equations. The first advantage of this procedure is that the boundary value problem for the initial equation has been transformed into the Cauchy problem of the system (4.10). Obviously under the condition $F(U,V) = 0$ the equation (4.9) yields from (4.10a) when suitable conditions at infinity are taken into account. It is actually the one-dimensional evolution equation that according to the scheme in Section 2 may be rewritten in the short form $R(v) = 0$. The influence of the active medium is similar to that of the dimensionality of the process. Exactly as in this case, $R(v) \neq 0$ any more but the derivative of $R(v)$ with respect to the coordinate ξ (measured from the wavefront) is equal to a function describing other than dynamical properties of the medium. In case of a two-dimensional wavebeam it is the diffractional expansion that must be taken into account (see the equation (2.1)). In case of the active medium it is the role of the additional processes in the medium. Such an equation must be of general interest for mathematicians, because its main properties could explain the characteristic features of wave distortion in complicated media. However, it must be pointed out that the equation (4.10a) was deduced from the initial equation of the second order. When the initial system has got another form, especially $F(U,V)$ vrs derivatives on the lefthand-side, then the result may be different (see further).

The pulse transmission in a nerve fibre is considered further in order to demonstrate the application of the evolution equation. There have been several attempts to overcome the shortcomings of the classical theory. One attempt was made by Zeeman [20] using the concepts of the catastrophe theory. His approach was criticized by Sussmann and Zahler [21]. Later Stewart and Woodcock [22] gave a detailed analysis of Zeeman's model and proved the critical statements [21] to be largely erroneous, but they pointed out some other weak points of his model. However, the question about the bounded solutions depending critically on the values of the velocity of a pulse remains open in this debate. Stewart and Woodcock suggested [22] to use Lieberstein's approach based on the hyperbolic cable equations [16] together with Zeeman's model. We shall use here the approach suggested earlier [17] for such a hyperbolic system to obtain an evolution equation for such a process. Beside the usual nonlinearity of the ion current, the nonlinearity in the transmission line itself is taken into account. The starting point is the system of equations given by Lieberstein [16] and used by Engelbrecht [17] but the following change is introduced: the charge q is not given by the usual linear relation but in the form

$$q = \pi a^2 (C_a v + \varepsilon_1 D_a v^2) \tag{4.11}$$

where a is the radius of an axon, C_a is its self-capacitance, v is the voltage across the membrane, ε_1 is the small parameter and D_a = const. Such a relation was proposed by Khokhlov [23] for electromagnetic shock waves. The main set of the equations reads now

$$\frac{\partial v}{\partial t} + \frac{1}{\pi a^2 C_a} \frac{\partial i_a}{\partial X} + \frac{1}{\pi a^2 C_a} j_i + \frac{2D_a \varepsilon_1}{C_a} v \frac{\partial v}{\partial t} = 0, \tag{4.12a}$$

$$\frac{\partial i_a}{\partial t} + \frac{\pi a^2}{L} \frac{\partial v}{\partial X} + \frac{r\pi a^2}{L} i_a = 0. \tag{4.12b}$$

Here i_a is the total longitudinal current flowing inside the membrane, r is the resistance, L is the axon specific self-inductance, j_i is the ion current. According to Khokhlov [23]

$$\varepsilon_1 D_a v << C_a \tag{4.13}$$

holds usually showing that the variable part of the self-capacitance is considerably smaller as compared to its constant part. The traditional second order equation for v is easily deduced from (4.12) consisting however the nonlinear terms. Further, the system of the first order equations is considered. The details of the ray method technique [4] that were also briefly described in [17] are not repeated here. The ray coordinates used for constructing the evolution equation are chosen

$$\xi = c_0 t - X, \quad \tau = \varepsilon X \tag{4.14}$$

where $c_0 = [a(2C_m L)^{-1}]^{1/2}$ and ε is a small parameter not necessarily being ε_1. However $\mathcal{O}(\varepsilon) \sim \mathcal{O}(\varepsilon_1)$ is assumed. Here C_m is the membrane capacitance. Following the scheme given in [17] the evolution equation is obtained in the following form:

$$\frac{\partial v_0}{\partial X} + a_{01} v_0 \frac{\partial v_0}{\partial \xi} + a_0 v_0 + b_0 j_i(v_0) = 0, \tag{4.15}$$

$$a_{01} = \varepsilon_1 D_a C_a^{-1}, \quad a_0 = \frac{1}{2} r\pi a(2aC_m L^{-1})^{1/2},$$

$$b_0 = (4\pi a C_m)^{-1}(2C_m L a^{-1})^{1/2}$$

where the amplitude parameter v_0 gives the voltage v in the first approximation. The difference between the previous model [17] and (4.15) is in the additional second term. As it is known from the wave mechanics [1] such a nonlinearity together with the dispersion mechanism may give rise to the stable wave profiles such as solitary waves, for example. As far as in a nerve fibre the information carrier is wave of a solitary type it is most intriguing to investigate the possibility of such cross--effects.

According to FitzHugh-Nagumo model [15]

$$j_i = 2\pi a[F(v_0) + R], \tag{4.16a}$$

$$F(v_0) = k_1 v_0 + k_3 v_0^3, \quad \frac{\partial R}{\partial t} = q_0(v_0 + q_1 + q_2 v_0) \tag{4.16b}$$

where k_1, k_3, q_0, q_1, q_2 are constants. The governing equation now takes the form

$$\frac{\partial^2 v_0}{\partial \xi \partial X} + C_1^2 \left(\frac{\partial v_0}{\partial \xi}\right)^2 + C_{02} v_0 \frac{\partial^2 v_0}{\partial \xi^2} + C_{01} v_0 \frac{\partial v_0}{\partial \xi} +$$

$$+ C_{21} v_0^2 \frac{\partial v_0}{\partial \xi} + C_1 \frac{\partial v_0}{\partial \xi} + D_1 \frac{\partial v_0}{\partial X} + C_0 v_0 + C_0^3 v_0^3 + C^0 = 0. \tag{4.17}$$

Here C_i, C_i^j, C_{ij}, D_i are constants. A simplified approach assuming $q_2 = 0$ gives

$$\frac{\partial^2 v_0}{\partial X \partial \xi} + C_1^2 \left(\frac{\partial v_0}{\partial \xi}\right)^2 + C_{02} v_0 \frac{\partial^2 v_0}{\partial \xi^2} + C_{21} v_0^2 \frac{\partial v_0}{\partial \xi} +$$

$$+ C_1 \frac{\partial v_0}{\partial \xi} + C_0 v_0 + C^0 = 0. \qquad (4.18)$$

The second and the third terms in this equation are due to the dynamical nonlinearity. We look for a steady-wave solution of (4.18) putting $v_0 = v_0(\eta)$, $\eta = X - \Theta\xi$. Here Θ is a dimensionless quantity that determines the change of the velocity. Note that actually $\eta = (1 - \Theta)X - \Theta c_0 t$.

We obtain

$$v_0'' (1 - C_{02}\Theta v_0) + (C_1 + C_{21} v_0^2 - C_1^2 \Theta^2 v_0') v_0' -$$

$$- C_0 \Theta^{-1} v_0 - C_0 \Theta^{-1} = 0. \qquad (4.19)$$

It is the Rayleigh equation [24] where the prime denotes the differentiation with respect to η. Its qualitative analysis shows the existence of the periodical solutions [24]. If the dynamical nonlinearity of (4.11) type is neglected, then the Liénard equation follows from (4.19). In both cases the "velocity" Θ is the unknown quantity determining actually the balance between the steepening of the wave-profile and the active processes in the medium. The situation is the same for the solitary waves modelled by the celebrated KdV equation where the balance between nonlinearity and the dispersion give rise to the soliton-type solutions.

5. Two-dimensional Evolution Equations

The two-dimensional evolution equations were analysed earlier by several authors, for example [2,4,25]. The solution of the nonlinear two-dimensional evolution equations (2.1) is a rather difficult mathematical problem [2,4], especially for arbitrary initial conditions. However, the semiself-similar solution may be constructed rather easily [25] giving a suitable description for ultra-scund wavebeams. The most universal methods are based on the numerical integration mainly using the method of finite differences. Recently for this purpose the procedure combining the universality of the method of the finite differences with the high accuracy of the Fast Fourier Transformation scheme [26] was proposed by T.Peipman. Let the operator R have the form

$$R = \frac{\partial v}{\partial \tau_1} + a_{01} v \frac{\partial v}{\partial \xi} + a_2 \frac{\partial^2 v}{\partial \xi^2} . \tag{5.1}$$

The Fourier transforms of the function v are given as $F_1 v$ and $F_2 v$ with respect of the variables ξ and τ_2, respectively, while the origins are determined as $F_1^{-1} v$ and $F_2^{-1} v$. The derivatives of v can be evaluated according to formula

$$\frac{\partial v}{\partial \xi} = F_1^{-1}(i\omega F_1 v), \tag{5.2a}$$

$$\frac{\partial^2 v}{\partial \xi^2} = -F_1^{-1}(\omega^2 F_1 v), \ldots \tag{5.2b}$$

where ω is the discrete spectrum. The final algorithm for finding the values of $\partial v/\partial \xi$ on the next layer $\tau_1 + \Delta\tau_1$ elaborated for the equation (2.1) with n = 0 and with the operator (5.1) is following:

$$\begin{aligned} &\partial v(\tau_1 + \Delta\tau_1, \tau_2, \xi)/\partial \xi - \partial v(\tau_1, \tau_2, \xi)/\partial \xi - \\ &- a_{01}\Delta\tau_1 v F_1^{-1}(\omega^2 F_1 v) + a_{01}\Delta\tau_1 [iF_1^{-1}(\omega F_1 v)]^2 + \\ &+ a_2 \Delta\tau_1 i F_1^{-1}(\omega^3 F_1 v) = -b\Delta\tau_1 F_2^{-1}(\omega^2 F_2 v). \end{aligned} \tag{5.3}$$

Similar algorithms for other types of operator R are easily deduced. As far as the convolution integrals in the Fourier space are calculated by simple multiplication, this approach is rather convenient for the operators R of the integro-differential type. The correctness of such an approach is demonstrated on the test problems.

References

1. Whitham, G.B.: Linear and nonlinear waves. New-York e.a.: John Wiley & Sons 1974.

2. Rudenko, O.V.; Soluyan, S.I.: Theoretical foundations of nonlinear acoustics. New-York: Plenum Press 1977.

3. Karpman, V.J.: Non-linear waves in dispersive media. Moscow: Nauka 1973 (in Russian).

4. Engelbrecht, J.K.; Nigul, U.K.: Nonlinear deformation waves. Moscow: Nauka 1981 (in Russian).

5. Taniuti, T.; Wei, C.C.: Reductive perturbation method in nonlinear wave propagation, I.J. Phys. Soc. Japan 24 (1968) 941-946.

6. Miura, R.M.: The Korteweg-de Vries equation and generalization. I: A remarkable explicit nonlinear transformation. Math. Phys. 9 (1968) 1202-1204.

7. Engelbrecht, J.: Theory of non-linear wave propagation with application to the interaction and inverse problems. Int. J. Non-Linear Mech. 12 (1977) 189-201.

8. Oikawa, M.; Yajima, N.: Generalization of the reductive perturbation method to multiwave systems. Progr. Theor. Phys. Suppl. No. 55 (1975) 36-51.

9. Leibowich, S.; Seebass, A.R. (eds.): Nonlinear waves. Ithaca, London: Cornell University Press 1974.

10. Engelbrecht, J.: On dispersion in two-dimensional wave processes. Acta et comm. Univ. Tartuensis, 564 (1981) 56-62 (in Russian).

11. Jeffrey, A.; Kakutani, T.: Weak nonlinear dispersive waves: A discussion centered around the Korteweg-de Vries equation. SIAM Rev. 14 (1972) 582-643.

12. Grekhova M.T. e.a. (eds.): Self-similar processes in diffusive systems. Gorky: Inst. of Appl. Phys. 1981.

13. Kolmogorov, A.N.; Petrovsky, I.G.; Piscounov, N.S.: Étude de l'équation de la diffusion avec croissance de la quantité de matière et son application á un problème biologique. Bull. Univ. Moscou (ser. int.) A1 (1937) 1-25.

14. Scott, A.C.: Neurophysics. New-York e.a.: John Wiley & Sons 1977.

15. Nagumo, J.; Arimoto, S.; Yoshizawa, S.: An active pulse transmission line simulating nerve axon. Proc. IRE, 50 (1962) 2061-2070.

16. Lieberstein, H.M.: On the Hodgkin-Huxley partial differential equation. Math. Biosci., 1 (1967) 45-69.

17. Engelbrecht, J.: On theory of pulse transmission in a nerve fibre. Proc. R. Soc. London, A 375 (1981) 195-209.

18. Engelbrecht, J.: The modes of the one-dimensional waves in a thermoelastic medium with the finite speed of heat. Proc. Estonian Acad. Sci. Phys.-Math., 22 (1973) 188-195 (in Russian).

19. Poston, T.; Stewart, I.N.: Catastrophe theory and its applications. London: Pitman 1978.

20. Zeeman, E.C.: Differential equations for the heartbeat and nerve impulse. In: Towards a theoretical biology (ed. by C.H. Waddington). Edinburgh University Press, 4 (1972) 8-67.

21. Sussman, J.H.; Zahler, R.S.: Catastrophe theory as applied to the social and biological sciences: critique. Synthese, 37 (1978) 117-126.

22. Stewart, I.; Woodcock, A.E.R.: On Zeeman's equations for the nerve impulse. Bull. Math. Biol. 43 (1981) 279-325.

23. Khokhlov, R.V.: On the theory of shock radiowaves in non-linear transmission lines. Proc. Univ. USSR: Radiotechnique and Electronics, No. 6 (1961) 917-925.

24. Reissig, R.; Sansone, G.; Conti, R.: Qualitative Theorie nichtlinearer Differentialgleichungen. Roma: Edizioni Cremonese 1963.

25. Engelbrecht, J.: Two-dimensional non-linear evolution equations: the derivation and the transient wave solution. Int. J. Non-Linear Mech. 16 (1981) 199-212.

26. Fornberg, B.; Whitham, G.B.: A numerical and theoretical study of certain nonlinear wave phenomena. Phil. Trans. Soc. London, A 289 (1978) 373-404.

An Approach to Constrained Thermoelastic Solids: Internal Constraint Responses and Wave Propagation

A. DONATO & D. FUSCO

Istituto di Matematica
Università di Messina
Via C. Battisti 90
98100 Messina, Italy

Summary

Thermomechanical restrictions on thermodynamic processes in a thermoelastic solid are studied within the thermodynamical framework proposed by I. Müller. By means of suitable assumptions on the form of the constitutive relations of the material, a class of constraints responses is characterized. Moreover the main features of wave propagation in constrained thermomechanical media are pointed out.

1. Introduction

In the last years some attention has been paid to work out a systematical study of thermomechanically constrained materials. Many Authors have elaborated on this subject by considering both different types of constraint equations and different methods of approach to characterize the constraint responses [1,...,5]. Particularly in [3] there have been studied thermomechanical restrictions on thermodynamic processes when the constraint is a scalar equation involving the present values of the strain, the absolute temperature and its gradient; such a type of constraint generalizes a classical one involving strain and absolute temperature only [6]. Then a constitutive theory of thermomechanically simple materials is obtained by using an extended formulation of the well known Truesdell & Noll's determinism principle leading to the mechanical theory of constrained simple materials [7]. In this paper an attempt is made in order to characterize a constitutive theory of thermoelastic constrained solids within the framework of the thermodynamical theory proposed by I. Müller [8]. The implications of such a theory (e. g. second sound effects, etc.) are well known and on this subject we refer to [9], [10], [11]. Let us stress here that in Müller's thermodynamical framework the system of

Nonlinear Deformation Waves
IUTAM Symposium Tallinn 1982
Editors: U. Nigul, J. Engelbrecht

the governing field equations may be hyperbolic [12],[13] and that only present values of the field quantities are involved in the constitutive equations All that allows one to study wave propagation problems in continuum thermodynamics in a more handful manner than other thermodynamical theories. Specifically for what concerns with the argument studied in the present paper, in Müller's theory context one can easy get the main features of wave propagation in thermomechanically constrained continua.

Let us briefly outline the contents of the paper. In Sect. 2 we consider a thermoelastic body [8] where thermomechanical restrictions are imposed on thermodynamic processes by a scalar equation involving the present values of the strain, the empirical temperature, the gradient and (time) rate of temperature. Such a type of constraint generalizes to the present framework the one considered in [3] . Thus by means of suitable assumptions on the form of the constitutive relations of the material, it is possible to characterize a class of constraint responses. Sect. 3 is devoted to point out the main features of wave propagation in a constrained thermoelastic body according to the general constitutive theory worked out in Sect. 2. First of all we state that, in general, wave propagation is not possible in thermomechanical media suffering more than three constraints. Moreover it is pointed out that, if in the constraint equation the temperature rate is involved, then it may exist a wave speed purely due to the constraint. However such a wave speed cannot exist in the one-dimensional case. Other implications are shortly discussed and propagation of " generalized longitudinal " [14] acceleration waves is considered.

In closing lét us recall that several results are known about acceleration waves in constrained elastic materials,(see [15] and refs. there quoted), and that wave propagation in some classical cases of thermomechanically constrained materials has been studied in [6] , [16] , [17] .

2. Constitutive theory for a constrained thermoelastic solid

If the body force (specific) and supply density of energy are assumed both vanishing, the standard balance laws of continuum thermodynamics, i. e. momentum, energy and mass conservation equations in material form are:

$$\rho_o \partial_t v_i - \partial_A T_{iA} = 0$$

$$\rho_o \partial_t (e + v^2/2) + \partial_A (q^A - v_i T_{iA}) = 0 \qquad (2.1)$$

$$\rho_o = \rho J$$

where t is time coordinate; $x_i = x_i(X_A, t)$ and X_A are, respectively, the spatial (current) coordinates and reference coordinates of a particle; ρ_o is the reference density; $v_i = \partial_t x_i(X_A, t)$; T_{iA} represents the first Piola-Kirchoff stress tensor; e denotes the specific internal energy; q^A is the heat flux; $J = \det \| F_{iA} \|$, $F_{iA} = \partial_A x_i$; $\partial_t = \partial / \partial t$, $\partial_A = \partial / \partial X_A$. Moreover throughout the paper lower case and upper case Latin indices take the values 1, 2, 3. We also employ the usual summation convention over repeated indices.

According to [8] , requiring the objectivity principle to be satisfied, the constitutive equations which we assume to characterize a homogeneous thermoelastic body are:

$$T_{AB} = T_{AB}(C_{CD}, \theta, \dot{\theta}, \theta_C)$$

$$q^A = q^A(C_{CD}, \theta, \dot{\theta}, \theta_C) \qquad (2.2)$$

$$e = e(C_{CD}, \theta, \dot{\theta}, \theta_C)$$

where $C_{AB} = F_{iA} F_{iB}$, $T_{iA} = T_{AB} F_{iB}$, $\dot{\theta} = \partial_t \theta$, $\theta_A = \partial_A \theta$, and $\theta = \theta(X_A, t)$ is the empirical temperature.

Restrictions on the response functions (2.2) are imposed by the entropy principle which we assume in the form proposed by I. Müller [8] , i. e.

$$\rho_o \partial_t \eta + \partial_A h^A \geqslant 0 \qquad (2.3)$$

where the specific entropy η and the entropy flux h^A are characterized by constitutive relations similar to (2.2). Let us notice, (see [18]), that the system (2.1) may be written as a symmetric system [19] which is also hyperbolic provided that $-\eta$ is a convex function of the field variables $\rho_o v_i$, F_{iA}, $\varepsilon = \rho_o(e + v^2/2)$; θ_A, θ .

We introduce thermomechanical constraints of the form

$$\gamma(F_{iA}, \theta, \dot{\theta}, \theta_A) = 0 \qquad (2.4)$$

By requiring the validity of the material frame-indifference principle for the constraint equation, (2.4) specializes to

$$\gamma(C_{AB}, \theta, \dot{\theta}, \theta_A) = 0 \tag{2.5}$$

A thermodynamic process [8], characterized by the quantities $x_i(X_A, t)$, θ, is said to be admissible in a constrained thermoelastic body if it obeys the balance equations (2.1), the constitutive relations (2.2) and the constraint equation (2.5).

By introducing suitable Lagrange multipliers as in [20], the following inequality must be satisfied

$$\Gamma^i\{\rho_o \partial_t v_i - \partial_A T_{iA}\} + \Gamma\{\rho_o \partial_t (e + v^2/2) + \partial_A (q^A - v_i T_{iA})\} + \nu \partial_t \gamma +$$

$$+ \nu^A \partial_A \gamma - \rho_o \partial_t \eta - \partial_A h^A \leqslant 0 . \tag{2.6}$$

Requiring the inequality (2.6) to hold for arbitrary values of v_i, F_{iA}, θ, $\dot{\theta}$, θ_A and any derivative of them, we get in the usual way, (see [20]) the following relations

$$\Gamma^i + \Gamma v_i = 0 \tag{2.7}$$

$$\Gamma\left(\rho_o \frac{\partial e}{\partial C_{AB}} - \frac{1}{2} T_{AB}\right) + \nu \frac{\partial \gamma}{\partial C_{AB}} - \rho_o \frac{\partial \eta}{\partial C_{AB}} = 0 \tag{2.8}$$

$$\rho_o\left(\Gamma \frac{\partial e}{\partial \theta_A} - \frac{\partial \eta}{\partial \theta_A}\right) + \Gamma \frac{\partial q^A}{\partial \dot{\theta}} - \frac{\partial h^A}{\partial \dot{\theta}} + \nu \frac{\partial \gamma}{\partial \theta_A} + \nu^A \frac{\partial \gamma}{\partial \dot{\theta}} = 0 \tag{2.9}$$

$$\rho_o\left(\Gamma \frac{\partial e}{\partial \dot{\theta}} - \frac{\partial h}{\partial \dot{\theta}}\right) + \nu \frac{\partial \gamma}{\partial \dot{\theta}} = 0 \tag{2.10}$$

$$\Gamma \frac{\partial q^A}{\partial \theta_B} + \nu^A \frac{\partial \gamma}{\partial \theta_B} - \frac{\partial h^A}{\partial \theta_B} = \omega_{[AB]} \text{ [1]} \tag{2.11}$$

$$\Gamma \frac{\partial q^A}{\partial C_{LB}} + \nu^A \frac{\partial \gamma}{\partial C_{LB}} - \frac{\partial h^A}{\partial C_{LB}} = \Omega^{A]}_{L[B} \tag{2.12}$$

$$\left(\Gamma \frac{\partial q^A}{\partial \theta} + \nu^A \frac{\partial \gamma}{\partial \theta} - \frac{\partial h^A}{\partial \theta}\right)\theta_A + \left\{\nu \frac{\partial \gamma}{\partial \theta} + \rho_o\left(\Gamma \frac{\partial e}{\partial \theta} - \frac{\partial \eta}{\partial \theta}\right)\right\}\dot{\theta} \leqslant 0 \tag{2.13}$$

---

[1] Indices in square brackets indicate that a tensor is skew symmetric with respect to these indices; $\omega_{[AB]}$ and $\Omega^{A]}_{L[B}$ are completely arbitrary tensors which appear because of the symmetries $\partial_A \theta_B = \partial_B \theta_A$; $\partial_A F_{iB} = \partial_B F_{iA}$.

In order to characterize the constraint responses we require the validity of the following relation between the heat flux and the entropy flux

$$h^A = \Gamma(\theta, \dot{\theta})\, q^A \tag{2.14}$$

holding in the unconstrained thermomechanical case for an isotropic thermoelastic body [8] . In addition to (2.14) we require Γ to be independent on the constraint, so, as in Müller's theory, it remains an universal function which at the equilibrium reduces to the inverse of absolute temperature.
By making use of the assumption (2.14), from (2.11) it easily follows that $\nu^A = 0$. Then, by introducing as in the unconstrained thermomechanical case [18] the function $L = \rho_o(\Gamma e - \eta)$, from (2.8), (2.9), (2.10) we get

$$T_{AB} = \bar{T}_{AB} + \nu\tilde{T}_{AB} \tag{2.15}$$

$$q^A = \bar{q}^A + \nu\tilde{q}^A \tag{2.16}$$

$$e = \bar{e} + \nu\tilde{e} \tag{2.17}$$

where $\bar{T}_{AB} = \frac{2}{\Gamma}\frac{\partial L}{\partial C_{AB}}$, $\tilde{T}_{AB} = \frac{2}{\Gamma}\frac{\partial \gamma}{\partial C_{AB}}$, $\bar{q}^A = (\frac{\partial \Gamma}{\partial \dot{\theta}})^{-1}\frac{\partial L}{\partial \theta_A}$, $\tilde{q}^A = (\frac{\partial \Gamma}{\partial \dot{\theta}})^{-1}\frac{\partial \gamma}{\partial \theta_A}$,

$\rho_o\bar{e} = (\frac{\partial \Gamma}{\partial \dot{\theta}})^{-1}\frac{\partial L}{\partial \dot{\theta}}$, $\rho_o\tilde{e} = (\frac{\partial \Gamma}{\partial \dot{\theta}})^{-1}\frac{\partial \gamma}{\partial \dot{\theta}}$.

Thus the assumption (2.14) and the **request** that Γ is independent on the constraint led us to get the constitutive relations (2.15), (2.16), (2.17) for T_{AB}, q^A, e which, like in other theories, [3], are composed of two summed parts; the first one is completely determined by the thermodynamical process and the second one by the constraint only to within an arbitrary parameter.
After (2.15) and (2.16) the residual inequality (2.13) reads

$$(\frac{\partial \Gamma}{\partial \theta}\rho_o\bar{e} - \frac{\partial L}{\partial \theta})\dot{\theta} + \frac{\partial \Gamma}{\partial \theta}\bar{q}^A\theta_A + \nu\{(\frac{\partial \Gamma}{\partial \theta}\rho_o\tilde{e} - \frac{\partial \gamma}{\partial \theta})\dot{\theta} + \frac{\partial \Gamma}{\partial \theta}\tilde{q}^A\theta_A\} \geq 0. \tag{2.18}$$

Let us now devote few words to the case involving $n \leq 6$ constraint equations like (2.5), i. e.

$$\gamma_\alpha(C_{AB}, \theta, \dot{\theta}, \theta_A) = 0 \qquad \alpha = 1,2,\dots,n \tag{2.19}$$

It is immediately seen that a class of constitutive relations, which generalizes our previous results to the case involving more than one thermomechani cal constraint, is the following

$$T_{AB} = \bar{T}_{AB} + \nu^{\alpha}\tilde{T}^{\alpha}_{AB}$$

$$q^{A} = \bar{q}^{A} + \nu^{\alpha}\tilde{q}^{A}_{\alpha}$$

$$e = \bar{e} + \nu^{\alpha}\tilde{e}_{\alpha}$$

where the meaning of the symbols is evident.

Finally we consider some examples of thermomechanical constraints belonging in the class (2.5) which represent a slight generalization of classical ones.

a) $B^{A}\partial_{A}\theta = 0$ (cf. [1]);

i. e. the temperature is uniform along the vector lines of the field of materials directions $\underline{B} = (B^{A})$.

b) incompressibility at a reference temperature θ_R (cf. [6]) :

$$\det\|C\| = f(\theta,\dot{\theta}) \qquad f_R = 1$$

c) inextensibility in the $\underline{B}$ direction at a reference temperature θ_R(cf.[6][7])

$$\underline{B}\,C\,\underline{B} = f(\theta,\dot{\theta}) \qquad f_R = 1$$

In all the cases a), b), c) it is straightforward to specialize the constraint responses appearing in (2.15), (2.16), (2.17). However let us notice that if in b) and c) $f = f(\Gamma(\theta,\dot{\theta}))$, then the coefficient of ν appearing in the residual entropy inequality (2.18) vanishes.

3. Wave propagation

Wave propagation of acceleration waves in thermoelastic materials in the framework of Müller constitutive theory has been studied by K. A. Lindsay & B. Straughan in [14] . The main aim of this Section is to point out some features of wave propagation occurring in a thermomechanically constrained medium according to the general constitutive theory exposed in Sect. 2.

Let us start by considering the case involving more than one constraint examined at the end of the previous Section. Let $\phi(X_A,t) = 0$ be a wave front, (singular surface), across which the discontinuities of the lowest order are the second derivatives of displacement and temperature and the first derivatives of ν^{α}. By defining [12] , [13]

$$\lambda = -\partial_t\emptyset/|\mathrm{Grad}\,\emptyset| \qquad \underline{N} = \mathrm{Grad}\,\emptyset/|\mathrm{Grad}\,\emptyset| \qquad \mathrm{Grad} = (\partial/\partial X_A)$$

$$\delta(.) = \left.\frac{\partial(.)}{\partial\emptyset}\right|_{\emptyset=0+} - \left.\frac{\partial(.)}{\partial\emptyset}\right|_{\emptyset=0-}$$

from the balance equations (2.1) in the standard manner it is possible to get the following compatibility relations

$$(Q_{ij} - \rho_o\lambda^2\delta_{ij})\,\delta v_j + \Omega_i\,\delta\dot{\theta} = \lambda k_i^{\alpha}\,\delta\nu^{\alpha} \tag{3.1}$$

$$\Gamma\Omega_i\,\delta v_i + \Gamma_{\dot{\theta}}\,R\,\delta\dot{\theta} = \lambda\Gamma_{\dot{\theta}}(\tilde{q}^A_{\alpha}N_A - \rho_o\lambda\tilde{e}_{\alpha})\delta\nu^{\alpha} \tag{3.2}$$

where

$$Q_{ij} = N_A N_B\frac{\hat{\partial}T_{iA}}{\partial F_{jB}}, \quad \Omega_i = N_A\left(N_B\frac{\hat{\partial}T_{iA}}{\partial\theta_B} - \lambda\frac{\hat{\partial}T_{iA}}{\partial\dot{\theta}}\right), \quad k_i^{\alpha} = \tilde{T}^{\alpha}_{iA}N_A$$

$$\Gamma_{\dot{\theta}} = \frac{\partial\Gamma}{\partial\dot{\theta}}, \quad R = \rho_o\lambda^2\frac{\hat{\partial}e}{\partial\dot{\theta}} - \lambda\left(\rho_o\frac{\hat{\partial}e}{\partial\theta_A} + \frac{\hat{\partial}q^A}{\partial\dot{\theta}}\right)N_A + \frac{\hat{\partial}q^A}{\partial\theta_B}N_A N_B$$

and if $p = \overline{p} + \nu^{\alpha}\tilde{p}^{\alpha}$ then $\dfrac{\hat{\partial}^m p}{\partial(.)^m} = \dfrac{\partial^m\overline{p}}{\partial(.)^m} + \nu^{\alpha}\dfrac{\partial^m\tilde{p}^{\alpha}}{\partial(.)^m}$.

By differentiating each constraint equation (2.19) with respect to time t and taking jumps we get that the quantities δv_i and $\delta\dot{\theta}$ in addition to (3.1) and (3.2) must satisfy the relations

$$\Gamma k_i^{\alpha}\,\delta v_i + \Gamma_{\dot{\theta}}(\tilde{q}^A N_A - \rho_o\lambda\tilde{e}_{\alpha})\,\delta\dot{\theta} = 0 \tag{3.3}$$

If the n equations (3.3) are independent, it is immediately seen that $n<4$; that is acceleration waves will not in general propagate in a thermomechanical medium suffering more than three constraints, otherwise some additional compatibility conditions must hold which restrict the class of materials. As a remark we notice that in the purely mechanical case N. Scott [15] showed that, in general, at most two constraints allow for wave propagation. In what follows we shall confine our attention to the case n = 1.

First of all we want to put in evidence the possibility to have a wave speed as a direct consequence of the constraint in the case where $\partial\gamma/\partial\dot{\theta} = \Gamma_{\dot{\theta}}\rho_o\tilde{e} \neq 0$. In fact if

$$w = \tilde{q}^A N_A - \rho_o\lambda\,\tilde{e} = 0 \tag{3.4}$$

from (3.1), (3.2) and (3.3) we gain

$$\delta v_i k_i = 0; \quad \delta\dot{\theta} = -(\Gamma/\Gamma_{\dot{\theta}})(\Omega_i\delta v_i); \quad \lambda\delta\nu = k_i\{Q_{ij} - (\Gamma/\Gamma_{\dot{\theta}})\Omega_i\Omega_j\}\,\delta v_j$$

However it is straightforward to realize that such a situation cannot occur when we consider one-dimensional wave propagation. Something like that happens by considering wave propagation in perfect fluids where the material wave, (i. e. $\lambda = \underline{v} \cdot \underline{n}$), cannot exist in the one dimensional case.

By assuming $w \neq 0$ and eliminating $\delta\dot{\theta}$ between (3.1), (3.2), (3.3) and $\lambda\,\delta\nu$ between (3.1) and (3.2) we get that the following relation must be satisfied:

$$\{w^2(Q_{ij} - \rho_o \lambda^2 \delta_{ij}) - \frac{\Gamma}{\Gamma_{\dot{\theta}}}(w(\Omega_i k_j + \Omega_j k_i) + R k_i k_j)\}\delta v_j = 0 \quad (3.5)$$ [2]

It is immediately seen that if $\partial\gamma/\partial\dot{\theta} \neq 0$, the characteristic polynomial, which comes out from (3.5), has the same degree as in the unconstrained thermomechanical case considered by K. A. Lindsay & B. Straughan in [14] . Of course the values of the wave speeds are different. Such a result qualitatively differs from the one obtained by N. Scott in [15] in the purely mechanical case, where a constraint present in the material implies a diminution of the number of the not vanishing wave speeds. In the present case it is easy to see that a situation similar to the Scott's one holds if $\partial\gamma/\partial\dot{\theta} = 0$. However it must be stressed that $\partial\gamma/\partial\dot{\theta} = 0$ does not allow the differential system for the quantities F_{iA}, v_i, θ, $\dot{\theta}$, θ_A, ν to be hyperbolic in the usual sense (see [12] , [13]); that is immediately seen by observing that $\partial\gamma/\partial\dot{\theta} = 0$ leads to absence of any time derivative of ν . Then we can conclude that a thermomechanical constraint involving even the rate of temperature implies an added internal energy $\tilde{e}$ which, in principle, allows one to have both a hyperbolic system for the quantities F_{iA}, v_i, θ, $\dot{\theta}$, θ_A, ν and a wave speed purely due to the constraint.

Let us recall now that in [14] , by considering propagation of a plane acceleration wave through an equilibrium state in a privileged direction, (generalized longitudinal wave), K. A. Lindsay & B. Straughan obtained a quartic equation for the wave speed λ which may admit two real positive solutions for λ^2. As well known such a type of situation does not occur for the quartic equation satisfied by the wave speeds of harmonic waves in a classical linear

---

[2] By eliminating $\lambda\delta\nu$ from (3.1) we get

$$(\delta_{ir} - k_i k^*_r)\{(Q_{rj} - \rho_o \lambda^2 \delta_{rj})\delta v_j + \Omega_r\, \delta\dot{\theta}\} = 0; \quad \underline{k}^* = \underline{k}/(\underline{k}\cdot\underline{k})$$

which in the purely mechanical limit reduces to (5.17) obtained in [15] .

elastic material;(see e. g. Sneddon [21] p. 43). In what follows we will work out an analysis strictly closed to the one considered in [14] in order to show that even in the case here studied, it is possible to get two real solutions for the square of the generalized longitudinal wave speeds, under the additional assumption that $\partial\gamma/\partial\theta_A|_E = 0 = \tilde{q}^A|_E$.[3]

Let us suppose that the material ahead of the wave is an isothermal region at rest in a fixed homogeneous configuration [14] and that the reference body has a centre of symmetry. Then in such a region we may assume

$$\left.\frac{\partial \hat{T}_{iA}}{\partial\theta_B}\right|_E = 0;\quad \left.\frac{\partial \hat{e}}{\partial\theta_A}\right|_E = 0;\quad \left.\frac{\partial \hat{q}^A}{\partial\dot{\theta}}\right|_E = 0;\quad \left.\frac{\partial \hat{q}^A}{\partial F_{iB}}\right|_E = 0;\quad \partial_t \nu = \partial_A \nu = 0 \tag{3.6}$$

Let $\Sigma(t)$ and $\sigma(t)$ be, respectively, the wave surfaces defined by $\emptyset(X_A,t) = 0$ and $\phi(x_i,t) = \emptyset(X_A(x_i,t),t) = 0$. By taking into account the relation, (see [22], (4.10))

$$N_A = F_{iA}\frac{|\text{grad}\,\sigma|}{|\text{Grad}\,\Sigma|}n_i \qquad\qquad \text{grad} = (\partial/\partial x_i)$$

we may express the acoustic tensor Q_{ij} as a function of $\underline{n} = (n_i)$ rather than $\underline{N}$. Then, by defining

$$T_{ij} = \frac{|\text{grad}\,\sigma|}{|\text{Grad}\,\Sigma|}\left(\frac{\partial \hat{T}_{iA}}{\partial\dot{\theta}}F_{jA}\right)\Big|_E \tag{3.7}$$

and by using the same arguments as in [14] , we can state that there is at least one direction $\underline{n}^*$ such that $T_{ij}n^*_j$ is an eigenvector of Q_{ij}. Since we are considering wave propagation in an isothermal region at rest, Q_{ij} is a constant matrix and so $T_{ij}n^*_j$ is fixed. Hence we may consider a plane acceleration wave propagating in the direction $\underline{n}^*$, with amplitude in the direction $T_{ij}n^*_j$;(generalized longitudinal wave). If $\underline{\mu}$ is the unit vector in the direction $T_{ij}n^*_j$ and $\delta\underline{v} = \alpha\,\underline{\mu}$, from (3.5) we get

$$\lambda^4 - \lambda^2(U_M^2 + m + T) + mU_T^2 = 0 \tag{3.8}$$

where

$$\rho_o U_M^2 = Q_{ij}\mu_i\mu_j\Big|_E ; \qquad\qquad U_T^2 = -\frac{\partial\hat{q}^A}{\partial\theta_B}N^*_A N^*_B/\rho_o\,\frac{\partial\hat{e}}{\partial\dot{\theta}}\Big|_E$$

---

[3] Subscript $|E$ means that a quantity is evaluated at the equilibrium state, specified by (3.6).

$$m = - \Gamma \frac{\hat{\partial} e}{\partial \dot{\theta}} (\underline{\mu} \cdot \underline{k})^2 / \Gamma_{\dot{\theta}} (\rho_o \tilde{e})^2 \Big|_E \qquad T = 2\Gamma T_{ij} \mu_i n^*_j \, (\underline{\mu} \cdot \underline{k}) / \Gamma_{\dot{\theta}} \rho_o^2 \, \tilde{e} \Big|_E$$

In the following we retain, as in the unconstrained thermomechanical case (see [8]), $\hat{\partial} e/\partial \dot{\theta}|_E > 0$, $\partial \Gamma/\partial \dot{\theta}|_E < 0$; hence $m > 0$.

It is immediately seen that equation (3.8) has real solutions if either

$$\text{i)} \quad T \leqslant U_M^2 + m - (4mU_T^2)^{1/2} \qquad \text{or} \qquad \text{ii)} \quad T \geqslant U_M^2 + m + (4mU_T^2)^{1/2} .$$

Because we are looking for positive roots λ^2 of (3.8), we are led to disregard the case ii). Then,if i) holds, we have two real positive values for λ^2.

In closing let us shortly deduce the growth equation for the amplitude of an acceleration wave propagating in a constrained thermomechanical medium. For the sake of simplicity we shall limit ourselves to the one-dimensional case; thus a thermodynamic process is characterized by $x = x(X,t)$ and $\theta = \theta(X,t)$. Specializing (2.1) and (2.4) to the case here considered and by taking jumps after differentiating (2.4) with respect to time t, we get the following algebraic system to be satisfied by $[\partial_X F]$ and $[\partial_X \nu]$;($[(.)] = (.)_{\emptyset=0+} - (.)_{\emptyset=0-}$; $F = \partial x/\partial X$)

$$M \, \underline{\Pi} = 0 \tag{3.9}$$

$$M = \begin{Vmatrix} \rho_o (U_C - \lambda)(\lambda^2 - U_M^2) - \dfrac{\Gamma \tilde{T}}{\Gamma_{\dot{\theta}} \rho_o \tilde{e}} \left(\lambda \dfrac{\hat{\partial} T}{\partial \dot{\theta}} - \dfrac{\hat{\partial} T}{\partial g}\right) & - \tilde{T} (U_C - \lambda) \\ \dfrac{\Gamma}{\Gamma_{\dot{\theta}}} \left\{ (U_C - \lambda)\left(\dfrac{\hat{\partial} T}{\partial g} - \lambda \dfrac{\hat{\partial} T}{\partial \dot{\theta}}\right) - \dfrac{\tilde{T} R}{\rho_o \tilde{e}} \right\} & \rho_o \tilde{e} (U_C - \lambda)^2 \end{Vmatrix} {}^{4} \qquad \underline{\Pi} = \begin{Vmatrix} [\partial_X F] \\ [\partial_X \nu] \end{Vmatrix}$$

where $U_C = \tilde{q}/\rho_o \tilde{e}$, $\rho_o U_M^2 = \hat{\partial} T/\partial F$, R has been defined above and $\det \| M \| = 0$ gives the characteristic polynomial. For the jumps of the second derivatives we easily get,in the usual way [22] [23], a non-homogeneous algebraic system of the form

$$M \, \underline{\Psi} = \underline{\omega} \tag{3.10}$$

where $\underline{\Psi}^T = \| [\partial^2 F/\partial X^2] \, , \, [\partial^2 \nu/\partial X^2] \|$,(superscript T refers to transposition), and $\underline{\omega}$ depends on the field and on its first derivatives jumps. When we consider a wave propagating into the equilibrium region specified above, forming the scalar product of (3.10) with the vector $\underline{\xi}$ such that $\underline{\xi} \, M = 0$, we obtain for the wave amplitude a differential equation of the form

[4] $g = \partial_X \theta$

$$a\frac{d\pi}{d\sigma} + b\,\pi^2 + c\,\pi = 0 \qquad\qquad \pi = [\partial_X F] \tag{3.11}$$

where $d/d\sigma = \partial_t + \lambda\,\partial_X$ and a, b, c are given by

$$a = 2\left(\rho_o^2\lambda^2\tilde{e} + \frac{\Gamma\,\tilde{T}^2}{\Gamma_{\dot\theta}\rho_o\tilde{e}\,\lambda^2}\,\frac{\hat\partial q}{\partial g}\right)$$

$$b = 3\frac{\rho_o^2\lambda\,\tilde{e}}{\tilde{T}}\{\lambda^2 - U_M^2 + \frac{\beta}{2}\}\{\rho_o U_M^{(c)^2} + \frac{\Gamma\,\tilde{T}}{\Gamma_{\dot\theta}\rho_o\tilde{e}}\left(\frac{R^{(c)}\tilde{T}}{\rho_o\tilde{e}\,\lambda^2} - 2\frac{\partial\tilde{T}}{\partial\dot\theta}\right)\} + \rho_o\lambda\,\tilde{T}\{\frac{\Gamma}{\Gamma_{\dot\theta}^2}\Gamma_{\dot\theta\dot\theta}\,\cdot$$

$$(\lambda^2 - U_M^2 - \frac{\beta}{2}) - 3(\lambda^2 - U_M^2)\} + \rho_o^2\tilde{e}\lambda\,\frac{\hat\partial}{\partial F}(U_M^2) + 3\,\frac{\Gamma}{\Gamma_{\dot\theta}}\rho_o\tilde{T}\{\frac{\Gamma\,\tilde{T}}{\Gamma_{\dot\theta}\,\rho_o^2\,\tilde{e}\lambda}\,(\underline{S}^T(\nabla\nabla^T T)\underline{S} -$$

$$-(\nabla(U_M^2)\cdot\underline{S})\} + \tilde{T}^3\left(\frac{\Gamma}{\Gamma_{\dot\theta}\rho_o\lambda\tilde{e}}\right)^2\{\,\underline{S}^T(\nabla\nabla^T q)\underline{S} - \rho_o\lambda\,\underline{S}^T(\nabla\nabla^T e)\underline{S}\}$$

$$c = \frac{\Gamma\,\tilde{T}}{\Gamma_{\dot\theta}^2\,\rho_o\lambda\tilde{e}}\,\{\frac{\tilde{T}}{\rho_o\lambda\,\tilde{e}}\,\frac{\partial\gamma}{\partial\theta}\,\frac{\hat\partial q}{\partial g} + \Gamma_{\dot\theta}\tilde{T}\left(\frac{\hat\partial q}{\partial\theta} - \rho_o\lambda\,\frac{\hat\partial e}{\partial\theta}\right) + \lambda\left(\frac{\partial\gamma}{\partial\dot\theta}\,\frac{\hat\partial T}{\partial\theta} - \frac{\partial\gamma}{\partial\theta}\,\frac{\hat\partial T}{\partial\dot\theta}\right)\}$$

$\Gamma_{\dot\theta\dot\theta} = \partial^2\Gamma/\partial\dot\theta^2$, $\nabla = \|\,\hat\partial/\partial g,\ \hat\partial/\partial\dot\theta\|$, $\underline{S}^T = \|\,-1,\ \lambda\|$ and

$$\beta = 2\,\frac{\Gamma\,\tilde{T}}{\Gamma_{\dot\theta}\,\rho_o^2\tilde{e}}\,\frac{\hat\partial T}{\partial\dot\theta}, \qquad \rho_o U_M^{(c)^2} = \frac{\partial\tilde{T}}{\partial F}, \qquad R^{(c)} = \rho_o\,\lambda^2\,\frac{\partial\tilde{e}}{\partial\dot\theta} + \frac{\partial\tilde{q}}{\partial g}\,.$$

For the sake of simplicity we omitted the $|_E$ notation in the quantities involved in a, b, c. Integration of (3.11) **yields:**

$$\pi = 1/\{\left(\frac{1}{\pi_o} + \frac{b}{c}\right)e^{c\,t/a} - \frac{b}{c}\} \tag{3.12}$$

where π_o denotes the initial wave amplitude i. e. $\pi_o = \pi|_{t=0}$.

In solving growth equations in nonlinear wave propagation problems a result like (3.12) is quite typical and its inspection may be carried on in the usual manner [12] , [13] , [23] . It is worth of mention here that p = -c/b represents the " critical amplitude " of the material for the prescribed state of strain and temperature compatible with the thermodynamical constraint (2.4). Moreover by retaining the assumptions $\hat\partial q/\partial g < 0$, $\Gamma_{\dot\theta} < 0$, [8] ,we get a >0;then from (3.12) it is easy to realize that if either

i) $c > 0,\ 0 < \frac{p}{\pi_o} < 1$ or   ii) $c < 0,\ \frac{p}{\pi_o} < 0$

there exists a " critical time " $t_c > 0$ defined by

$$t_c = -\frac{a}{c}\log(1-\frac{p}{\pi_o})$$

at which the wave amplitude becomes unbounded. As well known at t_c a shock wave formation may occur [12][13] .

References

1. Green, A.E.; Naghdi, P.M.; Trapp, A.J.: Thermodynamics of a continuum with internal constraints. Int. J. Engng. Sci. 8 (1970) 891-908.

2. Trapp, A.J.: Reinforced materials with thermomechanical constraints. Int. J. Engng. Sci. 9 (1971) 757-773.

3. Andreussi, F.; Podio Guidugli, P.: Thermomechanical constraints in simple materials. Bull. Acad. Pol. Sci., Sér. Sci. Tech. XXI 4 (1973) 199-205.

4. Gurtin, M.E.; Podio Guidugli, P.: The thermodynamics of constrained materials. Arch. Rational Mech. Anal. 51 (1973) 192-208.

5. Alts, Th.: Thermodynamics of thermoelastic bodies with kinematic constraints. Fiber reinforced materials. Arch. Rational Mech. Anal. 61 (1976) 253-289.

6. Manacorda, T.: Introduzione alla termomeccanica dei continui. Quaderni dell'Unione Matematica Italiana 14. Bologna: Pitagora Ed. 1979.

7. Truesdell, C.; Noll, W.:Encyclopedia of Physics III/3 Flugge, S.(Ed.) Berlin, Heidelberg, New York: Springer 1965.

8. Müller, I.: The coldness, a universal function in thermoelastic bodies. Arch. Rational Mech. Anal. 41 (1971) 319-332.

9. Green, A.E.; Laws, N.: On a global entropy production inequality. Quart. J. Mech. and Applied Math. 25 (1972) 1-11.

10. Green, A.E.; Lindsay, K.A.: Thermoelasticity. J. Elasticity 2 (1972) 1-7

11. Hutter, K.: The foundations of Thermodynamics, its basic postulates and implications. A review of modern Thermodynamics. Acta Mechanica 27 (1977) 1-54.

12. Boillat, G.: La propagation des ondes. Gauthier-Villars Paris 1965.

13. Jeffrey, A.: Quasilinear hyperbolic systems and waves. Res. Notes in Math. 5, London: Pitman 1976.

14. Lindsay, K.A.; Straughan, B.: Propagation of mechanical and temperature acceleration waves in thermoelastic materials. J. Applied Math. Phys. (ZAMP) 30 (1979) 477-490.

15. Scott, N.: Acceleration waves in constrained elastic materials. Arch. Rational Mech. Anal. 58 (1975) 57-75.

16. Manacorda, T.: On the propagation of discontinuity waves in thermoe lastic incompressible solids. Arch. Mech. Stos. 24 2 (1971) 277-285.

17. Amendola, G.: Acceleration waves in incompressible materials. Atti Sem. Mat. Fis. Modena 24 (1975) 381-395.

18. Donato, A.: On a supplementary conservation law for the balance laws in thermoelastic bodies. (To be published).

19. Friedrichs, K.; Lax, P.D.: Systems of conservation equations with a convex extension. Proc. Nat. Acad. Sci. U. S. A. 68 (1971) 1686-1688.

20. Liu, I-Shih: Method of Lagrange multipliers for exploitation of the entropy principle. Arch. Rational Mech. Anal. 46 (1972) 131-148.

21. Sneddon, I.N.: The linear theory of thermoelasticity. Wien, New York: Springer 1974.

22. Chen, P.J.: Growth and decay of waves in solids. Handbuch der Physik VIa/3; Berlin, Heidelberg, New York: Springer 1973.

23 McCarthy, M.F.: Singular surfaces and waves. Continuum Physics Vol. II, Eringen,A.C. (Ed); New York, San Francisco, London: Academic Press, 1975.

Acknowledgement: This work was supported by the C. N. R. through the Gruppo Nazionale per la Fisica Matematica.

Asymptotic Approach to Some Nonlinear Wave Problems

V.M.Babič[1], I.A.Molotkov[2], S.A.Vakulenko[3]

[1]Leningrad Branch of Mathematical Institute of Academy of Sciences USSR, Leningrad

[2]Physical Faculty of Leningrad University, Leningrad

[3]Leningrad Agricultural Institute, Leningrad

Summary

The subject of our paper is obtaining of asymptotic expansions of solutions of some wave problems. Our methods are based on introducing some special independent variables: these are "slow" and "quick" times in the first part of our paper, and stretched variables in the second part.
We consider some equations of non-linear oscillation theory and obtain their asymptotic solutions in the first part of the paper. We justify our expansions by the method of succesive approximations.
We consider the equation $\Delta u + \omega^2 W(|u|^2, x, y) u = 0$.
in the second part. Solutions of the equations concentrated in a small neighbourhood of a curve are found there.

1. One dimensional problems

A number of asymptotic methods in theory of wave propagation are connected with quasi-stationary phenomena. Quasi-stationary wave processes are "quick" wave processes, which parameters are not constants, but slowly varying functions of time and other variables. On this fact several asymptotic methods are based. We shall consider here the well-known one-frequency problems of the theory of nonlinear oscillation. According to our opinion our approach is more direct and simpler than the traditional one.

Let us begin from the equation*)

$$\frac{d}{dt}\left(m(\tau)\frac{dx}{dt}\right) + n(\tau)x = \varepsilon f\left(\tau, x, \frac{dx}{dt}\right) \tag{1.1}$$

*) There is another approach to the equation (1.1) in the book [1]

Nonlinear Deformation Waves
IUTAM Symposium Tallinn 1982
Editors: U. Nigul, J. Engelbrecht

(ε - is a small parameter, $\tau = \varepsilon t$, $m(\tau)$, $n(\tau)$ - are smooth positive functions). It is possible to obtain an asymptotic solution of the equation (1.1) in the form

$$x \sim \sum_{j=0}^{\infty} x_j(t_1,\tau)\varepsilon^j; \quad x_j(t_1+2\pi,\tau)=x_j(t_1,\tau);\ t_1=\frac{\Lambda(\tau)}{\varepsilon}. \qquad (1.2)$$

The series (1.2) is two-timing expansion: τ - is slow time, t_1 - is the quick one. The method of introducing of quick time is not new - see the paper [2]. The expansion (1.2) is "ansatz" of our problem. On the basis of it we obtain the first approximation defined by the formulae:

$$\Lambda(\tau)=\int_0^{\tau}\sqrt{\frac{n}{m}}\,d\tau, \quad x_0=A_0(\tau)\cos(t_1+B_0(\tau)) \qquad (1.3)$$

(A_0, B_0 - are unknown functions). The equation for calculation of the second approximation is the following

$$n\left(\frac{\partial^2 x_1}{\partial t_1^2}+x_1\right)+\left(2m\frac{\partial^2 x_0}{\partial t_1\partial\tau}+\frac{\partial m}{\partial\tau}\frac{\partial x_0}{\partial t_1}\right)\sqrt{\frac{n}{m}}+m\frac{\partial x_0}{\partial t_1}\left(\frac{\sqrt{n}}{m}\right)'=f\left(\tau,x_0,\frac{\partial x_0}{\partial t_1},\sqrt{\frac{n}{m}}\right).$$

(I.4)

The equality (1.4) is the second order differential equation for unknown 2π -periodical function t_1 . The periodical solution x_1 of the equation (1.4) exists, if

$$\int_0^{2\pi}\left[-\left(2m\frac{\partial^2 x_0}{\partial t_1\partial\tau}+\frac{dm}{d\tau}\frac{\partial x_0}{\partial t_1}\right)\sqrt{\frac{n}{m}}+m\frac{\partial x_0}{\partial t_1}\left(\sqrt{\frac{n}{m}}\right)'+f\right]\sin(t_1+B_0)\,dt_1=0$$

$$\int_0^{2\pi}\left[-\left(2m\frac{\partial^2 x_0}{\partial t_1\partial\tau}+\frac{dm}{d\tau}\frac{\partial x_0}{\partial t_1}\right)\sqrt{\frac{n}{m}}+m\frac{\partial x_0}{\partial t_1}\left(\sqrt{\frac{n}{m}}\right)'+f\right]\cos(t_1+B_0)\,dt_1=0.$$

(1.5)

Here $x_0=A_0\cos(t_1+B_0)$.

These formulae are equivalent to

$$\frac{d}{d\tau}\left(\sqrt{mn}\,A_0^2\right)=\frac{-A_0}{\pi}\int_0^{2\pi} f\left(\tau, A_0\cos\varphi, -\sqrt{\frac{n}{m}}\,A_0\sin\varphi\right)\sin\varphi\, d\varphi$$

$$2\pi\sqrt{mn}\,A_0B_0'=-\int_0^{2\pi} f\left(\tau, A_0\cos\varphi, -\sqrt{\frac{n}{m}}\,A_0\sin\varphi\right)\cos\varphi\, d\varphi. \qquad (1.6)$$

It is possible to continue this process of obtaining x_j, $j=1,2,\ldots$

Let some initial data for equation (1.1) are given when t=0 and $0\leqslant\tau\leqslant\tau_0<+\infty$ is the segment, where the corresponding solution of equation (1.6) exists. Then

1) the solution of the initial a value problem for equation (1.1) exists, if $0\leqslant\varepsilon t\leqslant\tau_0$, if ε is small enough,

2) the expansion (1.2) is the asymptotic expansion of the solution, if $0\leqslant t\leqslant\frac{\tau_0}{\varepsilon}$.

It is possible to prove this theorem, using classical successive approximation method.

Another example of the two-timing expansion method application arises in connection with the well known system *)

$$\frac{\partial\varphi}{dt}=\omega(I)+\varepsilon f(\varphi, I,\varepsilon)$$

$$\frac{dI}{dt}=\varepsilon g(\varphi, I,\varepsilon) \qquad 0<\varepsilon\ll 1$$

$$f(\varphi+2\pi), I,\varepsilon)\equiv f(\varphi, I,\varepsilon);\quad g(\varphi+2\pi, I,\varepsilon)\equiv g(\varphi, I,\varepsilon);\ \omega(I)>0. \quad (1.7)$$

*) See another approach in the book [3].

The ansatz has the form:

$$\varphi = t_1 + \sum_{j=0}^{\infty} \varphi_j(t_1,\tau)\varepsilon^j; \quad I = \sum_{j=0}^{\infty} I_j(t_1,\tau)\varepsilon^j. \tag{1.8}$$

Here $\tau = \varepsilon t$ is slow time, $t_1 = \frac{\varphi_{-1}(\tau)}{\varepsilon} + \varphi_0(\tau)$ is quick time of our problem; $\varphi_1(t_1,\tau)$, $I_j(t_1,\tau)$ are the functions, which are to be calculated.

The first approximation gives the classical equation of the averaging method.

Let the Cauchy data for the system (1.7) to be given at t=0 and the correspondent solution of the equation of the averaging method exist if $0 \leqslant \tau \leqslant \tau_0 < +\infty$, then the solution of the Cauchy problem for the system (1.7) exists, if

$$0 \leqslant t \leqslant \frac{\tau_0}{\varepsilon}$$

(ε is small enough), and expansion (1.8) give the asymptotic of the solution. We have proved this theorem, using the succesive approximation method. Some of this results were obtained with collaboration with V.D.Azhotkin.

2. Waves in nonlinear inhomogeneous medium, which are concentratedin the vicinity of a given curve

In this part (see also [4]) we deal with equation

$$\Delta u + \omega^2 W(|u|^2, s, n)u = 0 \tag{2.1}$$

in the vicinity of a preassigned plane and sufficiently smooth curve S . Here (s,n) are ordinary coordinates in the vicinity of S (length of arc and normal), $\omega > 0$ is frequency,real function $W(|u|^2, s, n)$ is squared nonlinear refractive index. We shall seek complex-valued asymptotic (for $\omega \to \infty$) solutions $u = u(s,n,\omega)$ of (2.1), which concentrate near S and have the meaning of waves propagating along S . Equation (2.1) and its special cases are very often encountered in wave-propagation problems and in other problems of mathematical physics.

In connection with determination of solutions concentrated near S we introduce the stretched normal $\nu = \omega n$ and consider two following problems for (2.1).

Problem 1. To find formal asymptotic for $\omega \to \infty$ solution (f.a.s.) of (2.1), which decreases on the both sides of S :

$$\lim_{\nu \to \pm\infty} u = 0 \tag{2.2}$$

(nonlinear analogy of the mode in internal guide).

Problem 2. To find f.a.s. of (2.1), which satisfies impedance boundary condition at S and decreases only on one side of S :

$$u_n + \omega \gamma(s) u \Big|_{n=0} = 0 \ (\gamma > 0), \quad \lim_{\nu \to +\infty} u = 0 \tag{2.3}$$

(nonlinear analogy of the undamped surface wave on the impedance boundary).

We shall make the following assumptions about the function $W(|u|^2, s, n)$ and its primitive $W(x,s,n) = \int_0^x w(t,s,n)\,dt$. In problem 1 we assume, that 1) $W \in C^\infty$, 2) difference $W(x,s,0) - \ell x$ for $\ell > w(0,s,0)$ has at least one positive zero, moreover the smallest of such zeros $x_0 = x_0(\ell, s)$ is simple. In problem 2 we presuppose in addition that $W(x,s,0)$ increase by x monotonically . The example of function W , satisfying all these conditions, is function

$$w(|u|^2, s, n) = \alpha(s,n)|u|^p + \beta(s,n), \ p > 0, \ \alpha(s,0) > 0. \tag{2.4}$$

Concentrated solutions of nonlinear equations have already been studied by asymptotic methods in [5]. Yet the equation (2.1) is different from those considered in [4], the method used below is also different.

We shall seek the solution of problem 1 in the form

$$u(s,n,\omega) = \mathcal{U}(s,u,\omega) \exp[i\omega Q(s,\nu,\omega)], \tag{2.5}$$

$$\mathcal{U}(s,\nu,\omega)=\sum_{j=0}^{\infty}\mathcal{U}_j(s,\nu)\omega^{-j};\quad Q(s,\nu,\omega)=\sum_{j=0}^{\infty}Q_j(s,\nu)\omega^{-j},$$

where $\mathcal{U}_j$ and Q_j are real, all $\mathcal{U}_j$ exponentially decrease for $\nu\to\pm\infty$, all Q_j don't have more than degree increase for $\nu\to\pm\infty$. We substitute (2.5) in (2.1) and separate the real and imaginary parts in the received equation:

$$\Delta\mathcal{U}+\omega^2\mathcal{U}[w(\mathcal{U}^2,s,\nu\omega^{-1})-(\nabla Q)^2]=0, \tag{2.6}$$

$$div(\mathcal{U}^2\nabla Q)=0. \tag{2.7}$$

We pass from (2.6), (2.7) to the recurrent sequence of equations, obtained by equating coefficients at every degree to zero. It is natural to comply with the two leading equations following from (2.7), if Q_0 and Q_1 are assumed to depend only on

$$Q_{0,1}(s)=\int_{s_0}^{s}q_{0,1}(s)ds,$$

where

$$q_0^2(s)>w(0,s,0). \tag{2.8}$$

The leading equation following from (2.6) is

$$\mathcal{U}_{0\nu\nu}-q_0^2\mathcal{U}_0+w(\mathcal{U}_0^2,s,0)\mathcal{U}_0=0 \tag{2.9}$$

and is integrated with respect to ν taking the behaviour for $\nu\to\pm\infty$ into account:

$$\nu+h(s)=\pm\int_{\mathcal{U}_0}^{\sqrt{x_0}}[q_0^2z^2-W(z^2,s,0)]^{-1/2}dz. \tag{2.10}$$

This formula defines the function $\mathcal{U}_0(s,\nu+h(s))$ decreasing exponentially for $\nu\to\pm\infty$. $\mathcal{U}_0$ is the even function of ν

for $h=0$. We should choose $h(s)\equiv 0$ in problem 1 which is being considered.

The subsequent equations following from (2.6) are

$$LU_m = U_0\left(\frac{\partial Q_m}{\partial \nu}\right)^2 + 2q_0 U_0 \frac{\partial Q_m}{\partial s} + F_m(U_0,\dots,U_{m-1},Q_0,\dots,Q_{m-1},s,\nu), \quad (m=1,2,\dots) \tag{2.11}$$

$$L = \partial^2/\partial\nu^2 + w(U_0^2,s,0) - q_0^2 + 2U_0^2 w'_{U_0^2}(U_0^2,s,0).$$

The laws of consevation

$$q_0(s) I_0(s) = c_0, \quad I_0(s) = \int_{-\infty}^{\infty} U_0^2 d\nu \tag{2.12}$$

$$\int_{-\infty}^{\infty} U_0^2 \frac{\partial Q_m}{\partial s} d\nu + 2q_0 \int_{-\infty}^{\infty} U_0 U_m d\nu + \int_{-\infty}^{\infty} G_m(U_0,\dots,U_{m-1},Q_0,\dots,Q_{m-1},s,\nu) d\nu = c_m \quad (m=1,2,\dots) \tag{2.13}$$

and equations

$$\frac{\partial Q_{m+1}}{\partial \nu} = H_{m+1}(U_0,\dots,U_{m-1},Q_0,\dots,Q_m,s,\nu), \; m=1,2,\dots \tag{2.14}$$

follow from equation (2.7). We obtain explicit expressions for $F_1, F_2, \dots, G_1, G_2, \dots, H_2, H_3, \dots$; $c_0, c_1, \dots$ are constant.

The linear nonhomogeneous equations (2.11) under the above mentioned conditions of decreasing for $\nu \to \pm\infty$ are solvable only if the right part of (2.11) is orthogonal to the solutions $A(s)U_{0\nu}(s,\nu)$ of the homogeneous problem:

$$2q_0 \int_{-\infty}^{\infty} U_0 U_{0\nu} \frac{\partial Q_m}{\partial s} d\nu + \int_{-\infty}^{\infty} \tilde{F}_m U_{0\nu} d\nu = 0, \quad m=1,2,\dots . \tag{2.15}$$

(2.14) was used when concluding (2.15), $\tilde{F}_m$ is analogous to F_m.

We'll assume that

$$\partial(q_0 I_0)/\partial q_0 \neq 0. \tag{2.16}$$

Then condition (2.12) determines the only function $q_0(s)$ for fixed value of constant c_0. Equality (2.15) when $m=1$ is as follows

$$\int_{-\infty}^{\infty} d\nu \int_0^{U_0^2} [\varkappa(s) W(x,s,0) + \frac{1}{2} W_n(x,s,0)] dx = 0, \tag{2.17}$$

$\varkappa(s)$ is curvature of S. Condition (2.17) gives the relation between $\varkappa(s)$, $W(U_0^2, s, 0)$, $W_n(U_0^2, s, 0)$ and generalizes the requirement of vanishing the effective curvature in the theory of linear Helmholtz equation (see [6]). These relations can't be fulfilled on the arbitrary S.

Let condition (2.17) be fulfilled. Then equations (2.11), (2.13)-(2.15) permit to find all the further terms of series U and Q. The function $U_1(s,\nu)$ vanishing for $\nu \to \pm\infty$ is obtained from (2.11) for $m=1$. Its even part $U_1^+ = q_1 \frac{\partial U_0}{\partial q_0}$ is expressed through q_1, q_0, U_0 and its odd part U_1^- - through the arbitrary function $A_1(s)$ (coefficient at $U_{0\nu}$), q_0 and U_0. The substitution of U_1 in (2.13) for m=1, when U_1^- disappears together A_1, gives the equation

$$q_1 \frac{\partial(q_0 I_0)}{\partial q_0} = c_1$$

determining $q_1(s)$ according to (2.16). Further we substitute U_1 in (2.15) for $m=2$. The evenness of $Q_2(s,\nu)$ with respect to ν, following from form H_2 in (2.14) leads to the disappearance of the first term in (2.15) $(m=2)$ and permits to determine $A_1(s)$ in the only way.

The next step determines U_2, Q_2 and occurs similarly. We find these functions from (2.11) $(m=2)$ and from (2.14) $(m=1)$. Even and odd parts of U_2 have the properties, which are similar to the properties of U_1^+ and U_1^-. The function $U_2(s,\nu)$ contains unknown coefficient $A_2(s)$, and

$Q_2(s,\nu)$ is determined to the accuracy of the unknown additive function $B_2(s)$. Function $B_2(s)$ is determined from (2.13) $(m=2)$. The latter equation does not contain $A_2(s)$ entering into U_2^-. Then condition (2.15) $(m=2)$ in general permits to find $A_2(s)$. The further steps determining U_m, Q_m $(m=3,4,\ldots)$ are the same as for $m=2$.

Constructed in the form of expansion (2.5) the solution of the problem 1 differs greatly from the solutions of the similar linear problem: 1) the amplitude in nonlinear case proves to be concentrated in a narrower vicinity ($Q(\omega^{-1})$ in width) of S instead of $O(\omega^{-1/2})$; 2) phase velocity of wave in nonlinear case proved to be dependent on the amplitude U_0 and contains arbitrary constants $C_0, C_1, \ldots$. In special case (2.4) the main terms of expansion (2.5) are very simple.

If condition (2.16) is not fulfilled then for given C_0 equation (2.12) can determine several different functions $q_0(s)$ corresponding to different dependences of phase velocity on s. In integrable case (2.4) it is possible only for $p>4$, i.e. for very strong nonlinearity. At bifurcation points of equation (2.12) the procedure of determining the terms of series from (2.5) is more complicated.

Now let us consider problem 2. Here we also find its solution in form (2.5) leading to equalities (2.6), (2.7) and its corollaries. Instead of (2.8) we now assume

$$q_0^2(s) > W(0,s,0) + \gamma^2(s) \tag{2.19}$$

in (2.10) take only the upper sign and choose $h(s) \neq 0$. The function $h(s)$ is determined from equation

$$U_{0\nu}(s,h) + \gamma(s)U_0(s,h) = 0 \tag{2.20}$$

according to (2.3). When making use of the assumption of monotonocity of $W(X,s,0)$ with respect to X and condition (2.19), equation (2.20) is always solvable according to function $h(s)$, which proves to be positive. Amplitude $U_0(s,\nu+h)$

is exponentially vanishing for ν just from the boundary S. In special case (2.4) from (2.20) we have

$$h(s)=\frac{1}{p\sqrt{q_0^2-\beta(s,0)}}\ln\frac{\sqrt{q_0^2-\beta(s,0)}+\gamma}{\sqrt{q_0^2-\beta(s,0)}-\gamma}.$$

As it is in problem 1 we obtain equation (2.11)-(2.14), but now interlation is fulfilled for $0\leqslant\nu<\infty$. The operator L on the semiaxis for the boundary condition (3) has not eigenfunctions. Orthogonality conditions in the problem 2 are absent. Corollary from this: the condition (2.17) disappears. Thus problem 2 is solvable for arbitrary smooth curve S as it is in linear case [7]. On the whole procedure of the determination of series - term in (2.5) is more simple then it is in the problem 1.

For definiteness we consider the periodical problem 1. Let conditions (2.16), (2.17) be fulfilled. Periodicity of U_j and Q_j $(j\geqslant 1)$ are investigated by means of equations (2.11) -(2.15). As a result we obtain quantification condition

$$\omega\int_0^l q_0(s,c_0)\,ds+\int_0^l q_1(s,c_1)\,ds+\sum_{m=2}^{\infty}\omega^{-m+1}\int_0^l B_m(s,c_m)\,ds=2\pi N. \qquad (2.21)$$

Here $B_m(s)$ $(m\geqslant 3)$ are similar to $B_2(s)$ and are contained in Q_m, N is integer, $N\gg 1$. It follows from (2.21) that

$$\int_0^l q_0(s,c_0)\,ds=\frac{2\pi N}{\omega}+O\left(\frac{1}{\omega}\right).$$

This condition is solvable with respect to c_0. It is possible to convince that condition (2.21) is satisfied for arbitrary ω at the expence of choice of $c_0, c_1, \dots$. In the periodical problem 2 we have similar situation.

References

1. Bogolyubov, N.N.; Mitropolskiy, Yu.A. Asymptotic methods in the theory of nonlinear oscillation. Moscow: Nauka 1974 (in Russian).

2. Kusmak, G.E.: Asymptotic solution of nonlinear equations of the second order with a variable coefficients (in Russian). Prikl.Math. and Mech. 23 (1959), N 3, 515-526.

3. Arnold, V.I. Additional chapters of the theory of the ordinary differential equations. Moscow: Nauka 1978 (in Russian).

4. Vakulenko, S.A.; Molotkov I.A.: Nonlinear waves concentrated in the vicinity of the given curve in unhomogeneous medium. Doklady AN of USSR, 262 (1982) N 3, 587-591 (in Russian).

5. Maslov, V.P. Complex WKB method for the nonlinear equations. Moscow: Nauka 1977

6. Babic, V.M., Buldyrev, V.S. Asymptotic methods in the diffraction problems for short waves. Moscow: Nauka 1972

7. Molotkov, I.A.: Surface-wave excitation in connection with diffraction at an impedance contour. Seminars in mathematics. V.A.Steklov Math.Institute, Leningrad, vol.17, 83-101. New York, London: Consultant Bureau 1972.

On Nonlinear Waves in Liquid-Filled Elastic Tubes

F. MAINARDI[1)] and H. BUGGISCH [2)]

1) Department of Mathematics, University of Bologna, I-40127 Bologna, Italy.

2) Department of Applied Mechanics, University of Karlsruhe, D-7500 Karlsruhe, FRG.

Summary

An analysis is developed to discuss the influence of viscosity on the propagation of non-linear waves in a liquid-filled elastic tube. The tube is assumed to be thin and uniform. The original system of equations of the one-dimensional theory is shown to be equivalent to a single non-linear wave equation with a Laplace convolution integral due to viscosity. This equation is considered in two different approximations (Poiseuille and boundary layer) to investigate shock waves running into a constant state. The boundary layer approximation is expected to be convenient to evaluate the non-ideal behaviour of shock tubes.

1. Introduction

The non-linear one-dimensional theory of wave propagation in liquid-filled elastic tubes has attracted considerable attention by researchers in view of the application to the arterial blood flow (see i.e.[1-42]). In spite of the many references that have already appeared on this topic the effect of the blood viscosity on the shock-wave formation is still poorly investigated. In this respect it should be noted that the effect of viscosity can be incorporated in a one-dimensional theory only after an analysis of the velocity profile has been separately performed (see i.e.[5,29,38]).

For flow with friction the work by Barnard et Al. [5] is noteworthy. They derived a one-dimensional theory based upon Navier-Stokes equations, which included effects due to a curved flow profile satisfying a no-slip boundary condition at the wall.

It is expected that the viscosity has a considerable influence on the shock-wave formation. Hoogstraten and Smit [35] obtained, by the use of wave-front expansions [43], explicit results about

Nonlinear Deformation Waves
IUTAM Symposium Tallinn 1982
Editors: U. Nigul, J. Engelbrecht

the circumstances under which shocks may form when the friction term is given by Poiseuille and Blasius formulae. However these formulae, suggested by Anliker et Al.[20] are in a strict sense only valid for steady flows at low and high Reynolds number respectively. For a reasonable picture of the effect of viscosity on shock-wave formation we believe that a time dependence in the friction term as derived from a boundary layer analysis is more convenient.

Starting from the basic equations by Barnard et Al.[5] we shall arrive at a single non-linear evolution equation which contains a Laplace convolution integral due to the viscosity. This equation enables us to obtain, in the steady-state regime, the results in [35] based on Poiseuille formula. In the relevant boundary-layer approximation our equation is similar to an equation derived by J.J.Keller[44] who considered the propagation of simple non-linear waves in gas-filled rigid tubes. All conclusions on shock-waves in [44] can be extended to our case.

2. Statement of the problem

The one-dimensional theory of blood flow derived in [5] is based on a number of assumptions which we recall hereafter. The blood is considered as an incompressible fluid, of density ρ and kinematic viscosity ν, and the artery as a straight, thin, elastic tube of circular cross section. Since arteries are constrained longitudinally, only axis-symmetric bulging motions of the (impermeable) tube wall are considered. Long-wave approximation is assumed, that is the typical axial wavelength of the flow is supposed to be much larger than the tube radius R. Hence only axial fluid acceleration and forces are taken into account.

In this model the independent variables are two: the axial coordinate x and the time t, and the dependent variables are three: the pressure difference across the tube wall p, the cross section area $A = \pi R^2$, and the averaged axial velocity U. Denoting by r the radial coordinate and $v = v(r,x,t)$ the local axial velocity, we get

$$U = \frac{2}{R^2}\int_0^R rv\,dr \, . \tag{2.1}$$

The equations for fluid motion are :
the continuity equation

$$A_t + (AU)_x = 0 , \qquad (2.2)$$

and the momentum equation

$$(AU)_t + (\chi AU^2)_x + Ap_x/\rho = \nu F , \qquad (2.3)$$

where index x and t denote partial derivatives with respect to x and t. In equation (2.3) χ and F are two functions defined by

$$\chi = \frac{2}{(RU)^2} \int_0^R rv^2 dr , \qquad (2.4)$$

$$F = 2\pi R \left.\frac{\partial v}{\partial r}\right|_R . \qquad (2.5)$$

Using the continuity equation, the momentum equation reads alternatively

$$U_t + UU_x + p_x/\rho = U(\chi-1)/A + \nu F/A . \qquad (2.6)$$

We note that for a flat profile ($\chi = 1$, $F = 0$) the R.H.S. of (2.6) reduces to zero and the usual momentum equation for the inviscid flow is obtained. In presence of viscosity it is necessary to know the variation of v with r (the "velocity profile") in order to evaluate χ and F. Since this variation cannot be properly determined without carrying out a full two-dimensional non-linear treatment of the problem, an approximate evaluation of χ and F will be discussed later in Section 3.

To complement equations (2.1),(2.2) we need a third equation. For a thin, tethered, elastic tube, in long-wave approximation, one usually takes the relation between p and A which holds under static conditions. We consider a uniform tube so that we write

$$p = \hat{p}(A), \qquad (2.7)$$

assuming $d\hat{p}/dA > 0$. Equation (2.7), also referred to as the wall equation, is to be used to eliminate one of the dependent variables.

Our basic equations are to be equipped with the appropriate initial and boundary conditions. We consider as in [35] a semi-infinite tube ($x \geqslant 0$), initially undisturbed and subjected at $x = 0$ to a continuous increase of the pressure. For the arterial flow this may describe the situation at the entrance of the aorta, where the aortic valve remains closed as long as the pressure inside the left ventricle of the heart is below p_0, and it opens suddenly when the ventricle pressure rapidly exceeds p_0. Thus

$$U(x,0) = 0, \quad p(x,0) = p_0, \quad A(x,0) = \hat{A}(p_0) = A_0, \tag{2.8}$$

and

$$p(0,t) = P(t), \quad P(0) = p_0. \tag{2.9}$$

3. Inviscid flow

Choosing A,U as dependent variables and introducing

$$c^2(A) = \frac{A}{\rho} \frac{d\hat{p}}{dA} , \tag{3.1}$$

we can eliminate the pressure gradient p_x in the momentum equation (2.3) or (2.6). We get

$$p_x/\rho = c^2(A) A_x/A . \tag{3.2}$$

Equations (2.2), (2.6), (3.2) provide our basic system

$$\underline{Y}_t + M \underline{Y}_x = 0, \tag{3.3}$$

where

$$\underline{Y} = \begin{pmatrix} A \\ U \end{pmatrix} = \underline{Y}(x,t) , \tag{3.4}$$

$$M = \begin{pmatrix} U & A \\ \frac{c^2(A)}{A} & U \end{pmatrix} = M(\underline{Y}) . \tag{3.5}$$

As a matter of fact the system (3.3) is hyperbolic resembling that of gasdynamics. The characteristics $\Gamma^{\pm}$ are described by the equations $\frac{dx}{dt} = \lambda^{\pm} = U \pm c(A)$.

In the linearized case about the undisturbed state one arrives at the wave-equation

$$\underline{Y}_{tt} - c_0^2\, \underline{Y}_{xx} = 0 \tag{3.6}$$

where

$$c_0 = c(A_0) = \sqrt{\frac{A_0}{\rho}\, \frac{d\hat{p}}{dA}\Big|_{A_0}} \,. \tag{3.7}$$

The linear solution to our problem is therefore a right-running wave propagating with velocity c_0. This velocity, first introduced by Young in 1808 but known as the Moens-Korteweg wave speed [38], reads

$$c_0 = \sqrt{\frac{Eh}{2\rho R_0}} \,, \tag{3.8}$$

where E is Young's modulus of the tube, and h, R_0 denote the wall thickness and the undisturbed radius, respectively.

Relevant examples of wall equations (2.7) are:

a) linear elastic material

$$p = \hat{p}(A) = p_0 + 2\rho c_0^2 \left\{(A/A_0)^{1/2} - 1\right\} ,$$

b) rubber-like material

$$p = \hat{p}(A) = p_0 + \tfrac{1}{2}\rho c_0^2 \left\{1 - (A/A_0)^{-2}\right\},$$

which provide respectively :

$$c(A) = c_0\, (A/A_0)^{1/4} \tag{3.9}$$

$$c(A) = c_0\, (A/A_0)^{-1} \tag{3.10}$$

The system (3.3) is known to admit an exact solution to our problem, the "simple wave", obtained by the Riemann invariants technique. The problem reduces to solve a kinematic wave - equation, obtained from the Γ^+ characteristic form, which reads

$$A_t + b(A)\, A_x = 0 \,, \tag{3.11}$$

where

$$b(A) = c(A) + \int_{A_0}^{A} \frac{c(A)}{A}\, dA \quad . \tag{3.12}$$

If we restrict our attention to the wave-front advancing into the undisturbed state, that is to the characteristic Γ_0^+ of equation $x = c_0 t$, it is easy to obtain the Bernoulli equation for the evolution of the weak discontinuities. Denoting by α the jump across Γ_0^+ we obtain

$$\frac{d\alpha}{dt} - \gamma\alpha^2 = 0 \ , \tag{3.13}$$

where

$$\gamma = b'(A_0) = c_0/A_0 + c'(A_0) \ , \tag{3.14}$$

A general equation ,in a vector form, ruling the weak discontinuities of the field variables can be obtained directly from (3.3) to (3.5) using the results by Boillat and Ruggeri[45]. From the solution of equation (3.13) a shock formation is expected at the critical time $t_s = 1/(\gamma\alpha_0)$ with $\alpha_0 = \alpha(0)$. For a linear elastic tube we get from (3.9), (3.14)

$$\gamma = \frac{5}{4}\, c_0/A_0 \ , \tag{3.15}$$

while, for a rubber-like tube we get from (3.10), (3.14)

$$\gamma = 0 \ , \tag{3.16}$$

which confirms that no shock will form in this case, in agreement with Olsen and Shapiro [10]. This case can be proved to be "exceptional" in Lax's sense [45].

From (3.11) to (3.14) we note that for the shock-wave formation on the wave-front it should be necessary to know the first order approximation $\tilde{b}(A)$ to $b(A)$, that is,

$$\tilde{b}(A) = c_0 + \gamma(A - A_0) \ . \tag{3.17}$$

The relative kinematic wave-equation

$$A_t + \tilde{b}(A)\, A_x = 0 \tag{3.18}$$

could be obtained directly from the basic system (3.3) working up to the second order of perturbation and concentrating on "relatively undistorted right-running waves" [46]

$$\underline{Y} = \begin{pmatrix} A \\ U \end{pmatrix} = \underline{Y}(\xi, \theta) \qquad (3.19)$$

with $\xi = x - c_0 t$, $\theta = t$, whereby $\underline{Y}$ depends only weakly on θ, $\underline{Y}_\theta << \underline{Y}_\xi$. Such a procedure will be generalized in order to treat the viscous flow where the Riemann invariants technique cannot be applied.

4. Viscous flow

Since for the arterial blood flow the effects of non-linearity and viscosity are expected to be relatively weak and comparable, in the present analysis we propose to utilize the linearized relations among pressure, area and velocity in order to evaluate the "friction term" νF entering the momentum equation.

A linearized analysis of the original two-dimensional field equations, based on the works by Womersley (see, i.e. [47]), allows one to obtain the velocity profile and to arrive at the following relations [48]

$$\nu F = c_0^2 A_x * \Phi(t) = - A_0 U_t * \Psi(t) \qquad (4.1)$$

where $*$ denotes the Laplace convolution, that is,

$$f * g = g * f = \int_0^t f(t-t')g(t')dt' .$$

In equations (4.1) $\Phi(t)$, $\Psi(t)$ denote two "memory functions" defined by their Laplace transforms

$$\bar{\Phi} = \bar{\Phi}(s\tau) = \frac{2}{\sqrt{s\tau}} \frac{I_1(\sqrt{s\tau})}{I_0(\sqrt{s\tau})}$$

$$\bar{\Psi} = \bar{\Psi}(s\tau) = \frac{2}{\sqrt{s\tau}} \frac{I_1(\sqrt{s\tau})}{I_2(\sqrt{s\tau})} \qquad (4.2)$$

where s is the complex Laplace parameter, $\tau = R_0^2/\nu = A_0/(\pi\nu)$, and I_0, I_1, I_2 denote the modified Bessel functions of order 0, 1, 2, respectively. It should be noted that in the linear case with friction one arrives at the following one-dimensional wave equations [48]

$$\underline{Y}_{tt} - c_0^2 \left[1 - \Phi(t)*\right] \underline{Y}_{xx} = 0,$$
$$\left[1 + \Psi(t)*\right] \underline{Y}_{tt} - c_0^2 \underline{Y}_{xx} = 0, \tag{4.3}$$

in analogy with the wave-equations of linear viscoelasticity, according to the "relaxation" and "creep" representations, respectively (see,i.e.[49]).

In the asymptotic limit $t/\tau \to \infty$, namely in the steady-state regime, we obtain

$$\Psi(t) \simeq \frac{8}{\tau} H(t) \tag{4.4}$$

where H(t) denotes the Heaviside step function, so that,from (4.1)

$$F \simeq -8\pi U . \tag{4.5}$$

This can be shown to correspond to a "parabolic" velocity profile according to Poiseuille law, that is,

$$v(r,x,t) = 2(1 - r^2/R^2)U(x,t) \tag{4.6}$$

and

$$\chi = 4/3 \tag{4.7}$$

The "boundary-layer" profile is obtained in the asymptotic limit $t/\tau \to 0$, so that

$$\Psi(t) \simeq \frac{2}{\sqrt{\pi}} (t\tau)^{-1/2} \tag{4.8}$$

and

$$\chi \simeq 1 . \tag{4.9}$$

Since the variation of χ with time is slow, in the present analysis we will assume χ constant.Then the basic equations for the viscous flow (2.2), (2.3) after using (3.2), (4.1) read

$$\begin{cases} A_t + (AU)_x = 0, \\ (AU)_t + \chi(AU^2)_x + c^2(A)\, A_x = -A_0 U_t * \Psi(t). \end{cases} \tag{4.10}$$

Looking for "relatively undistorted right-running waves" and working up to the second order of perturbation, we arrive at the following evolution equation

$$A_t + \tilde{b}_\chi(A)A_x = -\frac{1}{2}A_t * \Psi(t) \qquad (4.11)$$

where

$$\tilde{b}_\chi(A) = c_0 + \left[c'(A_0) + \chi c_0/A_0\right](A-A_0). \qquad (4.12)$$

In Poiseuille approximation the equation (4.11) reduces to

$$A_t + \tilde{b}_\chi(A)A_x + (4/\tau)(A-A_0) = 0, \qquad (4.13)$$

which implies the following Bernoulli equation for the amplitude of the "acceleration waves"

$$\frac{d\alpha}{dt} - \tilde{b}'_\chi(A_0)\,\alpha^2 + \frac{4}{\tau}\alpha = 0. \qquad (4.14)$$

This equation is similar to the corresponding one in [35] and, except for the different value of χ,† it predicts for the shock formation the same critical pressure rate (at $x = 0$) and the same critical time. However one has to interpret this result carefully since the steady-state assumption is of course no longer correct if the flow changes rapidly as in a shock transition.

In order to have more insight into the shock-wave formation we should consider the short-time behaviour, that is viscous effects have to be taken into account in the "boundary-layer" approximation. In this situation the equation (4.11) explicitly reads

$$A_t + \tilde{b}(A)A_x = \frac{1}{\sqrt{\pi\tau}}\int_0^{t-t_s} \frac{\partial}{\partial t'}A(x,t-t')\,\frac{dt'}{\sqrt{t'}} \qquad (4\text{-}15)$$

where $t_s = t_s(x)$ is the time at which the wave-front, advancing into a constant state, reaches the point x. This equation is of the same form as equation (11) in the paper [44] by Keller who considered non-linear simple waves in gas-filled rigid tubes, taking into account viscous effects in "boundary-layer"

† It should be noted that most authors (as [35]) who consider Poiseuille flow take erroneously $\chi = 1$.

approximation. All conclusions obtained in that paper are also true for our case. It can be shown that both continuous and discontinuous asymptotic waves exist for which the non linear distorsion and the viscous effect owing to the boundary-layer just balance. The continuous solution is given by $c_0 = c_s$, and

$$A = A(t-x/c_0) = \sqrt{\pi/\tau}\ \left\{c_0/\tilde{b}'(A_0)\right\}(t-x/c_0)^{1/2}, \qquad (4.16)$$

which is an exact solution of equation (4.15). Following [50], it can be shown that (4.15) is the unique continuous, non-trivial solution of (4.15). If we assume that a shock at the wave-front is moving into the undisturbed state at a constant speed c_s, we find the following relation between the shock-amplitude ΔA and c_s:

$$\Delta A = \frac{2}{\tilde{b}'(A_0)}(c_s - c_0). \qquad (4.17)$$

Introducing a dimensional variable

$$z = \frac{4}{\pi\tau}\left(\frac{c_s}{c_s-c_0}\right)^2 (t - x/c_s) \qquad (4.18)$$

the shock-wave is found to have the following asymptotic expressions in terms of z

$$A(z) \simeq \Delta A\left\{1 + z^{1/2} - (1-\pi/4)z\right\}, \quad z \to 0, \qquad (4.19)$$

$$A(z) \simeq \Delta A\ \pi/4\ \ z^{1/2}, \quad z \to \infty. \qquad (4.20)$$

We note that in order to keep a stationary wave of this type running, the cross-section area and consequently the pressure must increase infinitely as $z^{1/2}$. We find that the limiting pressure rate at $x = 0$ to keep this shock running is proportional to $t^{-1/2}$. Since the pressure must remain finite, any jump will decay to zero after a finite time.

Furthermore, as expected, the asymptote of (4.20) agrees with the exact solution (4.16). This means that both types of solutions can be collapsed into one normalized asymptotic solution [44]. A more detailed analysis will be presented in a next paper.

References

1. Lambert, J.W.: Fluid flow in a non rigid tube, Ph.D.Thesis, Purdue University ,1956.

2. Lambert, J.W. : On the nonlinearities of fluid flow in non rigid tubes, J.Franklin Inst., 266 (1958) 83-102.

3. Miekisz, S.: Non-linear theory of viscous flow in elastic tubes, Phys.Med.Biol. 6 (1961) 103-109.

4. Streeter, V.L.; Keitze, W.F. and Bohr, D.F. :Pulsatile pressure and flow through distensible vessels, Circ.Res., 13 (1963) 3-21.

5. Barnard, A.C.L.; Hunt, W.A.; Timlake, W.P. and Varley,E.: A theory of fluid flow in compliant tubes, Biophys.J., 6 (1966) 717-724.

6. Barnard, A.C.L.; Hunt, W.A.; Timlake, W.P. and Varley, E.: Peaking of the pressure pulse in fluid-filled tubes of spatially varying compliance, Biophys.J. 6 (1966) 735-746.

7. Rudinger,G.: Review of current mathematical methods for the analysis of blood flow, in Biomedical Flüid Mechanics Symposium, ASME : New-York, 1-33 (1966).

8. Skalak, R.: Wave propagation in blood flow, in Fung,Y.C. (editor): Biomechanics, ASME: New-York, 20-40 (1966).

9. Olsen, J.H.: Waves in fluid-filled elastic tubes, Ph.D.Thesis, M.I.T., Cambridge (USA),1966

10. Olsen, J.H.; Shapiro,A.H.: Large amplitude unsteady flow in liquid-filled tubes, J.Fluid Mech., 29 (1967), 513-538.

11. Nicholson, H.W.;Heiser,W.H.;Olsen,J.H.: Wave propagation in liquid-filled elastic tubes, Bull.Mech.Eng.Educ. 6 (1967) 371-376.

12. Seymour,B.R.; Varley,E., High frequency pulsatile flow in a tapering viscoelastic tube of varying stiffness, in Copley, A.L. (editor):Hemorheology, Oxford,New-York: Pergamon Press 1967, 131-142.

13. Rockwell, R.L.: Non linear analysis of pressure and shock waves in blood vessels, Ph.D. Thesis, Stanford University, 1969.

14. Seymour,B.R.: Pulsatile flow in flexible tubes, Ph.D.Thesis, University of Nottingham, 1969.

15. Rudinger, G.: Shock waves in mathematical models of the aorta, J.Appl.Mech., 37 (1970) 34-37.

16. Seymour,B.R.: Discussion of 'Shock waves in mathematical models of the aorta' by Rudinger,G., J.Appl.Mech., 37 (1970) 1202-1204.

17. Johnson,R.S.: A non linear equation incorporating damping and dispersion, J.Fluid Mech. 42 (1970) 49-60.

18. Johnson,R.S.: Non-linear waves in fluid-filled elastic tubes and related problems, Ph.D.Thesis, University of London,1971.

19. Bodley,W.E.: The nonlinearities of arterial blood flow, Phys.Med.Biol. 16 (1971) 663-672.

20. Anliker,M.;Rockwell,R.L.;Ogden,E. :Nonlinear analysis of flow pulses and shock waves in arteries,I,II, Z.Angew.Math. Phys.,22(1971), 217-246, 563-581.

21. Anliker,M.: Toward a nontraumatic study of the circulatory-system, in Fung;Y.C.;Perrone,N. and Anliker,M (eds): Biomechanics, its foundations and objectives, Englewood Cliffs, New Jersey: Prentice-Hall 1972.

22. Kenner,T.: Flow and pressure in arteries, in Fung,Y.C.;Perrone,N and Anliker,M. (eds): Biomechanics, its foundations and objectives. Englewood Cliffs,New Jersey : Prentice-Hall 1972.

23. Ling,S.C.;Atabek,H.B.:A non linear analysis of pulsatile flow in arteries, J.Fluid Mech.,55 (1972), 493-511.

24. Hughes,T.J.R.;Lubliner,J.: On the one-dimensional theory of blood flow in the larger vessels, Math.Bioscience, 18 (1973) 161-170.

25. Hughes,T.J.R.: A study of the one-dimensional theory of arterial pulse propagation, Report n.SESM 74-13, Structural Engineering Lab.,Univ.of Calif.,Berkeley,Dec.1974.

26. McDonald,D.A.: Blood flow in arteries. London: Edward Arnold 1974, 2nd ed.

27. Kivity,Y;Collins,R.:Non linear wave propagation in viscoelastic tubes: application to aortic rupture, J.Biomech., 7 (1974) 67-76.

28. Kivity,Y;Collins,R.: Steady state fluid flow in viscoelastic tubes. Application to blood flow in human arteries, Arch. Mech. 26 (1974) 921-931.

29. Lighthill,M.J.:Mathematical Biofluidodynamics, Philadelphia: Soc. for Industrial&Appl.Maths (SIAM), 1975, Chap .12.

30. Seymour,B.R.: Unsteady flow in flexible tubes: a modulated simple wave, Int.J.Engng Sci. 13 (1975) 579-594.

31. Rogers,C.;Clements,D.L.: Wave propagation in fluid filled elastic tubes, Acta Mechanica,22 (1975) 1-9.

32. Collins,R;Flaud,P;Geiger,D;Kivity,Y; Oddou,C. : Propagation of shock-like waves in long viscoelastic tubes, Biomechanics, Sofia, 1976.

33. Moodie,T.B.;Haddow,J.B.: Waves in thin-walled elastic tubes containing an incompressible inviscid fluid, Int.J.Non-Linear Mechanics, 12 (1977) 223-231.

334. Shapiro,A.H.: Steady flow in collapsible tubes, J.Biomech. Eng. 99 (1977) 126-147.

35. Hoogstraten,H.W.;Smit,C.H.: A mathematical theory of shock wave formation in arterial blood flow, Acta Mechanica, 30 (1978) 145-155.

36. Ravindran,R; Prasad,P. : A mathematical analysis of nonlinear waves in a fluid filled visco-elastic tube, Acta Mechanica, 31 (1979) 253-280.

37. Chen,P.J.;McCarthy,M.F.: Unsteady flow of incompressible fluids in deformable tubes, Acta Mechanica, 33 (1979)189-197.

38. Pedley,T.J.: The fluid mechanics of large blood vessels, Cambridge: University Press, 1980.

39. Flaud,P;Geiger,D.;Oddou,C: Large amplitude wave propagation in viscoelastic collapsible tubes, Paper presented at Euromech 127, Taormina, Italy, April 14-18,1980.

40. Barclay,D.W.;Moodie,T.B.; Madan,V.P.: Linear and non linear pulse propagation in fluid-filled compliant tubes. Meccanica 16 (1981) 3-9.

41. Cowley,S.J.: High Reynolds number flow through distorted channels and flexible tubes, Ph.D.Thesis,University of Cambridge (UK), 1981.

42. Cowley,S.J.: Elastic jumps on fluid-filled elastic tubes, J.Fluid Mech. 116 (1982) 459-473.

43. Whitham,G.B.:Linear and non linear waves, Wiley-Interscience: New-York 1974.

44. Keller,J.J.: Propagation of simple non-linear waves in gas filled tubes with friction, Z.Angew.Math.Phys. 32 (1981) 170-181.

45. Boillat,G; Ruggeri,T.: On the evolution of weak discontinuities for hyperbolic quasi-linear systems, Wave Motion, 1 (1979) 149-152.

46. Buggisch,H. : Nonlinear waves, in Rimrott,F.P.J.; Tabarrok, B.(eds), Proc. XV Int.Congress of Theor. and Appl.Mech., Amsterdam: North-Holland, 1980, 57-65.

47. Womersley,J.R.: An elastic tube theory of pulse transmission and oscillatory flow in mammalian arteries, WADC Tech.Report TR 56-614, Wright Air.Development Center, Jan.1957.

48. Mainardi,F: A linear viscoelastic model for distorsion of the arterial pulse, Preprint IFUB 1982, Proceedings of

III Int.Conf.on Mechanics in Medicine and Biol. (ICMMB), Compiègne, July 10-13,1982.

49. Graffi,D. : Mathematical models and waves in linear visco-elasticity, in Mainardi,F (editor): Wave Propagation in viscoelastic media, Boston,London: Pitman, 1982, 1-27.

50. Schneider,W.R.: The general solution of a non-linear integral equation of convolution type, Z.Angew. Math.Phys. 33 (1982)

Nonlinear Waves in Gas-Filled Elastic Tubes

H.W. BUGGISCH

Institut für Mechanische Verfahrenstechnik und Mechanik
Universität Karlsruhe
FR Germany

Summary

A model equation for the propagation of one-dimensional long nonlinear waves in elastic fluid-filled tubes will be derived. Essential assumptions are, that the fluid is compressible and inviscid, and that the wall deformation remains small, so that linear shell theory is applicable. The resulting wave equation for the gas pressure and the dispersion law will be discussed, and solutions of shock-wave-type will be obtained.

Introduction

In the past hundred years much research work has been done concerning the problem of linear and nonlinear wave propagation in elastic fluid-filled tubes. Usually, the fluid has been supposed to be incompressible. This assumption, which particularly underlies all nonlinear theories known to the author (with the exception of [1]), makes only sense, if the sound speed of the liquid is large compared to the wave speed. The effect of the finite compressibility of the fluid has been considered, however, in linear theories. Among the first significant contributions to the linear problem are those of Korteweg (1878), Lamb (1898), Joukowsky (1900), Allievi (1909), and Braun (1909) [2 - 6]. In the present paper a nonlinear theory of wave propagation in gas-filled elastic tubes is presented, where the nonlinear effects are mainly due to the nonlinearity of the pressure-density-relation of the gas. A similar theory has been put forward by the author in [1]. The following presentation is, however, more general and more detailed: Whereas in [1] the derivation of the model equation has only been sketched roughly, the present paper contains a detailed discussion of the basic equations; further, it is no longer assumed, that Poissons ratio of the wall material is zero. Finally, the present paper includes a discussion of the dispersion law.

The theory is based on the following simplifying assumptions:

Nonlinear Deformation Waves
IUTAM Symposium Tallinn 1982
Editors: U. Nigul, J. Engelbrecht

(a) The deformation of the tube wall remains small, so that linear theory suffices to include its effect. Flügge's shell theory [7,8] will be used for the wall.

(b) Viscosity and heat-conductivity of the gas are neglected. The gas-pressure p is supposed to depend on the gas-density ρ only.

(c) Waves are assumed to be sufficiently long, so that the gas flow is essentially in axial direction and can be treated by one-dimensional stream-tube-theory.

These assumptions suggest the following perturbation method: In first approximation the tube is assumed to be rigid, and a nonlinear pressure wave is propagated, the solution for the pressure being known from classical gas-dynamics of one-dimensional wave propagation (e.g. [9]). Let us denote this solution by $p_0(x,t)$. In a second step, the wall motion $r_0(x,t)$, due to $p_0(x,t)$, can be determined. Finally, the reaction of the gas pressure on the wall motion can be taken into account. This method, which is straight-forward, will not be pursued further in this paper; it works for relatively strong shock waves which propagate with a speed U significantly different from the sound speeds in front and behind the shock, respectively. Such solutions are presented in [1]. The most interesting feature of the structure of strong shocks is the occurence of a high-frequency precursor, due to the bending stiffness of the tube. The described procedure leads, however, into difficulties for weak nonlinear waves. The reason is, that in the small amplitude case $p_0(x,t)$ and consequently the wall motion $r_0(x,t)$ represent waves, traveling nearly with sound speed. This causes resonance phenomena for the pressure correction. Therefore, weak nonlinear waves have to be treated by a different appraoch: In this paper a simplified nonlinear differential equation for long small-amplitude pressure waves will be derived. Finally, solutions of this model equation will be discussed.

Basic equations for the wall

We introduce the following notations:

- h: thickness of the tube wall
- R: inner radius of the undeformed wall
- r: inner radius of the deformed wall
- s: axial displacement of a wall element

ρ_w: mass density of the wall material
E: Youngs modulus
ν: Poissons ratio
Δp: excess pressure in the tube

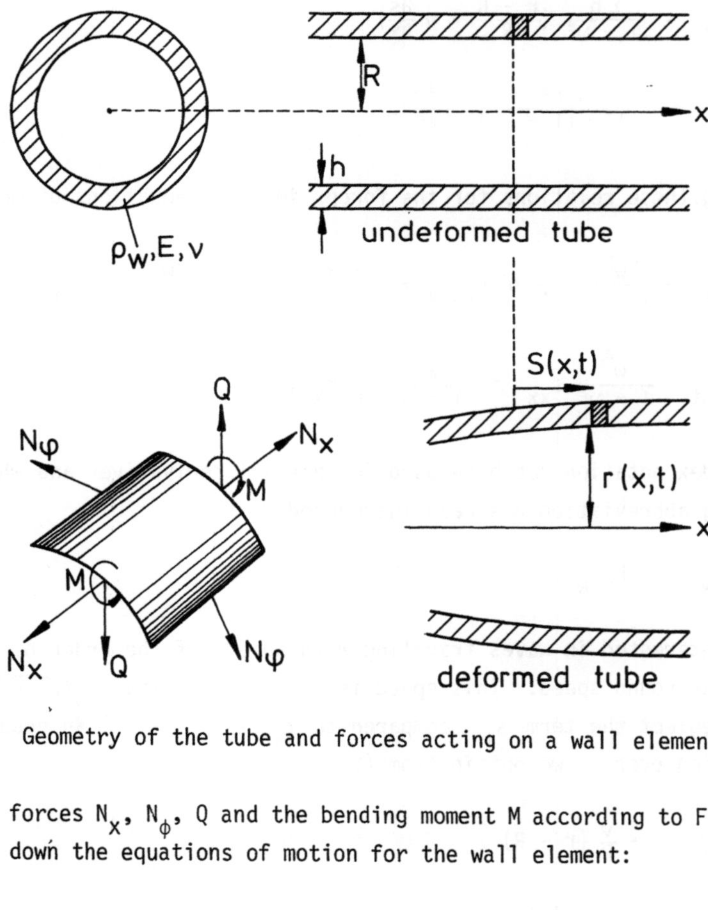

Fig. 1. Geometry of the tube and forces acting on a wall element

Defining the forces N_x, N_ϕ, Q and the bending moment M according to Fig. 1, we can write down the equations of motion for the wall element:

$$\rho_w h\, s_{tt} = \partial N_x/\partial x$$

$$\rho_w h\, r_{tt} = \partial Q/\partial x - N_\phi/R + \Delta p \; .$$

As usual, we shall neglect the rotational inertia of the wall element; therefore the equilibrium condition

$$Q = \partial M/\partial x$$

must be satisfied. The forces are related to the deformation of the tube according to Flügge's shell theory [7,8] by:

$$N_x = \frac{E\,h}{1-\nu^2}\left(\frac{\partial s}{\partial x} + \frac{\nu}{R}(r-R)\right)$$

$$N_\phi = \frac{E\,h}{1-\nu^2}\left(\frac{r-R}{R} + \nu\frac{\partial s}{\partial x}\right)$$

$$M = -\frac{E\,h^3}{12\cdot(1-\nu^2)}\,\frac{\partial^2 r}{\partial x^2}\,.$$

Inserting these relations for the forces into the equations of motion we get

$$r_{tt} + \frac{c_w^{\,2}}{1-\nu^2}\left\{\frac{1}{12}h^2\, r_{xxxx} + \frac{r-R}{R^2}\right\} = \frac{\Delta p}{\rho_w h} - \frac{c_w^{\,2}}{1-\nu^2}\,\frac{\nu}{R}\, s_x \qquad (1)$$

$$s_{tt} - \frac{c_w^{\,2}}{1-\nu^2}\, s_{xx} = \frac{c_w^{\,2}}{1-\nu^2}\,\frac{\nu}{R}\, r_x\,, \qquad (2)$$

where index notation has been used for partial derivatives and where the following abbreviation has been introduced

$$c_w^{\,2} = E/\rho_w\,.$$

We are interested in waves traveling with speeds of the order of magnitude of the gas sound speed. This speed is much smaller than $c_w/\sqrt{1-\nu^2}$. Hence we can neglect the term s_{tt} compared to $c_w^{\,2}\, s_{xx}/(1-\nu^2)$ in equation (2). Integrating over x we obtain from (2):

$$s_x = -\frac{\nu}{R}(r-R)\,.$$

Inserting this into (1) and, finally, adding a damping term $\lambda\cdot r_t$, which may be due to internal damping (or to external damping, as indicated in fig. 2), we arrive at the following equation for the radial motion of the wall:

$$\frac{c_w^{\,2}}{1-\nu^2}\,\frac{1}{12}h^2\, r_{xxxx} + r_{tt} + \lambda\, r_t + c_w^{\,2}\,\frac{r-R}{R^2} = \frac{\Delta p}{\rho_w h} \qquad (3)$$

This equation can not be simplified further, since the relative order of magnitude of the different terms depends on the wavelength. For very long waves on the left hand side only the term $c_W^2 (r - R)/R^2$ is important, i.e. bending stiffness, inertia and the damping term are neglegible. For shorter waves the next important term is due to the damping. For still shorter waves also the effects of inertia and finally of the bending stiffness play a certain role.

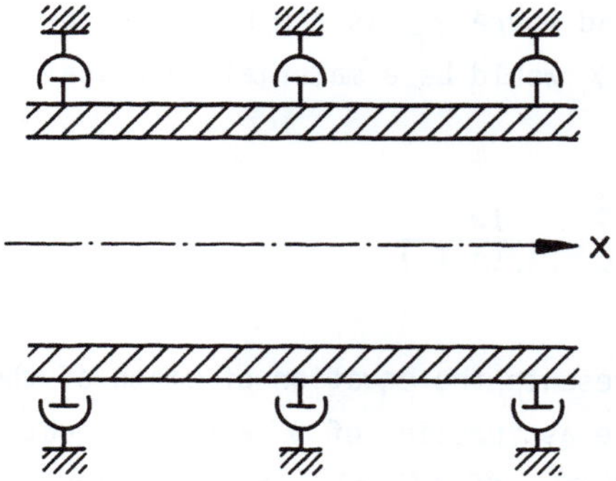

Fig. 2. Model for linear external damping of the wall

Basic equations for the gas

We introduce the following notations:

$p = p_0 + \Delta p$: gas-pressure in the wave

ρ: gas-density

u: speed of gas flow in axial direction

$D/Dt = \partial/\partial t + u\partial/\partial x$: material derivative

Momentum balance is expressed by the equation

$$\rho\, Du/Dt = - p_x .$$

The mass of the gas between cross-sections $x = 0$ and x, which is given by $m(x,t) = \int_0^x \rho\pi r^2\, dx$, obeys the mass balance equation

$$\frac{\partial m(x,t)}{\partial t} = \frac{\partial}{\partial t}\int_0^x \rho\pi\, r^2\, dx = \rho\pi r^2 u\Big|_{x=0} - \rho\pi r^2 u\Big|_x .$$

Differentiating this with respect to x we obtain

$$\frac{D\rho}{Dt} + \rho u_x = - \frac{2\rho}{r}\frac{Dr}{Dt} \tag{4}$$

The following transformation of the coordinate x turns out to be helpful:

$$X = \frac{1}{\rho_o} \int_{x_A(t)}^{x} \rho(\alpha,t)\, d\alpha \tag{5}$$

where $x_A(t)$ is the coordinate of a material point of the gas (e.g. the position of a piston), and where ρ_o is the reference density of the gas. Note that in a rigid tube X would be a material coordinate. We can solve (5) for x:

$$x = x_A + \rho_o \int_{o}^{X} \frac{d\alpha}{\hat{\rho}(p(\alpha,t))} \tag{5a}$$

where $\rho = \hat{\rho}(p)$ represents the equation of state of the gas, and where $p(X,t)$ is the pressure as function of X and t. Equation (5a) can be used to obtain finally $x(X,t)$, if $p(X,t)$ has been computed.

Consider a function $f(x,t)$, which in terms of X and t becomes $F(X,t)$. Obviously:

$$\partial f/\partial x = \partial F/\partial X \cdot \partial X/\partial x = \partial F/\partial X \cdot \rho/\rho_o \tag{6}$$

$$\frac{Df}{Dt} = \frac{\partial F}{\partial t} - 2\frac{\partial F}{\partial X} \cdot \int_{x_A(t)}^{x} \frac{\rho}{\rho_o r} \frac{Dr}{Dt}\, dx \tag{7}$$

From now on we suppose that the nondimensional pressure amplitude be $\delta_g \ll 1$, and that the nondimensional amplitude of the wall motion, δ_w, is small compared to δ_g. We will neglect terms of order $\delta_g \cdot \delta_w$, whereas terms of order δ_g^2 will be taken into account. Therefore, if F is the gas-speed or the gas-pressure or density, approximately

$$\frac{Df}{Dt} = \frac{\partial F}{\partial t} + O(\delta_g \cdot \delta_w) \approx \frac{\partial F}{\partial t} \tag{7a}$$

With the same order of accuracy we can replace $1/r \cdot Dr/Dt$ by $1/R \cdot \partial r/\partial t$. Hence, with help of (6) and (7a), momentum- and mass-balance can be written in the simpler form

$$\rho_o \, \partial u/\partial t \; = \; - \, \partial p/\partial X \tag{8}$$

$$- \rho_o \, \partial(1/\rho)/\partial t + \partial u/\partial X \; = \; - \, (2/R) \cdot \partial r/\partial t \, . \tag{9}$$

We add the equation of state

$$\rho \; = \; \hat{\rho}(p) \tag{10}$$

and eliminate from equations (8), (9) and (10) the density ρ and the flow-speed u. This leads to the following wave equation for the gas-pressure

$$(p_t/B^2(p))_t - p_{XX} \; = \; - \, 2 \, \rho_o/R \; r_{tt} \tag{11}$$

where

$$B^2 \; = \; (\hat{\rho}(p)/\rho_o)^2 \cdot (d\hat{\rho}/dp)^{-1} \tag{12}$$

is the sound speed with respect to the coordinate X.

<u>Nondimensional wave equations for the coupled system</u>

Applying the coordinate transformation (5) also for the equation (3), and again neglecting terms of order $\delta_g \cdot \delta_w$, we can replace in (3) r_{xxxx} by r_{XXXX}. Before proceeding further we finally introduce nondimensional quantities:

$$X^* \; = \; X/R \qquad\qquad t^* \; = \; t \, B(p_o)/R \; = \; t \, B_o/R$$

$$p^* \; = \; p/(\rho_o \, B_o{}^2) \qquad r^* \; = \; 1 + \Delta r^* \; = \; r/R$$

$$B^* \; = \; B/B_o$$

Making use of the abbreviations

$$\alpha \; = \; \frac{1}{12} \, \frac{h^2}{R^2} \cdot \frac{1}{1 - \nu^2} \quad ; \qquad \beta \; = \; B_o/c_w$$

$$\eta \; = \; \beta^2 \, \lambda \, R/B_o \qquad ; \qquad \delta \; = \; R \, \rho_o \, \beta^2/(h \, \rho_w)$$

the equations (3) and (11) can be written finally in the form

$$L_w \Delta r^* = \delta \cdot \Delta p^* \tag{13}$$

$$N_g \{\Delta p^*\} = -2\,\partial^2 \Delta r^*/\partial t^{*2} \tag{14}$$

where L_w is the "linear wall wave operator"

$$L_w = \alpha\,\partial^4/\partial X^{*4} + \beta^2\,\partial^2/\partial t^{*2} + \eta\,\partial/\partial t^* + 1 \tag{15}$$

and N_g is the "nonlinear gas wave operator"

$$N_g \{\Delta p^*\} = \frac{\partial}{\partial t^*}\left[\frac{\partial\,\Delta p^*/\partial t^*}{B^{*2}(p^*)}\right] - \frac{\partial^2\,\Delta p^*}{\partial X^{*2}} \,. \tag{16}$$

For simplicity, from now on we shall drop the symbol *; i.e. we shall e.g. denote the nondimensional pressure by p, the nondimensional radius by r, the nondimensional time by t and the nondimensional coordinate by X.

For some purpose it is useful to eliminate Δr from the system (13), (14). This is easily done by applying L_w to (14):

$$L_w\, N_g \{\Delta p\} = -2\,\delta\,\partial^2 p/\partial t^2 \,. \tag{17}$$

This equation also can be written in the following form:

$$(L_w - 1)\,\{(p_t/B^2)_t - p_{XX}\} + \{(p_t/A^2)_t - p_{XX}\} = 0 \tag{18}$$

where

$$A^{-2} = B^{-2} \cdot (1 + 2\,\delta) \,.$$

Equation (13) shows, that the "coupling parameter" δ is a measure for the influence of the pressure wave on the wall motion. If $\delta = 0$ (this occurs if the tube is rigid, or if the tube mass per unit length is infinite), the pressure wave is not able to put the wall into motion. Hence, $\Delta r \equiv 0$ is a possible solution of the problem in this case. The amplitude of the wall motion, δ_w, is related to the amplitude of the pressure wave, δ_g, by $\delta_w = \delta \cdot \delta_g$. Hence, the assumption $\delta_w \ll \delta_g$ is satisfied, if the coupling parameter is small.

<u>Dispersion law for linear waves</u>

In order to get some insight into the problem of pressure wave propagation in gas-filled tubes, we shall next treat linear harmonic waves. Note that in linear approximation X can be identified with x. The subsequent discussion is based on equation (17), in which the gas wave operator can be linearized for small amplitude waves:

$$L_w \, L_g \, p \;=\; -\,2\,\delta\,\partial^2 p/\partial t^2 \tag{19}$$

where $L_g = \partial^2/\partial t^2 - \partial^2/\partial X^2$ is the linearized nondimensional gas wave operator. We are interested in harmonic waves, $\Delta p = \delta_g \exp\{\,i\,k(x - ct)\,\}$. Inserting this ansatz into (19) we obtain the dispersion relation

$$\{\alpha k^4 - \beta^2 k^2 c^2 - i\,\eta\,k\,c + 1\}\{c^2 - 1\} \;=\; -\,2\,\delta\,c^2 \tag{20}$$

If the coupling parameter $\delta = 0$, this can be satisfied either by $c^2 = 1$, or by making the first bracket zero. This means, that either undisturbed gas-pressure waves are propagated in the tube, which remains undeformed, or that the pressure wave is enforced by the wall motion (which in turn is not influenced by the pressure). Solving (20) with $\delta = 0$ for c, we get the following result for rightrunning waves

$$c \;=\; \begin{cases} 1 \\ \dfrac{1}{\beta k}\left\{1 + \alpha k^4 - \dfrac{\eta^2}{4\beta^2}\right\}^{1/2} - \dfrac{i\,\eta}{2k\beta^2} \end{cases} \tag{21}$$

The first line represents the "gas-wave-branch", the second the "wall wave branch" in the uncoupled ($\delta = 0$) case. In fig. 3(a) both branches are sketched.

The main aim of this chapter is, to study the influence of weak coupling ($\delta \ll 1$) on the gas wave branch. For this purpose we try to solve (20) by a naive perturbation procedure: we put

$$c \;=\; 1 + \delta \cdot \Delta c_g \tag{22}$$

and neglect higher order terms in δ. The result is

$$\Delta c_g \;=\; -\,(\alpha k^4 - \beta^2 k^2 + 1 - i\,\eta\,k)^{-1}\,. \tag{23}$$

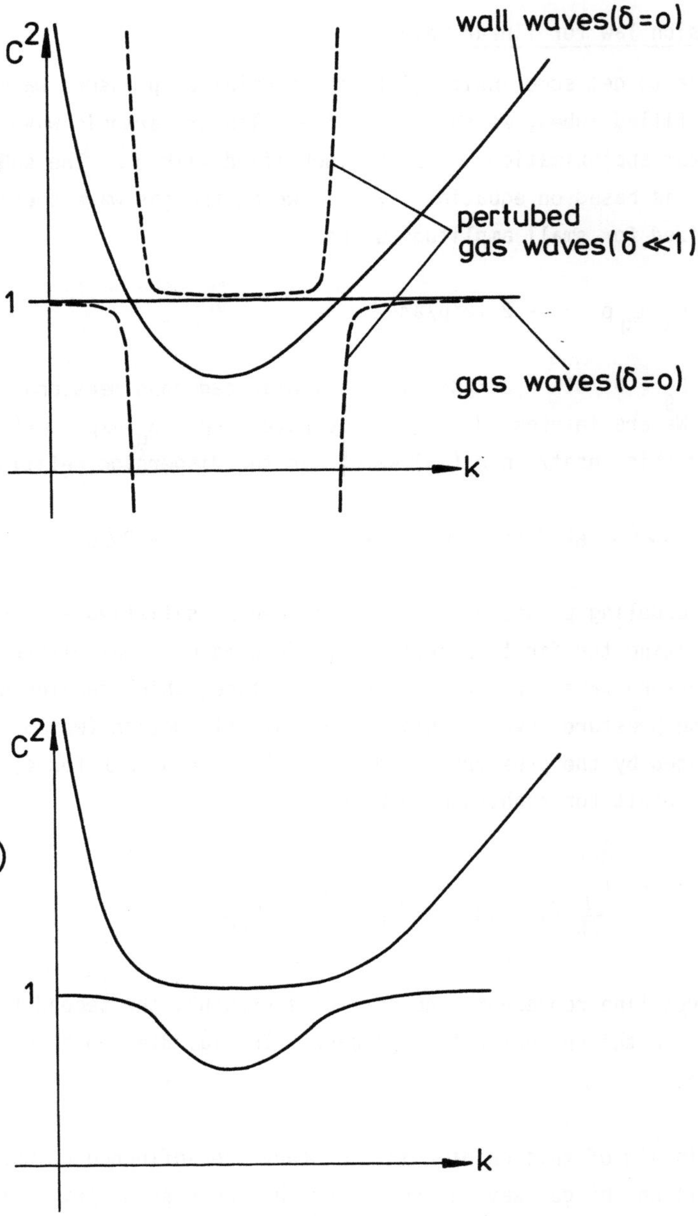

Fig. 3. Dispersion relation in the undamped case ($\eta = 0$)

(a) gas and wall-waves for $\delta = 0$ (solid lines) and perturbed gas waves, according to (22) and (23) (dotted lines)

(b) gas and wall-waves, according to the exact solution of (20)

This represents a valid approximation for all k, if either η is sufficiently large, or if $\beta^4 < 4\alpha$. If, however, $\eta = 0$ and $\beta^4 > 4\alpha$ (see fig. 4a, dotted line) the right hand side of (23) becomes infinite for $k_{1,2}^2 = \beta^2/(2\alpha) \mp \{\beta^2/(4\alpha^2) - 1/\alpha\}^{1/2}$. Therefore, for values of k close to k_1 or k_2, the above regular perturbation approach breaks down because of internal resonance of the system. Fortunately, if $\eta = 0$, equation (20) can be solved exactly for c^2. In fig. 3(b) the exact result is sketched. It becomes clear from the figure, why the regular perturbation procedure fails in the vicinity of $k_{1,2}$: There exists no connection between the perturbed gas-wave solutions in the intervalls $k < k_1$, $k_1 < k < k_2$ and $k > k_2$, respectively. Instead, in the vicinity of $k_{1,2}$, the perturbed gas wave branch is linked to the perturbed wall wave branch and vice versa.

Governing equations for long nonlinear waves

The subsequent discussion is based on equation (18). A measure for the nondimensional length scale, on which the pressure is changing, is $\Lambda = (\partial p/\partial X_{max})^{-1}$. If $\Lambda \gg 1$, in (18) only the lowest derivatives are important. Hence, very long nonlinear waves are governed by

$$(p_t/A^2(p))_t - p_{XX} = 0 . \qquad (24)$$

Such waves are therefore propagated with the pressure dependent speed

$$A(p) = B(p) \cdot (1 - \delta) + O(\delta^2) . \qquad (25)$$

The physical interpretation of this result is obvious: In long waves the tube wall, which may be treated as a membrane, is in static equilibrium with the gas pressure.

The result (24), which holds true asymptoticly if terms $\delta\delta_g/\Lambda^3$ are neglected, can be utilized to obtain from (18) the next higher approximation, in which only $\delta\delta_g/\Lambda^4$ and $\delta\delta_g^2/\Lambda^3$ are neglected. Replacing $(p_t/B^2)_t - p_{XX}$ by $(-2\delta) \cdot p_{XX}$ we get if $\eta \neq 0$:

$$(p_t/A^2(p))_t - p_{XX} - 2\eta\delta\, p_{XXt} = 0 \qquad (26)$$

If $\eta = 0$, the next higher approximation is

$$(p_t/A^2(p))_t - p_{XX} - 2\beta^2 \delta p_{xxtt} \cdot \qquad (27)$$

Whereas (26) leads to Burgers equation for unidirectional waves, (27) will lead to Korteweg - de Vries - equation [1,10]. For small values of η one should combine (26) and (27) and work with the model equation

$$(p_t/A^2(p))_t - p_{XX} - 2\delta(\eta p_{XXt} + \beta^2 p_{xxtt}) = 0 \qquad (28)$$

Structure of shock waves

Since the structure of shock waves, based on equation (18), has been discussed in detail by the author elsewhere [1], I shall give only a short summary here.

Stationary rightrunning shocks are defined as solutions of (18) with the following properties:

(a) $p - 1 = \Delta p(x - Ut) = \Delta p(z)$ with $U = \text{const}$

(b) $\Delta p \to 0$ for $z \to \infty$

(c) $\Delta p \to \delta_g$ for $z \to -\infty$.

Inserting the ansatz (a) into (18) and integrating twice over z from $-\infty$ to $+\infty$ one obtains, making use of (b) and (c), the following result for the nondimensional shock speed U:

$$U^{-2} = \delta_g^{-1} \int_1^{1+\delta_g} dp^*/A^2(p^*) \, . \qquad (29)$$

This is in accordance with the well known statement of relaxation gas-dynamics, that the speed of stationary shocks depends only on the low-frequency limit of sound speed.

If the shock strength is sufficiently large, the shock-structure contains an unsteady subshock, which obeys the same jump-conditions as it would in a rigid tube, since the cross section varies continuously. The direct perturbation method, described in the introduction, leads to shock-structures sketched in fig. 4.(a) and (b); (see [1]).

Weak shocks are long stretched continuous wave trains. Therefore, their structure can be discussed on basis of equation (28). According to [10] (see also [1]) the qualitative behavior of the solutions depends on a parameter γ, which in our case is defined by

$$\gamma = \sqrt{\frac{2}{\delta}} \frac{\beta}{\eta} \sqrt{\frac{dB}{dp}} \Bigg|_{p=1} \cdot \sqrt{\delta_g}$$

where B(p) is the nondimensional sound speed as function of the nondimensional pressure.

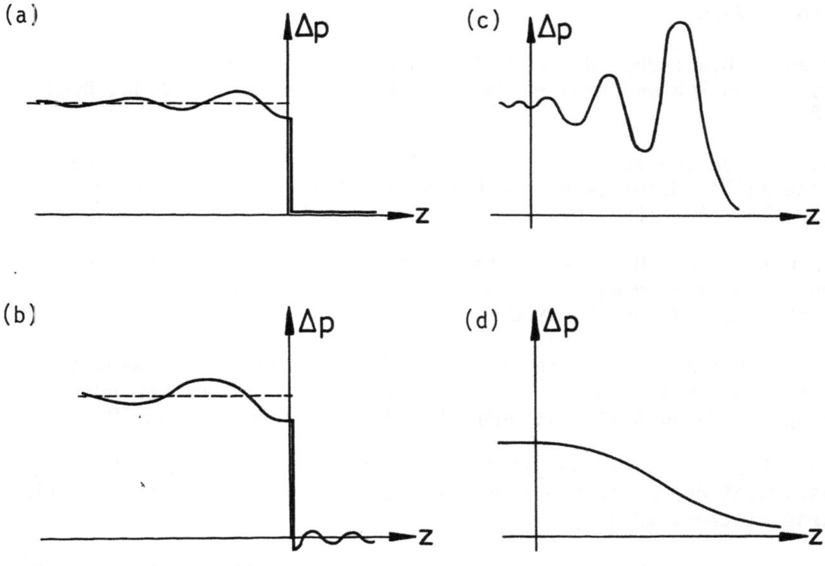

Fig. 4. Shock-structures of stationary shocks

(a) strong shock, membrane theory ($\alpha = 0$)
(b) strong shock, shell theory ($\alpha \neq 0$)
(c) weak shock, $\gamma > 1$
(d) weak shock, $\gamma < 1$

If $\gamma > 1$, the pressure oscillates around its asymptotic value $p = 1 + \delta_g$, whereas, if $\gamma < 1$, the pressure is a monotonously decreasing function of z (see [1] and [10]). If $\gamma \ll 1$, one obtains the well known Taylor-shock-structure

$$\Delta p = \frac{1}{2} \delta_g (1 - \mathrm{Tanh} \frac{z}{\Lambda})$$

where

$$\Lambda = 4 \eta \delta (dA/dp)^{-1} \cdot \delta_g^{-1}$$

is the characteristic length of the fully dispersed wave.

References

1. Buggisch, H.: Nonlinear Waves. Proceedings of the 15th International Congress of Theoretical and Applied Mechanics, Toronto 1980, Rimrott, F.P.J. and Tabarrok, B. (eds.), North Holland Publishing Company (1980), 57-66.

2. Korteweg, D.J.: Über die Fortpflanzungsgeschwindigkeit des Schalles in elastischen Rohren. Annalen der Physik und Chemie, 5. Folge, Band 5, 1878.

3. Lamb, H.: On the Velocity of Sound in a Tube as Affected by the Elasticity of the Walls. Memoires of the Manchester Literary and Philosophical Society, Vol. 42, 1898, p.1.

4. Joukowsky, N.E.: Über den hydraulischen Stoss in Wasserleitungsrohren. Memoires de l'Academie Imperiale des Sciences de St. Petersburgh, 8 serie, Vol. 9, No. 5, 1900.

5. Allievi, L.: Allgemeine Theorie über die veränderliche Bewegung des Wassers in Rohrleitungen, in: Allgemeine Theorie über veränderliche Bewegung des Wassers in Leitungen. Berlin, Springer-Verlag, 1909.

6. Braun, E.: Druckschwankungen in Rohrleitungen mit Berücksichtigung der Elastizität der Flüssigkeit und des Rohrmaterials. Stuttgart, Verlag von Konrad Wittwer, 1909.

7. Flügge, W.: Statik und Dynamik der Schalen. Berlin, Springer-Verlag, 1962.

8. Flügge, W.: Stresses in Shells. Berlin-Heidelberg-New York, Springer-Verlag, 1973.

9. Becker, E.: Gasdynamik. Stuttgart, Teubner-Verlag, 1965.

10. Karpmann, V.I.: Nichtlineare Wellen in dispersiven Medien. Braunschweig, Vieweg, 1977.

Propagation of Non-Uniform Shock Waves Over Large Distances

D. G. CRIGHTON

Department of Applied Mathematical Studies
University of Leeds
Leeds. LS2 9JT England

Summary

Model equations of nonlinear acoustics are analysed by Matched Asymptotic Expansions to predict the development over arbitrarily large ranges and times of flows containing weak shock waves. The equations include generalisations of Burgers' equation to include geometrical effects and density stratification and a Modified Burgers' equation with cubic nonlinearity for visco-elastic behaviour. Examined in detail are the different ways in which the structure of the shock wave breaks down at large distances.

1. Introduction

The aim of this paper is to advertise a study in theoretical nonlinear acoustics which continues to be carried out at the University of Leeds by the author and his students; some of the results have already been published [1,2] while further papers are sub judice [3,4]. This study involves the systematic use of Matched Asymptotic Expansions (MAE) to analyse model equations [5] which are typically versions of Burgers' equation generalised by the inclusion of linear terms representing geometrical spreading effects or modified in nonlinearity to describe viscoelastic behaviour.

As first shown in detail by Lighthill [6] , Burgers' equation partially substantiates a well-known and widely used procedure for dealing with flows containing relatively weak shock waves; in that procedure - "Weak Shock Theory", lossless solutions are joined by discontinuous shocks whose location is fixed by an "equal areas" rule, or equivalently (for quadratic nonlinearity) whose velocity is the mean of the velocities on either side of it. The treatment of Burgers' equation in [6] legitimises that procedure to a certain extent - though not perhaps to the extent

Nonlinear Deformation Waves
IUTAM Symposium Tallinn 1982
Editors: U. Nigul, J. Engelbrecht

hoped for by Lighthill - and shows also that the interior of of the shocks in a thermoviscous fluid is identical with the *steady* weak shock structure first analysed by Taylor [7].

When other linear mechanisms are present, however, causing the flow outside the shock to change partly in accordance with the demands of geometry or density stratification, for example, or giving rise to dissipation from relaxation of internal degrees of freedom or from tube wall attenuation, the situation is not the same. Lighthill (pp 322-323) argued that it *was* the same, at any rate as far as geometrical effects were concerned, setting out a procedure whereby a *Generalised Burgers' equation* could be derived in which, however, the coefficient of the viscous second derivative varied strongly with time or range; he argued, nonetheless, that it could be taken as constant, and the analytical advantages of Burgers' equation retained, because the viscous term was only ever needed in thin regions where shocks were present. What was not recognised in the argument, however, was the possibility that in general the shock wave might not be able to retain its steady Taylor structure, nor could the argument cope with the case in which the shock wave thickness had become comparable with some macroscopic scale.

Now the whole basis for use of a term like "shock wave" relies on the smallness of some appropriate inverse Reynolds number ε, and in order to examine, in a wide class of circumstances, the evolution, over large times or ranges, of flows containing weak shock waves, we have undertaken a systematic analysis of model equations in the limit $\varepsilon \to 0$. And since space-time then presumably breaks up into asymptotically disparate (but overlapping) regions in which different balances of physical mechanisms obtain, the use of MAE is clearly indicated. (One might, of course, hope to obtain exact solutions of the model equations; progress in that direction continues to be very slow, however, and we are not encouraged by our proof [8] that Burgers' equation itself is, together with the linear heat equation, the *only* parabolic equation which has a Bäcklund transformation either onto itself or onto another parabolic equation - this proof allowing arbitrary (x,t) dependence both in the BT and in the coefficients of the parabolic equations.) Our study can be regarded as bringing together the previously distinct ideas of shock dynamics and weak shock propagation (Whitham [9], Chapters 8 and 9) and as adding to those a detailed description of the effect which internal shock structure has on the lossless exterior flow, and vice-versa. In what follows, an outline will be given of the main issues at stake when one tries to follow a pattern of weak shock waves over arbitrarily large times and ranges.

2. Generalised Burgers' equations, with geometrical effects

Consider unidirectional waves, propagating down a tube of slowly varying area containing a thermoviscous fluid. It is well-known ([5], [6], [10] that the appropriate modification to Burgers' equation to account for area changes lies in the replacement of $\partial u/\partial t$ by $\partial u/\partial t + (\frac{1}{2} c_o u) \frac{d}{dx} \ln A(x)$, and that the resulting equation can always be put in the form

$$\frac{\partial V}{\partial s} - V \frac{\partial V}{\partial \theta} = \varepsilon\, g(s) \frac{\partial^2 V}{\partial \theta^2} \qquad (2.1)$$

in which $V \propto A^{\frac{1}{2}}u$, s is a time-like variable (often a range), θ a retarded time or linear phase variable, ε an inverse Reynolds number based on a wave length, a signal amplitude and the "diffusivity of sound" [6] , and the function g(s) represents the effects of tube area variation. The Generalised Burgers' equation (2.1) is to be solved in $s \geq 0$, $-\infty < \theta < +\infty$, (or in one period, if appropriate) subject to $V(0,\theta) = F(\theta)$ say. Specific cases of particular interest are $g(s) = 1$, $\frac{1}{2}s + s_o$, $\exp(s/s_o)$ for plane, cylindrical and spherical waves; these emphasise the strong variability of g(s), and point up the explicit appearance in the differential equation (though not then in the boundary condition) of a second dimensionless parameter s_o, essentially the product of Mach number and Helmholtz number at the source $s = 0$ $(r = r_o)$ In the first place we are interested in (2.1) as $\varepsilon \to 0$, s_o fixed. For many purposes that is not sufficient, however; for example, if, as is common, the variation in ε is brought about by a change in source amplitude or source frequency, then s_o will change in a definite way with ε and $s_o = O(1)$ will not be relevant. Ideally one would like to explore the whole (ε, s_o) parameter plane, but this is extremely demanding and has so far only been carried out for cylindrical and spherical sinusoidal waves ($F(\theta) = \sin\theta$) (see Scott [2]).

The lossless solution of (2.1) is the first term of an expansion

$$V(s,\theta,\varepsilon) = V_o(s,\theta) + \varepsilon V_1(s,\theta) + \ldots \tag{2.2}$$

in which the lossless equation for V_o has the characteristic solution

$$V_o = F(p) \quad , \quad p = \theta + sF(p) \tag{2.3}$$

while

$$V = \frac{1}{1-sF'(p)} \int_0^s g(\tau)\, \frac{\partial}{\partial p} \left[\frac{F'(p)}{1-\tau F'(p)} \right] d\tau \; . \tag{2.4}$$

At the shock formation range, $s = 1$ say, $\partial p/\partial\theta$ and $\partial V_o/\partial\theta$ become infinite, and for $s > 1$ the solution V_o is triple-valued for some ranges of θ. Resolve this in the first instance by inserting shock discontinuities at phase locations $\theta(s)$, as predicted by weak shock theory ([6], [9], [10]). Consider a particular shock, at $\theta = \Theta(s)$, say, and as an example suppose that for $\theta = \Theta(s) - 0$, $V_o = 0$, while for $\theta = \Theta(s) + 0$, $V_o = J(s)$. We analyse the shock structure by writing $\theta^* = (\theta - \Theta(s))/\varepsilon$ and expanding in the form

$$V(s,\theta^*,\varepsilon) = V_o^*(s,\theta^*) + \varepsilon V_1^*(s,\theta^*) + \ldots \; . \tag{2.5}$$

The equation for V_o^* balances convection and diffusion, and has a solution

$$V_o^* = \tfrac{1}{2}J(s) \; \{ \; 1 + \tanh \left[\frac{J(s)\,(\theta^*-A(s))}{4g(s)} \right] \} \tag{2.6}$$

after matching to the lossless solution, this requiring $V_o^* \to J(s)$ as $\theta^* \to +\infty$, $V_o^* \to 0$ as $\theta^* \to -\infty$. A constant $A(s)$ of integration is undetermined by this matching; it locates the centre of the shock in the sense that $V_o^* = \tfrac{1}{2}J(s)$ where $\theta^*=A(s)$, i.e. $\theta = \Theta(s) + \varepsilon A(s)$. This shift of the shock centre location from the prediction of weak shock theory is called by Lighthill [6] the "displacement due to diffusivity". As part of the matching leading to (2.6), one obtains also $d\Theta/ds = -\tfrac{1}{2}J(s)$, which of course is precisely the weak shock theory prediction.

Equation (2.6) is the steady Taylor shock solution, with amplitude $J(s)$ and thickness $\varepsilon g(s)/J(s)$ (with respect to θ).

To complete the $O(1)$ picture, the function $A(s)$ must be determined. This can sometimes be very efficiently achieved if a suitable integral conservation law is known, into which one inserts a uniformly valid composite expression for V, composed from V_o and V_o^* (see [1]). A more general procedure, which has further advantages, is to calculate the next term in the shock expansion (2.5), and to perform a (1,1) matching of the shock solution $V_o^* + \varepsilon V_1^*$ to the lossless solution $V_o + \varepsilon V_1$; this leads to an equation for dA/ds in terms of $J(s)$ and $g(s)$ which can be solved by quadrature to give $A(s)$ up to an additive constant. The general result is complicated and will be presented elsewhere; it simplifies very considerably for the case in which the initial function $F(\theta)$ is an N-wave, and correspondingly $J(s) \propto s^{-\frac{1}{2}}$. Then, aside from multiplicative constants depending upon numerical details of the various normalisations, the result is (cf [1])

$$A(s) = s^{\frac{1}{2}} \int_1^s \frac{g(\sigma)}{\sigma} d\sigma + \text{constant} , \tag{2.7}$$

so that for large s we have $A(s) \sim s^{\frac{1}{2}} \ln s$, $A(s) \sim s^{3/2}$, $A(s) \sim s^{-\frac{1}{2}} \exp s$ for plane, cylindrical and spherical N-waves. Of these, the first was obtained by Lighthill [6] who, however, regarded it as leading only to a small correction to the weak shock location. In any event, the function $A(s)$ is determined up to an additive constant whose value is to be determined by matching both the lossless solution V_o and the Taylor shock solution V_o^* to a solution $\hat{V}_o$ valid in the "embryo shock region". This is a region around the range, s=1 say, at which each shock first forms; the extent $|s-1|$ and the range of θ , with respect to ε, depend on the initial function $F(\theta)$ (for example, $|s-1| = O(\varepsilon^{\frac{1}{2}})$, $|\theta - 2n\pi| = O(\varepsilon^{\frac{3}{4}})$, $V = O(\varepsilon^{\frac{1}{4}})$ are the appropriate scalings for $F(\theta) = \sin\theta$) but whatever the scalings the leading order velocity $\hat{V}_o$ satisfies the *plane* Burgers' equation

$$\frac{\partial \hat{V}_o}{\partial \hat{s}} - \hat{V}_o \frac{\partial \hat{V}_o}{\partial \hat{\theta}} = g(1) \frac{\partial^2 \hat{V}_o}{\partial \hat{\theta}^2} \quad .$$

Details of the solution by the Cole-Hopf transformation and

matching to the shock and lossless fields are tricky [1], but the upshot is a complete determination of A(s). As to the correction $\varepsilon V_1{}^*$ to the Taylor solution, this is determined by the (1,1) matching, except for a function B(s) which represents a further correction to the shock location, i.e. the shock "centre" is at $\Theta(s) + \varepsilon A(s) + \varepsilon^2 B(s) + O(\varepsilon^3)$. This function can be determined either by still higher order matching (and very laborious calculation of the terms V_2, $V_2{}^*$) or [1] by the use of an integral conservation law.

We are now in a position to examine the consistency of the separation of the flow into lossless regions separated by thin Taylor shocks at the weak shock theory locations (aside from the embryo shock region in which the shocks are not yet recognisable as such). The consistency of this separation evidently requires that

I: the shock displacement due to diffusivity remains small,

II: the shocks remain thin compared with the scale of the lossless flow,

III: the Taylor shock solution itself remains valid in the shock regions.

We have found cases so far which lead to the violation of each of I-III at large values of s (occasionally at some finite value of s), and the breakdown of structure leads to different large s behaviour depending upon which violation occurs first. We give examples to illustrate the way in which the breakdown is diagnosed and the subsequent structure of the wave form established; we are interested, in particular, in whether the wave eventually matures into "old-age" - a phase in which non-linear effects are everywhere negligible and the wave dies under linear dissipative and geometrical mechanisms.

Example I: Plane N-waves [1]

Here weak shock theory gives $J(s) = s^{-\frac{1}{2}}$, while $g(s) = 1$, $A(s) = s^{\frac{1}{2}} \ln s$. Also, the overall width of the N-wave is $s^{\frac{1}{2}}$, so that the shocks remain thin compared with the overall scale, but because of diffusive effects drift far from their weak shock location by the time $s = O(\ln \varepsilon^{-1})$. Further, if the ratio $\varepsilon V_1{}^*/V_0{}^*$ is estimated as $s \to \infty$, it is found to be O(1) when $s = O(\ln \varepsilon^{-1})$. Thus the lossless solution appears to hold except within a thin region - whose location is unknown when s is as large as $\ln \varepsilon^{-1}$ - within which the steady Taylor solution can no longer hold. It is hard to analyse this "translational nonuniformity" - even when one has the exact solution in integral form via the Cole-Hopf transformation. The best we have been able to do is construct a smooth solution of the complete Burgers' equation, describing the whole wave form and holding for $s = O(\ln \varepsilon^{-1})$, this solution matching (as s decreases) the N-wave plus Taylor shock solution already described. (Although this solution satisfies the complete Burgers' equation, it does not *satisfy* the initial N-wave conditions; it is a simple solution which uniformly approximates to the complicated exact solution everywhere except in the embryo shock region which immediately follows the initial establishment of an N-wave.) Naturally, since the new solution is smooth, it fails to describe in detail what happens when the thin Taylor shocks drift far from their "lossless" location, and the analysis of the "translational nonuniformity" could still be pursued further. As s increases beyond $O(\ln \varepsilon^{-1})$, the new solution takes the form of a dipole solution to the linearised equation, and the old-age phase of linear decay has been reached.

Example 2. Plane harmonic waves [11]

Here $F(\theta) = \sin\theta$, the shock formation range is $s = 1$ and by symmetry the shocks remain centred at $\theta = 2n\pi$ for all s. Thus $A(s) = 0$, and it is also found that the Taylor shock solution remains valid for all s in the sense that V_1^*/V_0^* is uniformly bounded for $1 < s < \infty$. On the other hand, the shocks do now become thick as $s \to \infty$, and in fact the thickness becomes $O(1)$ at $s = O(\varepsilon^{-1})$. The nonuniformity involves the whole of the wave, and for $s = O(\varepsilon^{-1})$ the leading order solution for the velocity satisfies the full Burgers' equation (obviously nothing less than this would be possible). The solution which matches the lossless wave and the Taylor shocks for $s \ll \varepsilon^{-1}$ (and for $\theta = O(1)$ and $\theta = O(\varepsilon)$ respectively) is the well-known Fay solution,

$$V \sim 2\varepsilon \sum_{n=1}^{\infty} \frac{\sin(n\theta)}{\sinh(n\varepsilon s)} . \tag{2.8}$$

Old-age decay follows at once, for $\varepsilon s \gg 1$, as

$$V \sim 4\varepsilon \exp(-\varepsilon s) \sin\theta \tag{2.9}$$

displaying the well-known phenomenon of amplitude saturation (i.e. in dimensional form, the amplitude of V is independent of the initial amplitude). The nonuniformity here is much more easily treated than that of ex I; the shocks remain at known locations and described by Taylor's solution, and lose their identity because of thickening only.

Example 3. Cylindrical N-waves [1]

Here $J(s) \sim s^{-\frac{1}{2}}$, the N-wave width $\sim s^{\frac{1}{2}}$, $g(s) \sim s$ and $A(s) \sim s^{3/2}$ as $s \to \infty$. Assumptions I, II, III are all violated at $s = O(\varepsilon^{-1})$ (and none before then). The nonuniformity involves the whole wave, and for $s = O(\varepsilon^{-1})$ the leading order solution for V satisfies the full cylindrical Burgers' equation

$$\frac{\partial V'_o}{\partial s'} - V_o'\frac{\partial V'_o}{\partial \theta'} = s'\frac{\partial^2 V'_o}{\partial \theta'^2} \tag{2.10}$$

(with suitable scalings of V, s and θ) together with matching conditions for $s \ll \varepsilon^{-1}$. As the only known solution to (2.10) is a similarity solution [10] which here is irrelevant, the ultimate fate of the wave cannot be followed in all detail. However, one can argue that for $s \gg \varepsilon^{-1}$ the nonlinear terms in (2.10) must be small over the whole wave, and the old-age solution will take the form of a dipole solution to the linear equation. The only part of this dipole solution which is undetermined is a purely numerical constant multiplier (independent of ε), so that the MAE approach supplies all the scalings throughout the evolution of the wave, despite our inability to solve (2.10).

Example 4. Spherical N-waves [1]

In this problem the first nonuniformity arises from a failure of III; the transition from $V \sim 0$ to $V \sim J(s)$ still occurs in a thin region centred on the weak shock location $\Theta = s^{\frac{1}{2}}$, but the transition is not one in which Taylor's solution, with its steady-state balance between convective steepening and viscous diffusion, can hold. The reason - as is clear from examination of the complicated expression for the correction $\varepsilon V_1{}^*$ to the Taylor solution $V_o{}^*$ ($V_1{}^*$ containing terms dg/ds, dJ/ds, representing the rate at which geometrical effects are causing the flow outside the shock to change) - is simply that the external lossless flow is changing so rapidly under convective nonlinear and geometrical linear mechanisms that the internal shock structure cannot adjust to maintain the Taylor form. This first *local* nonuniformity arises at a range $S_1(\varepsilon)$ defined implicitly by $\varepsilon S_1(\varepsilon) \exp(S_1(\varepsilon)) = 1$ (such that $\varepsilon V_1^*/V_o^* = O(1)$ for $\theta^*=O(1)$) and at such ranges the new "evolutionary" shock is governed by the full spherical Burgers' equation

$$\frac{\partial V'_o}{\partial s'} - V_o'\frac{\partial V'_o}{\partial \theta'} = (\exp s')\, \frac{\partial^2 V'_o}{\partial \theta'^2} \tag{2.11}$$

in scaled variables (the scalings here being highly unusual). As in the cylindrical case, this equation cannot be solved exactly; however, even for ranges $s \gg S_1(\varepsilon)$ there is no failure in anything but the internal shock structure, and so the solution to (2.11) satisfies matching to the outer lossless N-wave both as $s \ll S_1$ and as $s \gg S_1$ (as well as matching to the Taylor tanh for $s \ll S_1$ and appropriate θ). Now as s increases beyond S_1, the nonlinear term in (2.11) becomes recessive, and there is a unique error function solution to the linear equation with the required matching properties. That solution, essentially

$$V'_o \sim \operatorname{erf}\,[\tfrac{1}{2}\theta'\, e^{-s'/2}]\,, \tag{2.12}$$

describes the large s behaviour of the evolutionary shock, and the width of this erf-shock can now be compared with that of the lossless N-wave. This comparison indicates a further nonuniformity - this time over the whole wave and arising from violation by (2.12) of Condition II - at a larger range $S_2(\varepsilon)$ defined implicitly by $\varepsilon\, S_2^{-1}(\varepsilon) \exp(S_2(\varepsilon)) = 1$. At ranges $O(S_2)$, and again with unusual scalings, the leading order solution V_o'' describing the whole wave satisfies the linear version of (2.11), namely

$$\frac{\partial V''_o}{\partial s''} = (\exp s'')\, \frac{\partial^2 V''_o}{\partial \theta''^2} \tag{2.13}$$

whose solution is to be matched to the lossless N-wave and to the erf-shock as $s \ll S_2$, for appropriate θ. Such a solution can be found, with all coefficients fully determined (despite the failure to solve (2.11) for the evolutionary shock), and in the ultimate old-age decay (with $s >> S_2$) the coefficient C in

$$V \sim C\, S_2^{\frac{1}{2}} \frac{\theta}{(\varepsilon e^s)^{3/2}} \exp\{-\frac{\theta^2}{4\varepsilon e^s}\} \tag{2.14}$$

is known exactly. This represents a triumph for the MAE approach - and one which in which no improvement in treatment or understanding will be achieved until the general solution of the full spherical wave Burgers' equation is found. A similar double nonuniformity in the transition from fully-developed shocks into old-age occurs also in the important problem of harmonic waves, and is resolved in a similar manner. Details, as far as they can be obtained, are summarised in [1] and [2] .

<u>Example 5. Spherical waves in an isothermal stratified atmosphere[3]</u>

To correspond with the notation of [3] , the variables (s,θ) are replaced by (z,τ), θ now denoting angle from the vertical in an atmosphere of constant temperature and exponential density stratification. The sound speed is constant, but the kinematic viscosity varies exponentially if it is assumed that the dynamic viscosity is a function of temperature alone. For a linear lossless spherical wave, $\rho^{-\frac{1}{2}}ru$ would be a function of phase only, and the phase has simple form, $\omega(t - (r-r_o)/c_o)$. In [3] we examine the sinusoidal wave problem, with $F(\tau) = \sin\tau$, introducting quadratic nonlinearity and thermoviscous diffusion with an exponentially varying diffusivity. The model equation can again be put in the form (2.1), with an implicit definition of $g(z)$; the essentials are that shock formation takes place at $z(r,\theta) \doteq 1$, that $g(z)$ and z both increase to infinity with r above the horizontal ($\theta < \pi/2$) and that below the horizontal $g(z) \to 0$ but $z \to$ a finite $z_\infty(\theta)$ as $r \to \infty$. A typical shock formation curve is shown in Figure 1; on a spherical expanding wave front there are shocks above the intersection of the front with $z = 1$, but not below. Above the horizontal, breakdown of the shock structure arises from violation of II and III when $z \sim Z_1(\varepsilon) = [\varepsilon^{\frac{1}{2}} \ln \varepsilon^{-1}]^{-1}$ and then the whole wave is described to leading order by the *cylindrical Burgers' equation* (for diverging waves) - i.e. by (2.10). Old-age follows as the nonlinear terms subside, amplitude saturation has already occurred, and the final linear decay of the wave can be calculated up to a multiplicative constant (which is a function of angle θ and of the ratio of wavelength to stratification scale height).

Below the horizontal the outer lossless flow tends to a *steady* state, $J(z) \to J(z_\infty)$, and the shocks get ever thinner (exponentially with r) at fixed locations. Again, however, the changes in external mechanisms are too rapid to allow the Taylor structure to survive, and condition III is violated at $r = O(\varepsilon^{-\frac{1}{2}})$, or $z - z_\infty = O(\varepsilon^{\frac{1}{2}} \exp(-1/\varepsilon^{\frac{1}{2}}))$. The new shocks are evolutionary, and in them we have

$$\frac{\partial V_o^+}{\partial z^+} + V_o^+ \frac{\partial V_o^+}{\partial \tau^+} + z^+ \frac{\partial^2 V_o^+}{\partial \tau^{+2}} = 0 \tag{2.15}$$

where $z^+ \to 0$ as $z \to z_\infty$. Equation (2.15) is again a type of cylindrical Burgers' equation, but the change of sign does not seem to permit any simple interpretation. Outside the shocks, the lossless solution persists, and no amplitude saturation occurs, since this requires z to become asymptotically large (at least as $O(\varepsilon^{-1})$). No solutions of (2.15) are known, beyond a similarity solution, discussed in [3] and shown to be irrelevant here. Propagation along horizontal rays is not covered in the discussion here; a much more complicated equation governs that propagation, as described in [3] . The particular interest of the present example is that it shows nonuniformity at large ranges (actually not all that large, $r = O(\varepsilon^{-\frac{1}{2}})$) in a flow where the shocks are at fixed phases and where their thickness gets ever smaller.

3. The Modified Burgers' Equation [4]

An equation of the form (again with a change of notation)

$$\frac{\partial V}{\partial x} + V^2 \frac{\partial V}{\partial \tau} = \varepsilon \frac{\partial^2 V}{\partial \tau^2} \tag{3.1}$$

was shown in [12] to govern shear waves in a certain model of viscoelastic behaviour and switch-on shocks in m.h.d. The paper [4] gives asymptotic solutions, with fascinating structure, to the N-wave and sinusoidal wave problems for (3.1). Figure 2(a), (b) shows the wave profiles, including shock waves; the nonlinear steepening is, of course, strange if one is used

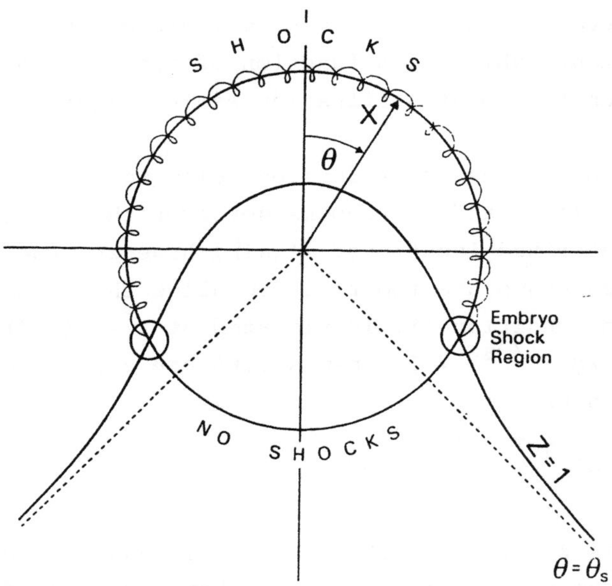

Figure 1. The shock formation curve z = 1 and a spherical wave front in an isothermal stratified atmosphere.

only to dealing with quadratic nonlinearity. One can find the first two terms of a **"lossless"** expansion, locate shock waves in accordance with an "equal areas" rule or the equivalent in differential form, and calculate two terms of an expansion describing the interior shock structure. (The "embryo shock region" for (3.1) is governed to leading order by the ordinary Burgers' equation.) The first term in the shock expansion may again be called a Taylor solution, balancing cubic nonlinearity against diffusion, with range x appearing only parametrically. All the solutions are a little more complicated for (3.1) than for the ordinary Burgers' equation but beyond that there appears at first to be no difficulty of principle in establishing the flow structure for finite ranges x. There is, in particular, an acceptable internal structure for the head shock of the N-wave problem for all finite x, this shock taking V from 0 for $\tau > c_h(x)$ to F(x) for $\tau < c_h(x)$.

The tail shock, however, provides a transition between values G(x) and H(x) of V (see Figure 2a), and because of the *cubic nonlinearity* and lack of any *quadratic* term in (3.1), any Taylor type of shock can only provide a transition between values of any two of $G(x)$, $H(x)$ and $-[H(x) + G(x)]$ which are not separated by the third of these. Initially a satisfactory transition exists, but at $x = X_1 = 10$ for the N-wave problem, and at $x = X_1 = 9.601$ for the sinusoidal wave, the values of H, G are such that $H + 2G = 0$. If it is assumed that on either side of the shock the values of V continue to be supplied from the initial data via lossless characteristics, then $H + 2G < 0$ for $x > X_1$ and the shock transition from G to H will have an unacceptable interior singularity. It is argued in [4] that an acceptable shock solution is mandatory for $x > X_1$, and that this then requires the outer flow to the right of the shock in Figure 2c to be related to that on the left by the condition $H + 2G = 0$ for all $x \geq X_1$. Data to the left of the shock come via characteristics, and determine $H(x)$; the characteristic emanating from $\tau = \sigma$ at $x = 0$ and carrying a value $V_o = H(\sigma)$ is "refracted" when it reaches the tail shock (for $x \geq X_1$), continuing on the right with gradient increased by a factor 4 and carrying now a value $V_o = -\frac{1}{2}H(\sigma)$. In this way, data for the lossless solution is supplied to the region (shaded in Figure 2c) above the shock path which would otherwise be unable to possess a lossless solution, lying as it does between the shock path and a characteristic which is tangential at X_1 to the shock path. Characteristics refracted through the tail shock may then (for $x > X_2 = 90$) impinge from the left on the head shock path, and then no acceptable solution will be possible for the head shock unless the condition $H + 2G = 0$ is also enforced there, with refraction through both shocks for $x > X_2$. For the sinusoidal initial distribution of V, characteristics propagate in this way from one half-cycle to the next indefinitely, so that at any finite x the values of V outside the shocks have to be traced through a finite number of refractions back to the initial data.

The analytical details to substantiate all this are complicated. There are also numerous other nonuniformities, some of a mild kind, arising both at finite x and as $x \to \infty$. For the N-wave problem, the shocks become thick and lose their Taylor form at $x = O(\varepsilon^{-2})$, and then the whole wave is governed to leading order by the *full* equation (3.1), followed by old-age decay

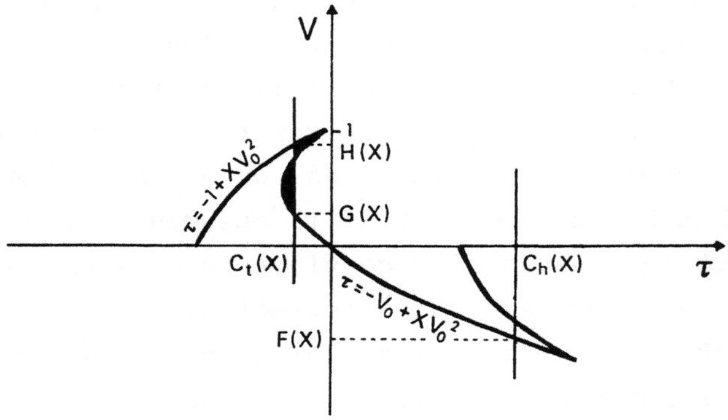

Figure 2(a). Evolution of an initial N-wave for the Modified Burgers' equation (MBE). The head shock path is given by $\tau = c_h(x)$, the tail shock path by $\tau = c_t(x)$. Shaded regions indicate satisfaction of an equal areas rule.

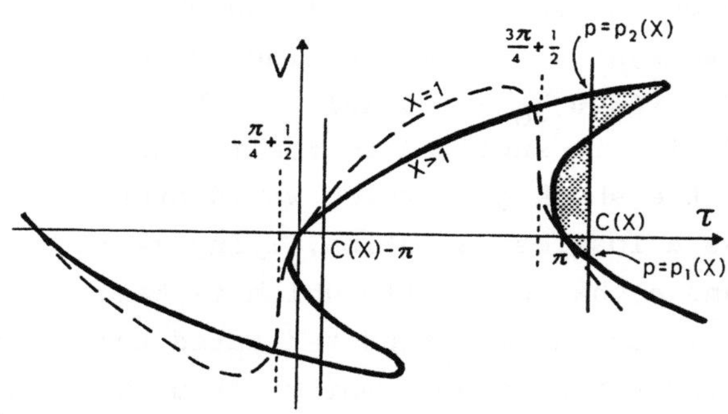

Figure 2(b) Evolution of an initial sinusoidal wave for the MBE, at the shock formation range $x = 1$ and at a larger range $x > 1$. The shocks form at $\tau = \frac{3\pi}{4} + \frac{1}{2}$ and $\tau = -\frac{\pi}{4} + \frac{1}{2}$ when $x = 1$, and are located at $\tau = c(x)$, $\tau = c(x) - \pi$ for $x > 1$.

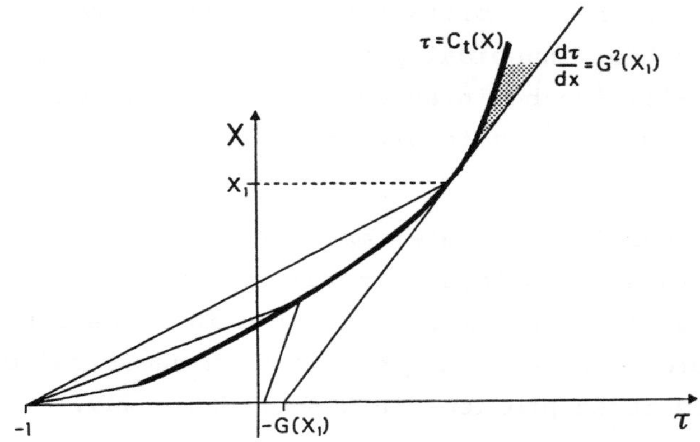

Figure 2(c). Tangency of the tail shock path and a characteristic for the N-wave MBE problem. The shaded region cannot be reached by characteristics emanating directly from x = 0.

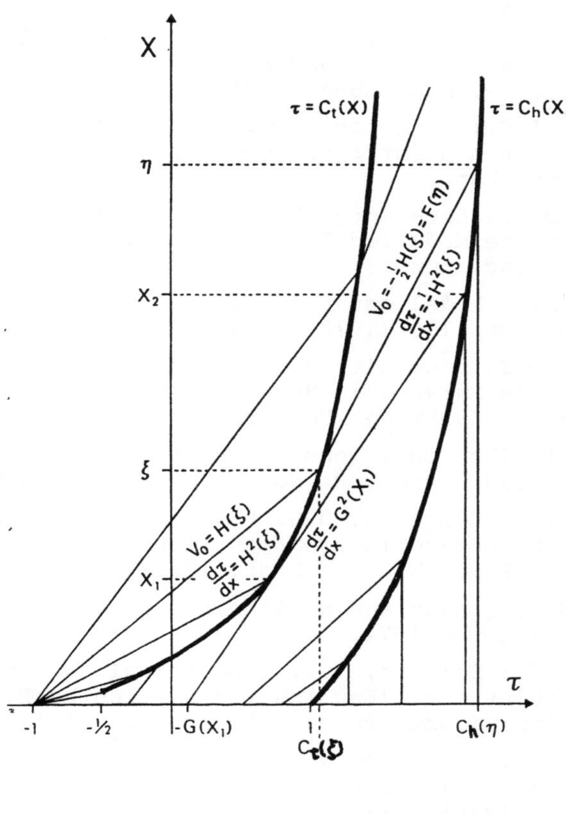

Figure 2(d). Refraction of characteristics through the tail shock and head shock.

(though the question of amplitude saturation has not been resolved). For the sinusoidal problem, all three criteria analogous to I-III can be investigated and all three are violated at $x = O(\varepsilon^{-1})$, again with the *full* (3.1) holding over the whole wave.

It should be clear from these examples that a wealth of interesting phenomena abounds in these generalised and modified forms of Burgers' equation, that many cases remain to be studied, and that MAE provides much insight and even numerical detail when applied to these problems in a systematic way.

References

1. Crighton, D.G.; Scott, J.F. Asymptotic solution of model equations in nonlinear acoustics. Phil. Trans. R. Soc. A292, (1979) 101-134.
2. Scott, J.F. Uniform asymptotics for spherical and cylindrical nonlinear acoustic waves generated by a sinusoidal source. Proc. R. Soc. Lond. A375, (1981), 211-230.
3. Lee-Bapty, I.P.; Crighton, D.G. Nonlinear spherical wave propagation in a stratified atmosphere. To be submitted to Proc. R. Soc. Lond.
4. Crighton, D.G.; Lee-Bapty, I.P. Nonlinear wave propagation governed by the Modified Burgers equation. To be submitted to Proc. R. Soc. Lond.
5. Crighton, D.G. Model equations of nonlinear acoustics. Ann. Rev. Fluid Mech. 11, (1979) 11-13.
6. Lighthill, M.J. Viscosity effects in sound waves of finite amplitude. In Surveys in Mechanics, eds. G.K. Batchelor & R.M. Davies, Cambridge University Press, 1956.
7. Taylor, G.I. The conditions necessary for discontinuous motion in gases. Proc. R. Soc. Lond. A84, (1910) 371-377.
8. Nimmo, J.J.C.; Crighton, D.G. Bäcklund transformations for nonlinear parabolic equations; the general results. Accepted for Proc. R. Soc. Lond. 1982/83.
9. Whitham, G.B. Linear and nonlinear waves. New York: Wiley-Interscience 1974.
10. Rudenko, O.V.; Soluyan, S.I. Theoretical foundations of nonlinear acoustics. (English translation by R.T. Beyer) New York: Consultants Bureau (Plenum Press) 1977.
11. Crighton, D.G. The plane wave problem using matched asymptotic expansions. Proc. Sixth ISNA, Vol.1. Moscow Univ. Press (1976) 19-29.
12. Nariboli, G.A.; Lin, W.C. A new type of Burgers' equation. ZAMM 53, (1973) 505-510.

The Propagation of Finite-Amplitude Edge Waves

T.R. AKYLAS

Department of Mechanical Engineering
Massachusetts Institute of Technology
Cambridge, Massachusetts 02139

Summary

A study is made of the temporal and spatial evolution of large scale modulations of weakly nonlinear edge waves on a uniformly sloping beach. It is found that a uniform train of free periodic edge waves is always unstable to large-scale variations, so that envelope solitons will develop; the resulting three-dimensional solitons are described in detail. In addition, it is shown that steady-state standing subharmonic edge waves, excited by incident wave trains on a long, mildly sloping beach, can be unstable to large-scale modulations. The physical significance of these findings is discussed.

Introduction

Edge waves are trapped wave modes, propagating along the coastline of a sloping beach and decaying in the offshore direction. Although the existence of a solution of the linear water-wave problem, representing edge waves on a uniformly sloping beach, has been known for quite some time [1], the physical significance of edge waves has been only recently recognized [2]. At present, it is generally believed that edge waves play an important role in the formation of coastal patterns [3], and in the generation of rip currents and periodic circulation cells in the nearshore region [4].

The main reason, for which very little physical relevance has been attached to edge waves for a long time, stems from the fact that such trapped wave disturbances are usually excited by nonlinear mechanisms. Using the shallow-water approximation, Guza & Davis [5] first pointed out that subharmonic edge waves can be generated by nonlinear interactions with incoming wave trains. In the further development of the theory, Whitham [6] showed that the shallow-water approximation is inadequate to describe the behavior of finite-amplitude edge waves uniformly in the offshore direction. Subsequently, Minzoni & Whitham [7] studied the excitation of monochromatic subharmonic edge waves by incident wave trains, using the full water-wave for-

Nonlinear Deformation Waves
IUTAM Symposium Tallinn 1982
Editors: U. Nigul, J. Engelbrecht

mulation; in the small beach angle limit, their results confirm the predictions of the shallow-water theory [8].

In the present paper, the theory of Whitham and Minzoni & Whitham is extended to interacting edge-wave packets. Two problems of physical interest are studied: first, the modulational instability of free periodic travelling edge waves is re-examined, and it is confirmed that the instability predicted by Whitham indeed leads to a series of envelope solitons; in addition, the three-dimensional structure of edge-wave envelope solitons is described in detail. Second, the stability to large-scale modulations of steady-state standing subharmonic edge waves is investigated. It is found that, in the small beach angle limit, large-scale modulations may grow, the instability increasing with the amplitude of the incoming wave train. It is suggested that this instability can play a role in explaining the origin of the long standing waves observed on natural beaches by Munk [9] and Tucker [10].

Edge-Wave Envelope Equations

Consider a uniformly sloping beach of angle α. A coordinate system is chosen such that x, y, and z are the offshore, vertical, and longshore coordinates respectively. In dimensionless variables, the classical gravity water-wave problem reads

$$\Phi_{xx} + \Phi_{yy} + \Phi_{zz} = 0 \ , \quad -x\tan\alpha < y < \varepsilon\eta \ , \tag{1}$$

$$\eta_t - \Phi_y + \varepsilon\left(\Phi_x \eta_x + \Phi_z \eta_z\right) = 0 \ , \quad y = \varepsilon\eta \ , \tag{2}$$

$$\Phi_t + \eta + \tfrac{1}{2}\varepsilon\left(\nabla\Phi\right)^2 = 0 \ , \quad y = \varepsilon\eta \ , \tag{3}$$

$$\Phi_x \sin\alpha + \Phi_y \cos\alpha = 0 \ , \quad y = -x\tan\alpha \ , \tag{4}$$

where $\Phi(x,y,z,t)$ is a velocity potential, $y = \varepsilon\eta(x,z,t)$ describes the free-surface elevation, and ε denotes a typical wave slope ($\varepsilon << 1$).

As noted by Stokes [1], the linear water-wave problem ($\varepsilon = 0$) admits a solution in the form of edge waves:

$$\Phi = F(x,y)\, e^{i(kz - \omega t)} + \text{c.c.} \ , \tag{5}$$

where

$$F(x,y) = \exp(-kx\cos\alpha + ky\sin\alpha) \ , \quad \omega^2 = k\sin\alpha \ , \tag{6}$$

and c.c. denotes the complex conjugate.

The main purpose of the present paper is to study the nonlinear interaction of an incident wave train of frequency ω, say, with two subharmonic edge-wave packets of frequency $\omega/2$, propagating in opposite directions along a beach. Accordingly, a suitable expansion for the three interacting wave packets is

$$\Phi(x,y,z,t) = \varepsilon^{-\frac{1}{2}}F(x,y)\Big\{(A_+(X,Y,Z,T)e^{i\theta_+} + c.c.)$$

$$+ (A_-(X,Y,Z,T)e^{i\theta_-} + c.c.)\Big\} + \Big\{S(x,y;X,Y,Z,T)e^{-i\omega t} + c.c.\Big\} + \ldots , \qquad (7)$$

where $\theta_\pm = \pm kz - \frac{1}{2}\omega t$, $\omega^2 = 4k\sin\alpha$, and $(X,Y,Z,T) = \mu(x,y,z,t)$ $(0 < \mu << 1)$ are long scales associated with the evolution of the wave envelopes. The $0(1)$ term in (7) represents the incident and reflected wave train, and will be made specific later. In addition, the dependence on the envelope scales remains unspecified at this stage; it will be determined by the requirement that (7) is uniformly valid in space and time.

Substitution of the proposed expansion into (1),..., (4) leads to a hierarchy of problems which determine S and the desired evolution equations for the envelopes $A_\pm$. Since the detailed derivation involves quite tedius algebraic manipulations, the appropriate evolution equations, for $\alpha = \pi/2N$ where N is an integer, are merely quoted (for details the reader is referred to Akylas [11]):

$$-\cos\alpha A_{\pm X} + \sin\alpha A_{\pm Y} \pm i A_{\pm Z} + 0(\mu) = 0 \ , \ -X\tan\alpha < Y < 0 \ , \qquad (8)$$

$$A_{\pm Y} - i\omega A_{\pm T} + \mu A_{\pm TT} = 0 \ , \quad Y = 0 \ , \qquad (9)$$

$$A_{\pm T} \pm \frac{\sin\alpha}{\omega} A_{\pm Z} = -i\frac{\mu}{\omega} A_{\pm TT} +$$

$$\frac{\varepsilon}{\mu}\Big\{-2i\frac{k^4}{\omega^2} A_\pm A_\pm^* + \gamma S_o A_\mp^* + \delta A_\pm A_\mp A_\mp^*\Big\} , \quad X = 0 \ , \quad Y = 0 \ , \qquad (10)$$

where γ and δ are certain constants depending on the beach angle α, and S_o is the amplitude of the incident wave at the shoreline. The nonlinear equations (10) determine the behavior of A_+ and A_- at the shoreline, and $A_\pm(X{=}0,Y{=}0,Z,T)$, together with (9), serve as boundary conditions for the determination of $A_\pm(X,Y,Z,T)$ through the linear equations (8).

Edge-Wave Envelope Solitons

The evolution equations for the envelope of a single free travelling edge-wave packet are obtained by setting $A_- = S = 0$ in (8), (9), (10). In addition, the balance $\mu = \varepsilon^{\frac{1}{2}}$ is adopted, so that the nonlinear effects enter at the same level as the dispersive effects. A special solution of these equations represents a uniform, periodic, finite-amplitude edge wave train; with the notation $A_+ \equiv A$, in a frame of reference moving with the group velocity, $c_g = \frac{\sin\alpha}{\omega}$, it reads

$$A = A_o \exp\left\{\frac{2\mu k^4}{\cos\alpha}|A_o|^2 (X\sin\alpha + Y\cos\alpha)\right\}\exp(-2i\mu\frac{k^4}{\omega}|A_o|^2 T), \qquad (11)$$

where $A_o = A(X=0,Y=0,T)$. The linear stability of (11) is readily determined: since the coefficients of A_{TT} and A^2A^* in (10) are of the same sign, linear perturbations to the special solution (11) tend to grow [12]. In fact, the instability gives rise to a series of envelope solitons [13].

The detailed structure of edge-wave envelope solitons is found to be [11]:

$$A = r\left\{\xi + i\kappa(X\cos\alpha - Y\sin\alpha)\right\}\exp\left\{i(\rho Z + \sigma\mu T) - \rho(X\cos\alpha - Y\sin\alpha)\right\} \qquad (12)$$

where
$$r(\xi) = \left(\frac{2\zeta}{\nu}\right)^{\frac{1}{2}}\operatorname{sech}\zeta^{\frac{1}{2}}\xi\ , \qquad (13)$$

and
$$\zeta = \frac{\beta\rho^2 - \sigma}{\beta\kappa^2}\ , \quad \nu = \frac{\gamma}{\beta\kappa^2}\ , \qquad (14)$$

with
$$\rho = \frac{\lambda}{2\kappa\beta}\ , \quad \beta = \frac{c_g^2}{\omega}\ , \quad \gamma = \frac{2k^4}{\omega}\ . \qquad (15)$$

It should be remarked that the decay rate in the seaward direction of an edge-wave packet with a soliton envelope is diminished due to nonlinear effects.

Stability of Standing Subharmonic Edge Waves

A uniform periodic wave train of frequency ω, normally incident and reflected on a beach, is unstable to standing edge-wave perturbations of frequency $\omega/2$ ([5],[7]). The growth due to the instability is exponential initially, but, as the edge-wave amplitude increases, the nonlinear terms due to edge-wave self-interactions become important, and eventually a steady state is reached. The stability of the resulting steady-state subharmonic standing edge waves to large-scale variations can be investigated, using the already developed envelope equations.

The steady-state standing finite-amplitude edge wave: $A_+ = A_- \equiv A$ is a special solution of (8), (9), (10):

$$A = \left(\frac{\gamma\omega S_o}{2ik^4 - \omega\delta}\right)^{\frac{1}{2}}\ . \qquad (16)$$

The linear stability of A is investigated by writing $A_\pm(X,Y,Z,T) = A + a_\pm(X,Y,Z,T)$ and linearizing for the perturbations $a_\pm$. Thus, (10) leads to four linear equations for the real and imaginary parts of

$a_{\pm} = a_{\pm r} + i\, a_{\pm i}$ at the shoreline. As is customary in linear stability analysis, it is assumed that

$$(a_{+r},\ a_{+i},\ a_{-r},\ a_{-i}) = (a_1,\ a_2,\ a_3,\ a_4)e^{iKZ + pT}, \qquad (17)$$

where K is taken to be real, and upon substitution of (17) to the equations for $a_{\pm r}$ and $a_{\pm i}$, a linear system of four homogeneous algebraic equations for the perturbation amplitudes a_1, a_2, a_3, a_4 is obtained. The solvability condition for this system determines the appropriate values of p, and thereby the linear stability of A can be decided depending on the sign of the real part of p.

The linear stability of A was investigated in the shallow water limit ($\alpha << 1$), assuming the balance $\mu = \varepsilon$. It was found that there always exists a range of real K for which p has a positive real part, implying that large-scale modulations to the steady-state waves tend to grow. In fact, the maximum growth rate is a linearly increasing function of the amplitude of the incident wave train at the shore. Furthermore, the corresponding imaginary parts of p turn out to be substantially smaller than the real parts, so that unstable disturbances represent nearly standing waves of growing amplitude.

Concluding Remarks

The present theory is based on a number of restrictive assumptions: inviscid, small-amplitude, non-breaking disturbances on perfectly reflective beaches are considered. Nevertheless one could speculate that the instability of standing subharmonic edge waves to large-scale variations could be of some importance for certain phenomena observed on natural beaches. The field observations of Munk [9] and Tucker [10] indicate that long standing waves can exist on beaches. Accordingly, the modulational instabilities of subharmonic edge waves could contribute to the generation of such long-period oscillations, in addition to the interaction mechanisms of Gallagher [14] and Foda & Mei [15]. Furthermore, the longshore variations, produced by the unstable edge-wave modulations, could play some role in the excitation of the observed widely spaced rip currents and circulation cells on natural beaches.

References

1. Stokes, C.G. 16th Meet. Brit. Ass. Adv. Sci., Southampton, pp. 1-20, London: John Murray 1846.
2. Munk, W.H., Snodgrass, F. & Carrier, G.F. Science 123 (1956) 127-132.

3. Guza, R.T. & Inman, D.L. J. Geophys. Res. 80 (1975) 2997-3011.
4. Bowen, A.J. & Inman, D.L. J. Geophys. Res. 74 (1969) 5479-5490.
5. Guza, R.T. & Davis, R.E. J. Geophys. Res. 79 (1974) 1285-1291.
6. Whitham, G.B. J. Fluid Mech. 74 (1976) 353-368.
7. Minzoni, A.A. & Whitham, G.B. J. Fluid Mech. 79 (1977) 273-287.
8. Guza, R.T. & Bowen, A.J. J. Mar. Res. 34 (1976) 269-293.
9. Munk, W.H. Trans, Am. Geophys. Union 30 (1949) 849-854.
10. Tucker, M.J. Proc. Roy. Soc. A202 (1950) 565-573.
11. Akylas, T.R. submitted for publication to J. Fluid Mech. (1982).
12. Benney, D.J. & Newell, A.C. J. Math. & Phys. 46 (1967) 133-139.
13. Zakharov, V.E. & Shabat, A.B. Sov. Phys. - JETP 34 (1972) 62.
14. Gallagher, B. J. Fluid Mech. 49 (1971) 1-20.
15. Foda, M.A. & Mei, C.C. J. Fluid Mech. 111 (1981) 319-345.

Iterated Compatibility Relations and Nonlinear Wave Propagation

H. COHEN

Dept. of Civil Engineering
University of Manitoba
Winnipeg, Canada

MARCELO EPSTEIN

Dept. of Mechanical Engineering
University of Calgary
Calgary, Canada

It has been suggested [1] that the connection between the singular surface technique, which provides information at the wave front only, and the global solution, can be revealed by finding the jumps of all derivatives of the deformation at the wave front In this way, a formal Taylor expansion around the wave-front should yield information for at least a short time after the passage of the wave. Such a process has been successfully effected in several recent papers [2,3] in the domain of linear plate and shell theory. One of the difficulties in executing such a scheme stems from the basic nature of the compatibility relations. Moreover, these relations when iterated for the higher order jumps are of such complexity as to render their application almost impossible, if in fact they are even available [4].

In this paper we present an extremely concise form of the iterated compatibility relations which are readily useful for finding the basic equations for the higher order waves. Their formulation involves an arbitrary choice of coordinates on the singular surface in space-time.

Assume that the moving surface S can be represented as

$$\phi(X^I) - t = 0 \tag{1}$$

where X^I ($I = 1,2,3$) and t are space and time coordinates, respectively. Eq. (1) can be also viewed as a fixed surface T in space-time. Any function $\Psi(X^I,t)$ can also be represented in

Nonlinear Deformation Waves
IUTAM Symposium Tallinn 1982
Editors· U. Nigul, J. Engelbrecht

terms of the variable change

$$\begin{cases} X^I = Y^I \\ t = \phi(Y^I) - \tau \end{cases} \tag{2}$$

as $$\Psi = \Psi(Y^I,\ \phi(Y^I) - \tau) = \hat{\Psi}(Y^I,\tau)\ . \tag{3}$$

By the chain rule

$$\Psi,_I = \hat{\Psi},_I + \dot{\hat{\Psi}}\ \phi,_I \tag{4}$$

$$\dot{\Psi} = -\dot{\hat{\Psi}} \tag{5}$$

where a comma denotes partial derivative with respect to X^I or Y^I and a dot, with respect to t or τ. If Ψ is singular across T, it follows from Hadamard's lemma that

$$[\hat{\Psi}],_I = [\hat{\Psi},_I]\ , \tag{6}$$

where brackets denote jumps. Denoting by a superimposed index, k say, the k-th t or τ derivative, it also follows that

$$[\overset{k}{\hat{\Psi}},_{I_1 \ldots I_n}] = [\overset{k}{\hat{\Psi}}],_{I_1 \ldots I_n}\ , \tag{7}$$

which may be referred to as the iterated superficial compatibility relations. These basic equations in conjunction with Eqs. (4), (5) can be used to obtain a concise form of the k-th iterated compatibility relations as

$$[\overset{k}{\Psi},_I] = (-1)^k\ ([\overset{k}{\hat{\Psi}}],_I + [\overset{k+1}{\hat{\Psi}}]\phi,_I) \tag{8}$$

and

$$\begin{aligned}[\overset{k}{\Psi},_{IJ}] = (-1)^k\ (&[\overset{k}{\hat{\Psi}}],_{IJ} + [\overset{k+1}{\hat{\Psi}}],_I\ \phi,_J \\ &+ [\overset{k+1}{\hat{\Psi}}],_J\ \phi,_I + [\overset{k+2}{\hat{\Psi}}]\ \phi,_I\ \phi,_J + [\overset{k+1}{\hat{\Psi}}]\ \phi,_{IJ})\end{aligned} \tag{9}$$

In the usual derivation of the geometric and kinematic compatibility relations [4], directional derivatives perpendicular and tangent to the wave front as well as the displacement time derivative are used. Here, the differentiations correspond to directions tangent to T only as prescribed by the arbitrary surface coordinates Y^I and transverse to T in the direction parallel to the time axis.

Consider, as an example, a homogeneous nonlinear elastic solid which, for simplicity, we assume initially undeformed and at rest. It follows that for any n quantities Ψ_i $(i = 1,\ldots n)$ discontinuous at the wave front (and vanishing ahead of it) the identity

$$[\prod_{i=1}^{n} \Psi_i] = (-1)^{n+1} \prod_{i=1}^{n} [\Psi_i] \tag{10}$$

holds, where π is the continuous product symbol.

In Cartesian coordinates, X_I and x_i, in reference configuration and in space, respectively, the equations of motion are given by

$$A_{KkQq}\, x_{q,KQ} + \rho_o\, f_k = \rho_o\, \ddot{x}_k\ , \tag{11}$$

where ρ_o is the reference density, f_k is the body force per unit mass and the functions $A_{KkQq}\ (x_{p,P})$ are the elasticities of the material. The n-th material time derivative of Eq. (11) is

$$\sum_{m=1}^{n} \sum_{p=1}^{m} \binom{n}{m} C_{mp} \overset{n-m}{x}_{q,KQ}\, A_{KkQq\, I_1\, i_1 \ldots I_p\, i_p}$$

$$\overset{m_1}{x}_{i_1,\, I_1}\, \overset{m_2}{x}_{i_2, I_2} \ldots \overset{m_p}{x}_{i_p,\, I_p} +$$

$$A_{KkQq}\, \overset{n}{x}_{q,KQ} = \rho_o\, \overset{n+2}{x}_K\ , \tag{12}$$

with

$$m_1 \geq m_2 \geq \ldots \geq m_p \geq 1 \tag{13}$$

and

$$m_1 + m_2 + \ldots + m_p = m \tag{14}$$

and where

$$A_{KkQqI_1i_1\ldots I_p\, i_p} = \frac{\partial^p A_{KkQq}}{\partial x_{i_1,I_1}\cdots\partial x_{i_p,I_p}} \,. \tag{15}$$

In writing Eq. (12) we have generalized to several variables the formula of Faá di Bruno [5], which provides a recipe for calculating the coefficients C_{mp}. We note here that $C_{m1} = 1$ for all m.

We focus our attention to acceleration waves, i.e.

$$[x_i] = [\dot{x}_i] = 0 \ , \quad [\ddot{x}_i] \neq 0 \tag{16}$$

and obtain, via Eqs. (8) and (9) the jump of Eq. (12) as

$$\begin{aligned}
&(A_{KkQq}\,\phi_{,K}\,\phi_{,Q} - \rho_o\,\delta_{kq})\,[\overset{n+2}{\hat{x}_q}] + \\
&(A_{KkQq}\,\phi_{,KQ} + A_{KkQqIi}\,\phi_{,K}\,\phi_{,Q}\,\phi_{,I}\,[\hat{\ddot{x}}_i])\,[\overset{n+1}{\hat{x}_q}] + \\
&(A_{KkQq} + A_{QkKq})\,\phi_{,K}\,[\overset{n+1}{\hat{x}_q}]_{,Q} + \overset{n}{R}_k = 0 \ ,
\end{aligned} \tag{17}$$

where we have lumped together all of the terms not containing jumps of order higher than n under the notation $\overset{n}{R}_k$. We refer to Eq. (13) as the general decay-induction equation. Note its essential features, namely, that except for n = 1 it is linear in $[\overset{n+1}{\hat{x}_q}]$ and that the coefficient of the highest order jump is exactly the one appearing in the eigenvalue problem of the propagation condition, used to determine ϕ in the first place. Indeed, for n = 0 Eq. (13) reduces to the propagation condition

$$(A_{KkPp}\,\phi_{,K}\,\phi_{,P} - \rho_o\,\delta_{kq})\,[\hat{\ddot{x}}_p] = 0 \ . \tag{18}$$

Let us assume that ϕ is one out of the three possible simple solutions of the characteristic determinantal equation of Eq. (18) and let e_q denote the corresponding unit eigenvector, viz

$$[\ddot{\hat{x}}_q] = a\ e_q \tag{19}$$

where a is the indeterminate magnitude of the acceleration jump. We wish to extract from Eq. (17) the decay and induced wave equations associated with the solution ϕ. To this end, we multiply Eq. (17) by e_k and we obtain, by virtue of the "filtering out" effected by the coefficient of the higher order jump, the decay equation

$$B_{Qq}\ [\overset{n+1}{\hat{x}}]_{,Q} + C_q\ [\overset{n+1}{\hat{x}}_q] + \overset{n}{D} = 0 \tag{20}$$

with

$$B_{Qq} = (A_{KkQq} + A_{QkKq})\ \phi_{,K}\ e_k\ ,$$

$$C_q = (A_{KkQq}\ \phi_{,KQ} + A_{KkQqIi}\ \phi_{,K}\ \phi_{,Q}\ \phi_{,I}\ a\ e_i)\ e_k$$

and

$$\overset{n}{D} = \overset{n}{R}_k\ e_k \tag{21}$$

Note that the characteristics of this first order PDE are independent of n.

To extract the induced waves behaviour we note that the operator $(A_{KkPp}\ \phi_{,K}\ \phi_{,P} - \rho_o\ \delta_{kp})$, when restricted to the two-dimensional subspace $e_k^{\perp}$ orthogonal to e_k is non-singular. Let P_{kq} denote the composition of this inverse preceded by projection onto $e_k^{\perp}$. Applying P_{kq} to Eq. (18) yields an algebraic equation for the determination of the part $[\overset{n+2}{\hat{x}}_q]^{\perp}$ orthogonal to e_k of the induced higher order waves for all $n \geq 0$.

References

1. Achenbach, J.D., "Wave Propagation in Elastic Solids", North Holland, Amsterdam (1973).

2. Cohen, H., Shah, A.H. and Thambiratnam, D.P., "Transient and Time Harmonic Waves in Elastic Plates", Int. J. of Solids and Structures, 15, 395-404 (1979).

3. Thambiratnam, D.P., Shah, A.H. and Cohen, H., "Axisymmetric

Transients in Shells of Revolution", Earthquake Engng. and Struct. Mech., 7, 369-382 (1979).

4. Truesdell, C. and Toupin, R., "The Classical Field Theories", in Handbuch der Physik, S. Flügge ed., Vol. III/1, Springer Verlag, Berlin (1960).

5. Roman, S., "The Formula of Faá di Bruno", American Math. Monthly, 87, 805-809 (1980).

Equations of Nonlinear Geometrical Acoustics

E.N.Pelinovskij, V.E.Fridman

Applied Physics Institute, AN SSSR, Gorky USSR
Radiophysical Research Institute, Gorky, USSR

Summary

An effect of nonlinearity on the character of wave propagation in smoothly inhomogeneous media with small dispersion and dissipation is studied. The transport equation for the wave variable is derived in the first approximation over the nonlinearity. Exact or approximate solution of this equation is found out, which defines the nonlinear correction to the velocity of the wave propagation along the ray as well as the relation between the amplitude and the ray tube cross section. Formulas obtained are used for closing the system of ray equations and their solution at the concrete wave process.

Ray Equations

In the most general case, the transformation of any phase surface of the wave field is described by kinematic equations for rays [1]. These equations connect the variation of the ray sloping angle with the variation of the propagation velocity and the ray-tube cross section both along and transverse the ray trajectory. For example, for a two-dimensional wave process they have the form ([1] p.283).

$$\frac{\partial \varphi}{\partial m} = \frac{\bar{C}}{\bar{S}} \frac{1}{C} \frac{\partial S}{\partial \ell} ; \quad \frac{\partial \varphi}{\partial \ell} = \frac{\bar{S}}{\bar{C}} \frac{1}{S} \frac{\partial C}{\partial \ell} \qquad \text{(I)}$$

Here C is the propagation velocity of a profile point with the fixed phase, S is the cross-section area of the ray tube, $\bar{C}$ and $\bar{S}$ are the values at the initial moment, φ is the ray sloping angle, ℓ and m are the ray and the transverse coordinates. For a linear wave in inhomogeneous medium the velocity $C(\vec{r})$ has been defined and system (I) is closed. Another situation is developed at the propagation of

Nonlinear Deformation Waves
IUTAM Symposium Tallinn 1982
Editors: U. Nigul, J. Engelbrecht

a nonlinear wave. In this case the velocity C is the field function, as a rule, and to close the system (I) one has to connect the field with parameters of ray tubes. The present paper is devoted to the problem of finding such a relation.

One-Wave Approximation

It is well known that propagation of waves with small but finite amplitude may be described by the nonlinear wave equation

$$\frac{\partial^2 p}{\partial t^2} - c_0^2(\varepsilon\vec{r})\Delta p = \varepsilon\hat{Q}(p;\varepsilon\vec{r}) \tag{2}$$

Here $\hat{Q}$ is, in general, the integro-differential nonlinear operator describing nonlinear, dissipative and dispersion properties of medium as well as smooth inhomogeneity. The parameter ε is assumed to be small, C_0 is the velocity of a wave propagation in the linear approximation. We are interested in the wave travelling in one direction, that is why, instead of coordinates $\vec{r}$ and time t new independent variables are introduced

$$\eta = \tau(\vec{r}) - t\ , \quad \vec{\rho} = \varepsilon\vec{r} \tag{3}$$

Here τ is the eikonal describing a moment of arrival of the wave profile fixed point. In these variables equation (2) is reduced to the form

$$\left[\frac{1}{C_0^2} - (\Delta\tau)^2\right]\frac{\partial^2 p}{\partial \eta^2} - \varepsilon\left[\frac{\partial}{\partial\eta}(2\nabla\tau\nabla_{\rho}p + \Delta\tau p) + \hat{Q}\right] - \varepsilon^2\Delta_{\rho}p = 0 \tag{4}$$

Here $\nabla_{\vec{\rho}} = \partial/\partial\vec{\rho}\ ,\ \Delta_{\vec{\rho}} = \partial^2/\partial\rho^2$. As it is made in geometrical methods we introduce the ray coordinate ℓ and the ray-tube cross section area S [2]

$$\Delta\tau = \frac{1}{S}\frac{d}{d\ell}\left(\frac{S}{C}\right)\ ;\ \nabla\tau\nabla_{\rho}p = \frac{1}{C}\frac{\partial p}{\partial \ell} \tag{5}$$

Here C is the velocity of propagation of the profile fixed point which differs from C_0. To define the value of C we shall find eikonal equation in the form

$$(\nabla\tau)^2 = \frac{1}{C_0^2} + \varepsilon\hat{\Phi} \tag{6}$$

where $\hat{\Phi}$ is the functional to be determined. Here

$$C = \left(\frac{d\tau}{d\ell}\right)^{-1} \simeq C_0\left(1 - \frac{\varepsilon}{2} C_0^2 \hat{\Phi}\right) \tag{7}$$

Substituting (6) in equation (4) and neglecting $\Delta_{\bar{\rho}} P$ we obtain the transport equation with the accuracy up to ε .

$$\frac{\partial P}{\partial \ell} + A_P + \frac{C}{2}\hat{\Phi}\frac{\partial P}{\partial \eta} + \hat{R}\left\{P;\ell\right\} = 0 \tag{8}$$

$$A = \frac{d}{d\ell}\ln q : q = (S/C_0)^{1/2} ; \ \hat{Q} = \partial\hat{R}/\partial\eta$$

The functional $\hat{R}$ contains only η derivatives and integrals. Since the functional $\hat{\Phi}$ is a correction to the velocity, it depends only on the ray coordinate ℓ . Then, the substitution of variables

$$P = \mathcal{P}/q ; \ \ \xi = \eta - \int \frac{C_0}{2}\hat{\Phi}\, d\ell \tag{9}$$

reduces equation (8) to the following form:

$$\frac{\partial \mathcal{P}}{\partial \ell} + \hat{M}\left\{\mathcal{P};\ell\right\} = 0 \tag{10}$$

where $M = q\hat{R}\{\mathcal{P};\ell\}$ contains only ξ derivatives and integrals.

<u>Algorithms of the Solution</u>

Variable coefficients $C_0(\ell)$, $S(\ell)$ and others (for example, the medium density or the depth) can not be defined without ray trajectories. Even so, there are physically important classes of solution for equation (10) which may be found without preliminary calculation of ray trajectories.

Primarily, this is the class of separating operators

$$\hat{M}\{\mathcal{P};c\} = \varphi(\ell)\hat{N}\{\mathcal{P}\} \tag{11}$$

describing Riemann wave in inhomogeneous medium, linear waves in media with automodel dispersion $(\omega \sim K^{\alpha})$ or dissipation $(\omega \sim iK^{\alpha})$ law, solitons in media with a special law of the inhomogeneity variation. In this case, solution of equation (10) found analytically or numerically has the form

$$\mathcal{P} = \mathcal{P}\left(\xi, \int \varphi(\ell)d\ell\right) \tag{12}$$

We connect lines of equal phases with fixed points of the profile $\mathcal{P}_m$ (for example, with the positions of maxima of the function $\mathcal{P}$).The equation of motion of the fixed point is easily obtained from the solution of (12).

$$\xi_m = \xi\left[\ell; S(\ell); \mathcal{P}_m\right] \tag{13}$$

Then, from (13) the formula for the correction to the eikonal is found

$$\xi_m = \text{const} - \int \frac{c_0}{2}\hat{\Phi}\,d\ell$$

or

$$\hat{\Phi}(\ell) = -\frac{2}{c_0}\frac{d\xi_m}{d\ell} \tag{14}$$

Substituting (14) into (7) an expression is obtained for the velocity in the form

$$c \simeq c_0\left(1 - \varepsilon c_0 \frac{d\xi_m}{d\ell}\right) \tag{15}$$

Another class of solutions for equation (10) may be found under the condition of slow variation of the medium parameters in comparison with the characteristic lengths of nonlinearity, dispersion, etc (these conditions are usually more rigid than geometrooptical conditions of smallness of the parameter variations by the wavelength). The presence of the small parameter μ (μ is quotient of the nonlinear length and the inhomogeneity scale) in equation (10) permits us to use the effective asymptotic methods (see, for example [1,3]). The equation (10) in the general case may be written in the operational form

$$\hat{T}\left\{\xi ; \ell ; \mathcal{P} ; \mu\right\} = 0 \tag{16}$$

In the zero μ approximation from equation (15) we find

$$\mathcal{P} = \mathcal{P}_0 \left(\xi, \ell, B_i\right) \tag{17}$$

where B_i is a set of constants defining the solution (amplitude, mean flow, etc). Here, parameters of the inhomogeneity, including the ray-tube cross section are considered constant. In this solution we may fix the point $\mathcal{P}_m$, define the corresponding value of the coordinate ξ_m , that permits us to find a correction to the velocity (see formula (15)). However, in contrast to (12), the function of (17) is not the solution of equation (16). To find the solution it is necessary to define a dependence of constants B_i on the ray coordinate ℓ . This dependence is specified by the condition of the absence of the correction increase to $\mathcal{P}_0$. The correction $\mathcal{P}_1$ is described by the equation in variational derivatives

$$\frac{\delta \hat{T}}{\delta \mathcal{P}_0} \mathcal{P}_1 = \frac{\delta \hat{T}}{\delta \mu} \tag{18}$$

The condition of the increase absence $\mathcal{P}_1$ is the condition of orthogonality $\delta \hat{T}/\delta \mu$ to the eigenfunctions Y of the operator conjugated with $\delta \hat{T}/\delta \mathcal{P}_0$

$$\left\langle Y \frac{\delta \hat{T}}{\delta \mu} \right\rangle = 0 \tag{19}$$

where $\langle\ \ \rangle$ denotes averaging for the wave period or over infinite limits depending on the form of the undisturbed solution. Since $\delta\hat{T}/\delta\mu$ contains derivatives from constants B_i, condition (19) is an equation estimating a relation between inhomogeneity parameters and the amplitude and its derivative. In particular cases, for example, for nondissipative media, the amplitude equation is degenerated into the law of the energy or the energy flux conservation. Thus, here the algorithm of the functional $\hat{\Phi}$ searching is described in the general form $\hat{\Phi} = \hat{\Phi}\{\ell; S(\ell); \mathcal{P}_m(\ell)\}$, a relation between $\mathcal{P}_m(\ell)$ $S(\ell)$ and ℓ is given by the amplitude equation which in a number of cases may be integrated. For closing ray equation (I) it is sufficient to know the functional $\hat{\Phi}$ together with the amplitude equation. The approach developed here proves self-refraction equations of shock waves and solutions used in papers [4-6] .

References

1. Whitham, G.B.; Linear and nonlinear waves. John Wiley & Sons.Inc.1974.
2. Babich,V.M.; Buldyrev,V.S.; Asymptotic methods in problems of short wave diffraction. Nauka, Moscow,1974.
3. Gorshkov,K.A.; Ostrovskij,L.A.; Pelinovskij,E.N.; Some problems of asymptotic theory of nonlinear waves.Proc. IEEE, N11,v.62 (1974) 1511-1517.
4. Ostrovskij,L.A.; Shrira,V.I.: Instability and self-refraction of solitons. ZhETF.N4,v.71 (1976) 1412-1420.
5. Miles,I.W.: Diffraction of solitary waves. Z.Angew.Math. and Phys. N5,v.25 (1977) 889-902.
6. Fridman,V.E.: Self-refraction of small amplitude shock waves. Wave Motion. N2,v.4 (1982) 151-161.

Some Exact Solutions of Korteweg- De Vries- Burgers Equation for Plane, Cylindrical and Spherical Waves

V.P.KOROBEINIKOV

Steklov Mathematical Institute,
Academy of Sciences,
Moscow, USSR

Summary

The problem of finding of new types of solutions is studied for the generalized Korteweg-de Vries-Burgers (KdVB) equation and for the cases of plane, cylindrical and spherical waves. All invariant solutions of the equation are shown by using group method theory. For the cylindrical case a type of particular solutions of KdV equation is considered. Concrete examples of the solutions and their properties and applications are also discussed.

1. Introduction

For the study of waves in cold plasma [1-4], magnetohydrodynamics [5], bubble filled liquid [6], nonlinear acoustics [7] and for the study of deformation waves in solid [8] generalized KdVB equation,

$$\frac{\partial u}{\partial t}+\frac{j}{2}\cdot\frac{u}{t}+\gamma u^{m}\frac{\partial u}{\partial x}-\mu\frac{\partial^{2}u}{\partial x^{2}}+\beta\frac{\partial^{3}u}{\partial x^{3}}=0 \qquad (1.1)$$

is used. Here $j=0,1,2$ for the plane, cylindrical and spherical symmetry correspondingly; m, γ, μ, β are some constants, $u(x,t)$ unknown function, x is a coordinate (space), t is the second coordinate (time).

The equation (1.1) is well studied for the cases $j=0$, $m=1$, $\gamma=1$, $\beta=0$ (plane Burgers (B) equation) and $j=0$, $m=1$, $\mu=0$ (plane KdV equation). For the general case equation (1.1) is less studied. The finding of new particular solutions is of definite interest for the theoretical aims and the different applications.

Nonlinear Deformation Waves
IUTAM Symposium Tallinn 1982
Editors U. Nigul, J. Engelbrecht

2. Method of invariant solution.

The full admissible group for (1.1) was found in [9]. The Ovsjannikov's algorithm for finding of the group operator algebra [10] was used.

For $m \neq 0$, $m \neq 1$, $j = 0$, equation (1.1) is invariant under translation groups with respect to t and x. The invariant solution has a form

$$u = U(t - \alpha x), \quad \alpha = const. \qquad (2.1)$$

The solution of type (2.1) was studied in [4]. Suppose now that $m = 1$.

3. Invariant solution of KdV equation.

Plane case, $m = 1$, $\gamma = 1$, $\mu = 0$, $j = 0$. For the plane case we have the following invariant solutions:

$$\begin{aligned} &u_1 = U(t), \quad u_2 = U(x - \alpha t), \quad u_3 = \frac{x}{t} + U(t), \\ &u_4 = \frac{x}{t} U(\lambda), \quad \lambda = x t^{-1/3}, \quad u_5 = \frac{\alpha}{2} + \frac{\frac{x}{t} U(\lambda)}{1 + 2\alpha\lambda t^{2/3}} \\ &\lambda = x t^{-1/3} - \frac{\alpha}{2} t^{2/3}, \quad u_6 = \frac{t + U(\lambda)}{\alpha}, \quad \lambda = x - \frac{t^2}{2\alpha} \end{aligned} \qquad (3.1)$$

The solution $u_4(x,t)$ is self-similar one [1]. We note that after substitution of the solutions (3.1) in equation (1.1), ordinary differential equations will be obtained for the functions $U(\lambda)$.

Let us consider in more details the solution $u_6(x,t)$. After substitution $u_6(x,t)$ in KdV equation we obtain

$$1 + \frac{U}{\alpha} U' + \beta U''' = 0$$

We can integrate this equation and after transformation

$$y=-\beta^{-1/5}(12\alpha)^{-3/5}U(\lambda), \quad z=(12\alpha\beta^2)^{-1/5}\lambda$$

obtain the result

$$y''=6y^2+z+C, \qquad C=const$$

The solution for this equation is known as Painleve function.

Cylindrical case, $j=1$. The KdV equation admits more poor transformation group. However, we have for this case the solution u_1 , u_4 from (3.1) and in addition we have

$u=\frac{x}{2t}+U(t)$

$$u(x,t)=\frac{x}{2t}+\frac{1}{t}U(\lambda), \quad \lambda=x/\sqrt{t} \tag{3.2}$$

We note that solution (3.2) has no analogy in the plane case.

Let us now consider the case $j=2$ for the KdV equation. The full collection of invariant solutions, constructing by means of nonsimilar subgroups is

$$u_1=U(t), \quad u_2=\frac{x}{t\ln t}+U(t), \quad u_3=\frac{x}{t}U(\lambda), \lambda=xt^{-1/3},$$

$$u_4=\frac{x}{t(\ln t+1)}+U(t), \quad u_5=t^{-2/3}U(\lambda), \ \lambda=\frac{x+1}{t^{1/3}} \tag{3.3}$$

$$u_6=t^{-2/3}U(\lambda)-\frac{1}{t}, \quad \lambda=(x+\ln t+3)/t^{1/3}$$

Besides self-similar solutions $u_3(x,t)$ we have here new interest solutions u_5 , u_6 . Obviously the self-similar solution u_3 has a "plane" asymptotic, investigated in [1]. Types of Cauchy problems which can be formulated for the solutions (3.3) prescribed by these forms.

4. <u>Invariant solutions of B equation.</u>

If $\gamma=1$, $\mu\neq 0$, $\beta=0$ the equation (1.1) is well known B equation. The admissible group of transformations was found in [9]. For the plane case B equation is reduced to well-known equation of conductivity for $\theta(x,t)$ by transformation $u=-2\mu\frac{\partial \ln\theta(x,t)}{\partial x}$. The large table of the B equation solutions is given in [12]. We note here only the solutions:

$$u_1=\frac{1}{\alpha}t+\frac{1}{\alpha}U\left(x-\frac{t^2}{2\alpha}\right),$$

$$u_2=\lambda+\frac{U(\lambda)}{t}, \quad \lambda=\frac{x}{t} \tag{4.1}$$

$$u_3=\frac{1}{t^{1/2}}U\left(\frac{x}{\sqrt{t}}\right).$$

The solution $u_3(x,t)$ corresponds to a group of stretching and it can be called as self-similar. This type of solution take place also for cylindrical and spherical symmetry. As an example we takes the point explosion problem for B equation with initial data

$$u(x,t)=E_0\delta(x), \quad \int_{-\infty}^{+\infty}u(x,0)\,dx=E_0 \tag{4.2}$$

$$t\neq 0$$

This problem is self-similar and its solution can be found in the form u_3 from (4.1). For the plane case ($j=0$) analytical solution of the problem is known [9, 13]. For cylindrical case it can be found using the results of the book [7] and for spherical case analytical solution is not known.

5. On the solutions of cylindrical KdV equation.

The analysis of the solution (3.2) showed, that after substitution the transformation (3.2) in equation (1.1) the member contained j-parameter disappears in an equation for invariant function $U(\lambda)$. This fact can be generalized and we have the next rule. For the case $j=1$ equation (1.1) by means of a substitution

$$u(x,t)=\frac{x}{2t}+\frac{1}{t}v\left(\frac{x}{\sqrt{t}},-\frac{2}{\sqrt{t}}\right) \tag{5.1}$$

reduces to the plane KdV equation

$$\frac{\partial v}{\partial \eta}+v\frac{\partial v}{\partial \lambda}+\beta\frac{\partial^3 v}{\partial \lambda^3}=0,\ \lambda=\frac{x}{\sqrt{t}},\quad \eta=-\frac{2}{\sqrt{t}} \tag{5.2}$$

The transformation (5.1) is the generalization of transformation (3.2)*.

If we have a solution $v(\lambda,\eta)$ for (5.2) we can find using (5.1) corresponding solution for KdV equation with $j=1$.
Examples:

a) $v=\frac{a_0}{2}=const,\quad u=\frac{x-a_0}{2t}$ (the invariant solution).

We can also take for $v(\lambda,\eta)$ invariant solutions from (3.1).

b) $v(\lambda,\eta)=3a\,ch^{-2}(\lambda-a\eta+b)$ (plane soliton),

then

$$u(x,t)=\frac{x}{2t}+\frac{3a}{t}ch^{-2}\frac{\sqrt{a}}{2}\left(\frac{x+2a}{\sqrt{t}}+b\right) \tag{5.3}$$

This is cylindrical "soliton" with variable amplitude and with linear in x background. The function $tu-x/2$ is similar to the ordinary soliton.

The multi-soliton solution $v(\lambda,\eta)$ will transfer by means of (5.1) to a new solution $u(x,t)$ with great numbers of maximum points in field of linear background.

<u>References</u>

1. Karpman,V.I.:Nonlinear wave in dispersive media, Novosibirsk university press, 1968(in Russian).
2. Maxon,S.; Viecelli, J.: Phys. Fluids, 1974, vol.17, No.8,

*) The transformation (5.1) was showed by a student of the Tomsk State University, V.F.Fal'mež (Mathematical Institute, 1979) during the fulfillment of his diploma work. The author found out at the Symposium (Tallinn, 1982) that a similar transformation has been proposed earlier by B .A.Lugovtsev from Novosibirsk.

3. Maxon, S.; Viecelli, J.: Phys. Rev. Letters, 1974, vol.32, No.1.

4. Jeffrey, A.; Kakutani, T.:SIAM Rev.,1972, vol.14, No.4.

5. Kakutani, T.; Kawanara, T.: J.Phys.Society Japan, 1970, vol.29, No.4.

6. Van Wijngaarden, L.: J. Fluid Mech., 1968, vol.33, pt.3.

7. Rudenko, O.V.; Solujan, S.I.: Theoretical foundation of nonlinear acoustic, "Nauka", Moscow, 1975 (in Russian).

8. Nigul, U.K.; Engelbrecht,Ju.K.: Nonlinear waves of deformation, Academy of Sci. of Est. SSR, Tallin, 1981 (in Russian).

9. Zakharov, N.S.; Korobeinikov, V.P.: PMM, 1980, vol.44, No. 5, p.943.

10. Ovsjannikov, L.V.: Group analysis of differential equations, "Nauka", Moscow, 1978 (in Russian).

11. Hopf, E.: Comm. Pure Appl. Math., 1950, vol.3, No.3, p. 201-230.

12. Benton, E.R.; Platzman, G.W.: Quart. Appl. Math., 1972, vol.30, No.2.

13. Whitham, G.B.: Linear and Nonlinear waves, John Wiley and sons, New York, 1974.

PART 2
WAVES IN HEREDITARY AND PLASTIC MEDIA

Deformation Waves in Nonlinear Hereditary Media

Yu.N.RABOTNOV, Yu.V.SUVOROVA, A.Ye.OSOKIN
Mechanical Engineering Research Institute, Moscow
Moscow Physico-Technical Institute, Moscow

Summary

The propagation of the one-dimensional deformation waves in nonlinear hereditary media described with the aid of the constitutive integral equation of Rabotnov type is considered. It is assumed that in a material under loading the process of damage accumulation can take place simultaneously with the process of viscous flow and both of them are of hereditary character. So, when a pulse of loading has passed through a section of a sample, this section might be damaged. We shall describe the influence of material properties and loading parameters on the pulse shape evolution and on the distribution of residual deformations due to damage accumulation. For obtaining the solutions of the problems under consideration a numerical method has been developed. The obtained solutions permit to treat the inverse problem, i.e. to calculate stress-strain diagrams for the material if the experimental data on pulse shape evolution are available.

For carrying out the analysis of the processes of wave propagation in nonlinear hereditary media, it is necessary, first of all, to choose a proper constitutive equation. As a rule that is done on the basis of the results of the dynamic experiments, but the difficulties in the test procedure and interpretation of the records often lead to the mistakes connected with the impossibility to distinguish the distortion of the pulse form due to material properties and due to the imperfections of the experiment (not quite one-dimensional waves, width of the gages etc.). One can use, of course, also the data of the quasistatic experiments for the formulation of the model, but it is not clear whether this model will be correct in the dynamic case. Therefore, the mathematical model is to be proved by a careful comparison of the theoretical and experimental results, concerning the pulse shape

Nonlinear Deformation Waves
IUTAM Symposium Tallinn 1982
Editors U Nigul, J Engelbrecht

evolution during its propagation.
For the description of the behaviour of inelastic materials the hereditary-type models are often used now, i.e. the constitutive equations are taken in the form of the Volterra type integral equations. These equations are known to give good results in case of the quasistatic problems. For solving the dynamic problems they have been comparatively rarely used due to some mathematical difficulties. In the literature, devoted to wave propagation, mainly linear constitutive equations are considered. In the linear case it is possible to apply the integral transform technique. This mathematical method is very powerful. Nevertheless, the mathematical difficulties are great, especially when the two- or three-dimensional problems are considered.

It is much more difficult to describe wave propagation in a nonlinear hereditary medium /1/. For this purpose the development of special methods is important. For example, the method /2/ of construction of successive approximations by integration of the linear equations should be mentioned. The application of the numerical methods is very perspective, too.

In this report we shall present some results of the numerical analysis of the one-dimensional wave propagation in semi-infinite and finite rods made of a material having hereditary properties (in particular, the kernel function with weak singularity) and essential nonlinearity.

We shall use the constitutive equation

$$\varphi(\varepsilon) = \sigma + \mathcal{L}^*\sigma + M^*\sigma \tag{1}$$

where σ - stress, ε - strain, $\varphi(\varepsilon)$ - the curve of the instantaneous deformation, $\mathcal{L}^*\sigma = \int_0^t \mathcal{L}(t-\tau)\,\sigma(\tau)\,d\tau$ - the operator which describes the process of reversible creep, and $M^*\sigma = \int_0^t M(t-\tau)\,\sigma(\tau)\,d\tau$ - the operator which describes the process of damage accumulation, that is the process of gradual failure of the material.

The reason of using of the two operators is connected with the necessity of separation of viscous and failure types of deformation. This is important especially when unloading processes are considered.

Failure deformation can't decrease under unloading, so the equation (1) for the processes $\dot{\varepsilon} < 0$ must be written in the form:

$$\varphi(\varepsilon) = \sigma + \mathcal{L}^{*}\sigma + M^{*}\sigma \Big|_{t=t_{*}} \qquad (2)$$

where t_{*} - the moment of the beginning of unloading.

The deformation, connected with the failure process, can't decrease, so the value of the third term in this equation can't vary under unloading. For this reason when there are N loading intervals, $t_i^- \leq t \leq t_i^+$, $1 \leq i \leq N$, the structure of the operator M^* takes the following form:

$$M^{*}\sigma = \sum_{i=1}^{N} \int_{t_i^-}^{t_i^+} M(t-\tau)\,\sigma(\tau)\,d\tau \qquad (3)$$

Advantages of such approach to the analysis of inelastic material behaviour in the quasistatic case have been described for example in /3,4/ where the problem of the choice of the kernels has also been considered. It was shown that the kernels might be used which differ from each other only by a coefficient. This simplifies the mathematical analysis of the problems, especially in case of the analytical approach. Although in the numerical calculation this assumption is not essential, we shall use it in our considerations.

We shall present some results of description of the wave propagation by making use of the constitutive equations (1) and (2).

The system of equation for the wave propagation process consists of the equation (1) and the field equations:

$$\rho \frac{\partial v(x,t)}{\partial t} = \frac{\partial \sigma(x,t)}{\partial x}$$
$$\frac{\partial \varepsilon(x,t)}{\partial t} = \frac{\partial v(x,t)}{\partial x} \qquad (4)$$

here ρ - density of material in the nondeformed state, x - space coordinate.

We must choose now an analytical form of the function $\varphi(\varepsilon)$. It is preferable to take it in the form:

$$\varphi(\varepsilon) = \begin{cases} E\varepsilon & \varepsilon \leqslant \varepsilon_0 \\ E\left\{\gamma\varepsilon + (1-\gamma)\left[2\varepsilon_0 - \frac{\varepsilon_0^2}{\varepsilon}\right]\right\}, & \varepsilon > \varepsilon_0 \end{cases} \qquad (5)$$

here E - modulus of elasticity, γ and ε_0 - nondimensional parameters of materials. If $\gamma = 1$ the equation (5) gives the Hooke's law $\varphi(\varepsilon) = E\varepsilon$, if $\gamma = 0$ it gives the Malvern's diagramm. So this parameter is the measure of nonlinearity of the instantaneous curve $\varphi(\varepsilon)$.

Let's introduce the following nondimensional variables

$$\bar{\sigma} = \frac{\sigma}{E\varepsilon_0} \; ; \qquad \bar{\varepsilon} = \frac{\varepsilon}{\varepsilon_0} \; ; \qquad \bar{v} = \frac{v}{c}$$
$$\bar{t} = \frac{t}{t_0} \; ; \qquad \bar{x} = \frac{x}{x_0} \qquad (6)$$
$$c = \sqrt{E/\rho} \; ; \qquad x_0 = c\,t_0$$

Later we take $x_0 = \ell$ if a rod with a finite length ℓ is considered, so $\bar{x}$ changes from 0 to 1. The bars above letters can be omitted since we shall deal only with nondimensional variables.

On the basis of (5) and (6) we obtain

$$
\varphi(\varepsilon)=\begin{cases} \varepsilon, & \varepsilon \leqslant 1 \\ \gamma\varepsilon+(1-\gamma)[2-1/\varepsilon], & \varepsilon>1 \end{cases}
$$

$$
c^2(\varepsilon)=\frac{d\varphi}{d\varepsilon}=\begin{cases} 1, & \varepsilon \leqslant 1 \\ \gamma+(1-\gamma)/\varepsilon^2, & \varepsilon>1 \end{cases} \tag{7}
$$

The parameter γ can take the value from the interval $0 \leqslant \gamma \leqslant 1$, and $o \leqslant c(\varepsilon) \leqslant 1$.

In the following calculations we must deal with the function $\psi(z)$, which is inverse to φ, i.e. $\psi[\varphi(\varepsilon)] = \varepsilon$.

From (7) we obtain

$$
\psi(z)=\begin{cases} z, & z \leqslant 1 \\ \dfrac{0{,}5z+(\gamma-1)}{\gamma}+\sqrt{\left[\dfrac{0{,}5z+(\gamma-1)}{\gamma}\right]^2+\dfrac{1-\gamma}{\gamma}}, & z>1 \end{cases} \tag{8}
$$

It is necessary to notice, that at $\varepsilon = 1$ both functions $\varphi(\varepsilon)$, $\psi(\varepsilon)$ and their first derivatives are continuous.

According to (1),(3),when knowing the intervals of loading and unloading during the process of waves propagation in the rod of finite length, we can write the following equation for the calculation of deformation:

$$
\varepsilon(t)=\psi\Big\{\sigma(t)+\int_0^t \mathcal{L}(t-\tau)\,\sigma(\tau)\,d\tau+
$$

$$+\sum_{i=1}^{N} \int_{t_i^-}^{t_i^+} M(t-\tau)\sigma(\tau)d\tau \}$$

In our numerical procedure we used this relation.

Let's consider now the problem of wave propagation in semi-infinite rod, made of material with the properties described by equation (1) and possesing weak singularity.

There is no necessity to separate processes of reversible creep and damage accumulation, so we shall write in the constitutive equation only one operator with the kernel $K(t-\tau)$:

$$\varphi[\varepsilon(t)] = \sigma(t) + \int_0^t K(t-\tau)\sigma(\tau)d\tau \qquad (9)$$

For construction of the stable, finite difference scheme, which is able to give correlated approximation of boundary value problems for the equations (4) and (9) let's introduce the regular kernels $K_\delta(t-\tau)$ approaching at $\delta \to 0$ the kernel $K(t-\tau)$.

For simplicity we shall consider the kernel of the Abel's type

$$K(t-\tau) = k(t-\tau)^{-\alpha}, \qquad 0 < \alpha < 1$$

The approximation of this kernel

$$K_\delta(z) = \begin{cases} k\delta^{-\alpha}, & 0 \leq z \leq \delta \\ kz^{-\alpha}, & \delta < z < \infty \end{cases}$$

is shown in fig. 1

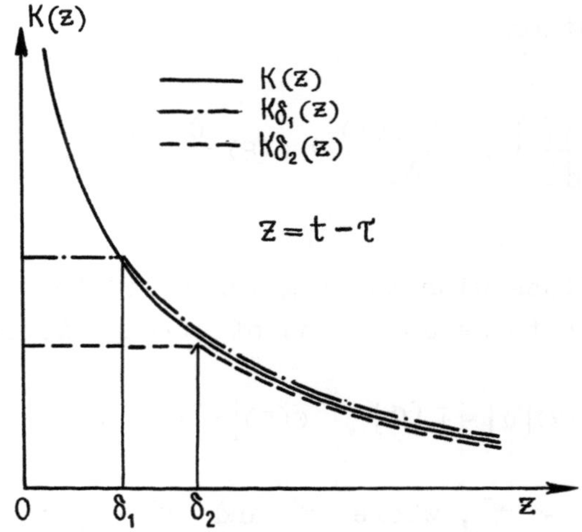

Fig. 1. Approximation of the Abel's kernel

Having chosen the kernel in such a form, let's reduce the system of equations (4) and (9) to the normal characteristic form. For this aim let's differentiate the equation (9) in respect to t:

$$c^2(\varepsilon)\frac{\partial \varepsilon}{\partial t} = \frac{\partial \sigma}{\partial t} + K_\delta(0)\sigma(t) + \int_0^t \frac{\partial K_\delta(t-\tau)}{\partial t}\,\sigma(\tau)\,d\tau \qquad (10)$$

We assume also that $d^2\varphi/d\varepsilon^2 < 0$ and generation of shock waves can't take place.

The next relations now can be written along the characteristics $dx = \pm c(\varepsilon)dt$;

$$\left\{\frac{dv}{dt}\right\}^{\pm} \mp \frac{1}{c(\varepsilon)}\left\{\frac{d\sigma}{dt}\right\}^{\pm} = \pm\frac{1}{c(\varepsilon)}\left\{K_\delta(0)\,\sigma(t) + \int_0^t \frac{\partial K_\delta(t-\tau)}{\partial t}\,\sigma(\tau)\,d\tau\right. \qquad (11)$$

here the notation

$$\left\{\frac{df}{dt}\right\}^{\pm} = \frac{\partial f(x,t)}{\partial t} \pm c(\varepsilon)\frac{\partial f(x,t)}{\partial x}$$

is used.

For the equations with the regular hereditary kernels the front is known to be as a line of strong discontinuity and

$$c(\varepsilon)[\vartheta] = \mp[\sigma], \quad c(\varepsilon)[\varepsilon] = \mp[\vartheta] \tag{12}$$

Here $[f] = f^{+} - f^{-}$, where f^{+} and f^{-} are the values of f at the left and right sides of the wave front.

For c = const it is easily to obtain from (11) and (12) the ordinary differential equations, describing the evolution of discontinuity of $[\sigma]$ or $[\vartheta]$ on the incident wave dx = cdt and reflected wave dx = -cdt. For c = 1 the equation for jump of σ have the form:

$$2K_\delta(0)^{-1}\frac{d[\sigma]}{dt} + [\sigma] = 0$$

Its solution is

$$[\sigma] = [\sigma]_0 \exp\left\{-\frac{K_\delta(0)}{2}(t - t_0)\right\} \tag{13}$$

It is clear, that the less δ (see fig. 1), the quicker damping occurres. For the equations with the singular kernels the propagation of strong discontinuities is impossible. To prove this fact let's consider the boundary value problem for difference of solutions, one of which possesses the regular kernel $K_\delta(t - \tau)$ and the other - irregular one $K(t - \tau)$. Let's introduce $\tilde{\vartheta} = \vartheta - \vartheta_\delta$, $\tilde{\sigma} = \sigma - \sigma_\delta$, $\tilde{\varepsilon} = \varepsilon - \varepsilon_\delta$.

For this difference the problem takes a form

$$\tilde{\vartheta}_t - \tilde{\varepsilon}_x = 0, \quad \tilde{\vartheta}_x - \tilde{\varepsilon}_t = 0, \quad \tilde{\varepsilon} = \tilde{\sigma} + K_\delta^{*}\tilde{\sigma} + q_\delta$$

$$\tilde{v}(0,x) = \tilde{\sigma}(0,x) = \tilde{\varepsilon}(0,x) = 0$$

$$\tilde{v}(t,0) = \tilde{\sigma}(t,0) = \tilde{\varepsilon}(t,0) = 0$$

$$q_\delta = \int_{t-\delta}^{t} \left\{ K(t-\tau) - K_\delta(0) \right\} \sigma(\tau)\, d\tau$$

We consider here only integrable singularities and bounded stresses, so the next estimate can be made for the term q_δ

$$q_\delta \leqslant \sigma_{max} \left\| K - K_\delta \right\|_{\mathcal{L}_1} = \frac{k}{1-\alpha} \left[t^{1-\alpha} - (t-\delta)^{1-\alpha} \right] - \delta^{1-\alpha} \longrightarrow 0$$

$$\text{for } \delta \longrightarrow 0$$

It means that for ($\tilde{v}$, $\tilde{\sigma}$, $\tilde{\varepsilon}$) the problem with arbitrary small disturbance and homogeneous initial and boundary conditions takes place. Therefore the solution approaches zero when $\delta \longrightarrow 0$.

Using this result we can replace the regular kernel in (13) by the irregular one. It is seen that a jump of stress must be equal to zero. Therefore the strong discontinuity can't propagate in the media with weak singularity and stress across the front is continuously differentiable. This result in a more detailed and precise form was proved in / 5 /. This fact was used for the realization of numerical procedure.

Using the theorem of commutability of convolution we obtain from (11):

$$\left\{ \frac{dv}{dt} \right\}^{\pm} \mp \frac{1}{c(\varepsilon)} \left\{ \frac{d\sigma}{dt} \right\}^{\pm} = \pm \frac{1}{c(\varepsilon)} \left\{ \int_0^t K(\tau) \frac{\partial \sigma(t-\tau)}{\partial t} d\tau \right\} \quad (14)$$

If we want to construct the difference scheme, we must make the approximation of the relations (14) using interpolation according to order of accuracy of the scheme for the characteristics passing, through the knots of the net. At the boundary we shall use only the relations on the characteristics directed inside the region of $x \sim t$ plane in which the solution is searched. It is necessary to add also the boundary conditions. Deformation is calculated with the help of the relation $\varepsilon(t) = \psi\{\sigma(t) + K^{*}\sigma\}$ where the function ψ is determined by (8). As a result we have the stable finite difference shceme, which can be considered as a modification of the scheme, developed in /6/.

The described method was used for the analysis of the waves propagation in the semi-infinite and finite length rods. The following problems were considered:

1. Construction of the pulse shapes and investigation of their evolution with propagation along the rods according to the material properties namely nonlinearity and heredity.

2. Analysis of damage accumulation during wave propagation and determination of residual deformation.

3. Solution of inverse problem and determination of dynamical stress-strain curves using information concerning the shape of the wave.

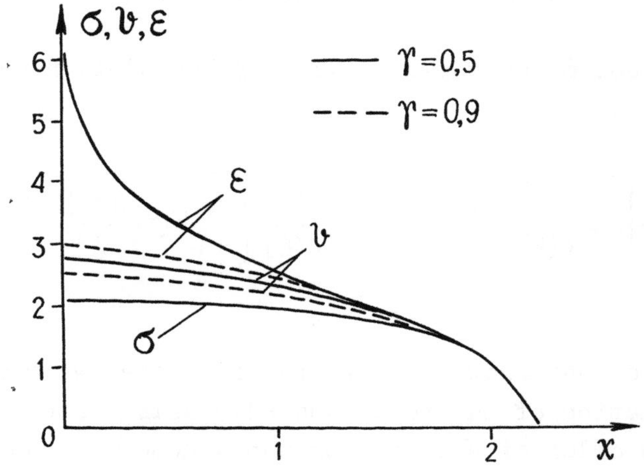

Fig. 2. Shapes of pulses propagating in the materials with the various degrees of nonlinearity. The condition at the end of the rod is: dv/dt = const

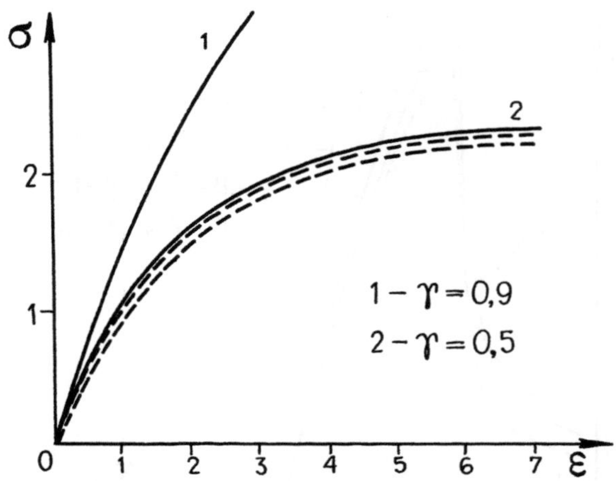

Fig. 3. Dynamical stress-strain curves and curves of instantaneous deformation for various degrees of nonlinearity

Fig. 2 shows the shape of pulses propagating in the materials with different degree of nonlinearity under the condition at the end of the rod dv/dt = const . The curves of instantaneous deformation for these materials are given in fig. 3. The curve 1 corresponds to value of $\gamma = 0{,}9$; curve 2 – $\gamma = 0{,}5$. The kernel of the constitutive equation (9) was chosen in the form of Abel's type $K(t-\tau) = k(t-\tau)^{-\alpha}$ (for material 1 $\alpha = 0{,}96$; $k = 0{,}099$; for material 2 $\alpha = 0{,}8$; $k = 0{,}05$). Fig. 2 demonstrates the influence of the nonlinearity on the growth of deformation behind the front: evidently, the rate of growth will be the higher the nearer curve $\varphi(\varepsilon)$ to the ideal plasticity type.

The fore part of the pulse doesn't disturb at a great extent due to the nonlinearity, in any case one couldn't notice this effect in fig. 2. It is more pronounced in fig. 4, which shows the shapes of the pulses for the various degrees of nonlinearity and hereditary effects. This example corresponds to the conditions of the loading by the pulse σ = const of finite duration, the kernels of the constitutive equation (1) were chosen in the form $\mathcal{L}(t-\tau) = \ell e^{-\beta(t-\tau)}$, $M(t-\tau) = m e^{-\beta(t-\tau)}$. The cases when $\ell = m = 0{,}005$ the parameter of heredity degree $\beta = 0{,}5$; 2; 10; and the degree of nonlinearity:

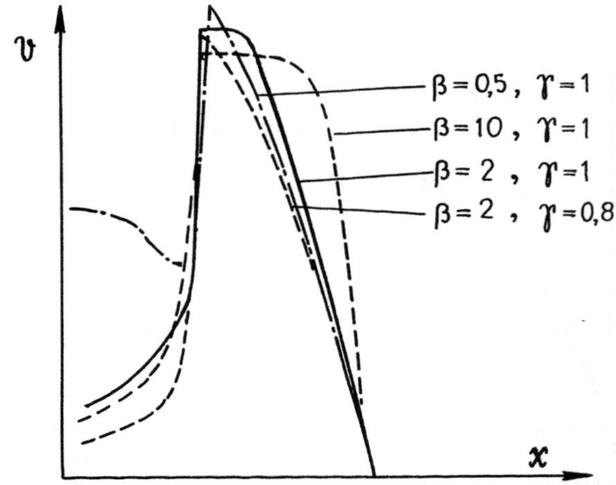

Fig. 4. Shapes of the pulses for the various degrees of nonlinearity (γ) and hereditary effects (β). The condition at the end of the rod is: σ = const, $0 \leqslant t \leqslant t_0$

γ = 1 and γ = 0,8 were considered. As can be seen the nonlinearity does not change the shape to a great extent. The parameter β appears much more essential than γ, the greater β , the nearer the pulse shape to the rectangular one corresponding to elastic material. The decrease of β leads to a greater rate of growth of deformation at the rear part of the pulse because of material's memory. This can be noticed in figure 4 for the case corresponding to β = 0,5. During propagation of the wave the unloading part of the pulse is also changing. But is is necessary to emphasise that if the distortion of the loading part is the consequence both viscous effects and damage accumulation, the distortion of unloading part depends only on viscosity.

As a result of wave propagation the material of the rod partly failures due to damage accumulation, therefore if wave of loading passes the rod many times, there occurs the residual deformation after the damping. Nonlinearity of the material properties influences on the distribution of the residual deformation and on its values. Fig. 5 shows the residual deformation after propagation of the finite duration pulse σ =const along the rigidly fixed rod for two cases γ = 1, γ = 0,5.

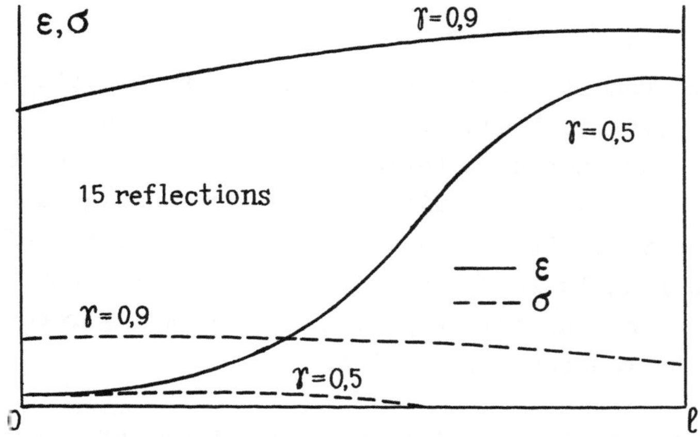

Fig. 5. Residual deformation in the rod after propagation of the finite duration pulse for various degrees of nonlinearity

It is seen that for γ = 0,5 the distribution of deformation much more irregular because of the less velocity of wave. The last fact is also the reason of quicker establishment of the numerical procedure for the nonlinear material. Fig. 5 corresponds to the 15 reflections of waves (1500 steps of calculation). One can notice that the maximum value of deformation for linear material is higher. It apparently can occur because of too little duration of calculations for linear material and this distribution doesn't yet correspond to the real value of residual deformation.

Having the programm for describing the pulse shape evoluation we have possibility to solve the inverse problem. Now if we have obtained in experiment some pictures of pulse for different sections of the sample, we are able to construct the dynamical stress-strain curves for a given material. Fig. 3 shows such diagramms, calculated using fig. 2 for the nonlinear material with γ = 0,5. These stress-strain curves are seen not to differ too much from the instantaneous curve. It allows to conclude, that the approach of Karman-Taylor-Rakhmatulin, based on the concept of the existence of only two diagramms - static and dynamical is valid for such a material and we can indentify the dynamical diagram with the instantaneous curve.

Moreover this results allow to make the other important conclusion. The dynamical experiments can be now the base for obtaining of the instant deformation diagramm. Up to now it was determined only by calculation procedure using data on creep or on stress-strain behaviour at various rates of loading / 3 /. Our analysis shows, that any of stress-strain curves, obtained in dynamical experiments can be regarded as curve of instantaneous deformation.

These results actually confirm the possibility of using for the dynamical problems the nonlinear hereditary-type equation, proposed previously / 7 / for the description of creep properties of the materials.

References

1. Rabotnov, Yu.N. Hereditary mechanics of solids. Moscow: Nauka 1977.

2. Nigul, U.K. Echo-sighals from elastic objects. Tallin: Valgus 1976.

3. Suvorova, J.V. Nonlinear effects at loading of hereditary media. Mech. Polym. 6 (1977) 976-980.

4. Suvorova, J.V.; Viktorova, I.V.; Mashinskaya G.P. Long--time failure of nonelastic composites. Mech. Comp. Mater. 5 (1979) 794-798.

5. Lokshin, A.A.; Suvorova, J.V. Mathematical theory of wave propagation in media with memory. Moscow: Moscow University Press 1982.

6. Suvorova, J.V.; Osokin, A.E. Propagation of one-dimensional waves in nonlinear-hereditary materials. Mech. Polym. 3 (1978) 425-429.

7. Rabotnov, Yu.N. Some problems in creep. Vestnik of Moscow University 10 (1948) 81-91.

The Analysis of the Strain Wave Processes and Fracture of the Structural Elastic-Viscoplastic Elements

A. G. Ugodchikov

Gorky State Lobachevsky University

Summary

Recent years witnessed a positive success in handling the nonlinear problems of the dynamic deformation and the media interaction, brought about by the application of modern numerical methods and high-speed computers. The present paper describes the methods and algorithms for the analysis of the wave deformational processes and fracture of the elastic-viscoplastic bodies in the axisymmetrical formulation under the pulse thermal loads and forces, including the real processes of elastoplastic deformation and fracture, as well as the large-scale distortion of the analysed body boundaries. The state equations used in the paper for viscoplastic media account for the temperature, strain-rate and plastic deformation level dependence of the material physical-mechanical parameters, and for the deformation anisotropy. A numerical analysis method for the thin-walled shell dynamic deformation, including large deflections and elastoplastic deformations, initiation and propagation of the shock waves in media, displacements and distortions of the contacting boundaries, and cavitation phenomena in liquids, is discussed. The analysis of the axisymmetric body deformation and fracture processes, and the axisymmetric shell-compressible fluid interaction is illustrated by a number of real studies.

Finite-Element Analysis of the Axisymmetric Body Dynamic Deformation and Fracture Processes [1-3, 8-10]

The nonlinear equation system, describing the transient deformation processes, is formulated as a two-velocity, four-stress system. The node velocities are found using the momentum equation in the integral form, after which the area integral is substituted by the boundary one. The resulting formulae for the natural approximation of the partial derivatives provide for the momentum conservation for any finite element set with the accuracy up to truncation errors. Discretization in time is carried out using an explicit, three-stage scheme, its stabi-

Nonlinear Deformation Waves
IUTAM Symposium Tallinn 1982
Editors U Nigul, J Engelbrecht

lity being confirmed by a number of numerical experiments and practically provided by the Courant condition with a constant factor.

The following equation system describes the axisymmetric stress wave propagation problem:

momentum equation

$$\bar{\sigma}_{11,1} + \sigma_{12,2} + \frac{\sigma_{12}}{x_2} = \rho \ddot{x}_1 ,$$

$$\sigma_{21,1} + \bar{\sigma}_{22,2} + \frac{\bar{\sigma}_{22} - \bar{\sigma}_{33}}{x_2} = \rho \ddot{x}_2 , \qquad (I)$$

$$\bar{\sigma}_{ii} = \sigma'_{ii} - (p+q)\,\delta_{ii} ,$$

continuity equation

$$\frac{\dot{V}}{V} = \dot{x}_{1,1} + \dot{x}_{2,2} + \frac{\dot{x}_2}{x_2} , \qquad (2)$$

energy equation

$$\dot{E} = -(p+q)\dot{V} + V(\sigma'_{11}\dot{e}_{11} + \sigma'_{22}\dot{e}_{22} + \sigma'_{33}\dot{e}_{33} + 2\sigma_{12}\dot{e}_{12}) + \dot{E}_B ; \qquad (3)$$

$$p = p(V, E)$$

Here x_1 is a rotation axis, p - hydrostatic pressure, q - artificial viscosity, V - relative volume, ρ - density, σ'_{ij} - stress deviator components, e_{ij} - strain tensor components, E - intrinsic energy per initial volume unit, and $\dot{E}_B$ - the intrinsic heat source power.

The elastoplastic response of the material is described in terms of the thermoplasticity theory with the kinematic and isotropic hardening[I].

The experimental studies show, that the fracture process under the pulse loads consists of the formation of numerous discontinuities, their growth and merging. Depending on the load conditions, the three stages have different importance for the structure service life, but in all cases the completion of the failure process takes a finite time interval, which depends on the stress-strain history. This necessitates the application

of the so-called time criteria and the damage accumulation theory, to evaluate the rigid body dynamic strength.
The works [2,3] , on the basis of [4] , develop the response models of the deforming media, incorporating the material defect initiation, growth and merging. The models use realistic defect structures: spheric pores in viscous failure, and penny-shaped microcracks in brittle fracture.
To account for a large number of cracks, the defect nucleation and growth processes are described statistically. To this end a crack parameter distribution function $f(q_1, q_2, \dots q_n)$ is used, where $q_i\,(i=1,2,\dots,n)$ are the internal crack coordinates, parameters, describing dimension, orientation and form, so that the expression $\int_{q_i}^{q_i+\Delta q_i} f(q_1', q_2', \dots q_n')\, dq_1' dq_2' dq_3' \dots dq_n$ is equal to the crack number in the interparameter interval $[q_i, q_i+\Delta q_i]$. The experimental studies [4,5] show, that the discontinuity distribution according to the dimensions is of the form

$$N(R) = N_0 \exp(-R/R_1) \quad , \tag{4}$$

where N is the total discontinuity number per unit volume, with the radius greater than R , N_0 - overall discontinuity number, and R_1 - a distribution parameter.
If the function (4) is known, one can evaluate the total void volume, integrating over the entire distribution; assuming the new nucleating voids to distribute still according to (4), one can find the discontinuity nucleation and growth rates. The analysis of the material deformation and failure behaviour is based on the porous body plasticity theory [2,3].
After the linear damage accumulation hypothesis, the evaluation of the time-to-failure value t_p under a random loading pattern $\sigma_э = S(t)$ can be reduced to the evaluation of the functional equation

$$\int_0^{t_p} \frac{dt}{t_p[S(t)]} = 1 \tag{5}$$

where $\sigma_э$ is the equivalent stress, and $t_p(S)$- time to failure at some basic test $S = const$.
If $t_p(S) = k/(\sigma_э - \sigma_0)^n$, then the equation (5) gives the well-known Tooler-Butcher criterion

$$\int_0^{t_p} (\sigma_э - \sigma_0)^n dt = k \quad , \tag{6}$$

where σ_0, n, and k are the experimentally derived parameters. The equation system (I)-(3), combined with the state equations and the equations, describing the material failure, boundary and initial conditions, make up a complete equation set, describing the axisymmetric viscoplastic body dynamic deformation and failure process.

The implementation of the state equations in practical calculations for realistic materials requires a development of the experimental-theoretical procedure for evaluating the state equation parameters on the basis of the basic experiment. This is done on the basis of the tension-compression tests of the cylindrical specimens at various (constant during each test) temperatures [I]. To define the strain rate dependence of the yeild surface, a construction of the dynamic deformation diagrams at various strain rates is required in the range of IO^0-IO^3 sec^{-I}. The described procedure for the dynamic deformation diagram construction is based on the well-known Kolsky method [IO] - the Hopkinson split pressure bar (HSPB) method.

The developed experimental procedure provides for the dynamic dependencies $\sigma = \sigma(\dot{e}, e)$ of various materials. Fig.Ia shows as an example, the dynamic deformation diagrams for the АД-0 aluminium at two strain rates, as compared to the static diagram. Fig. I illustrates the strain rate variation of the deforming specimen material.

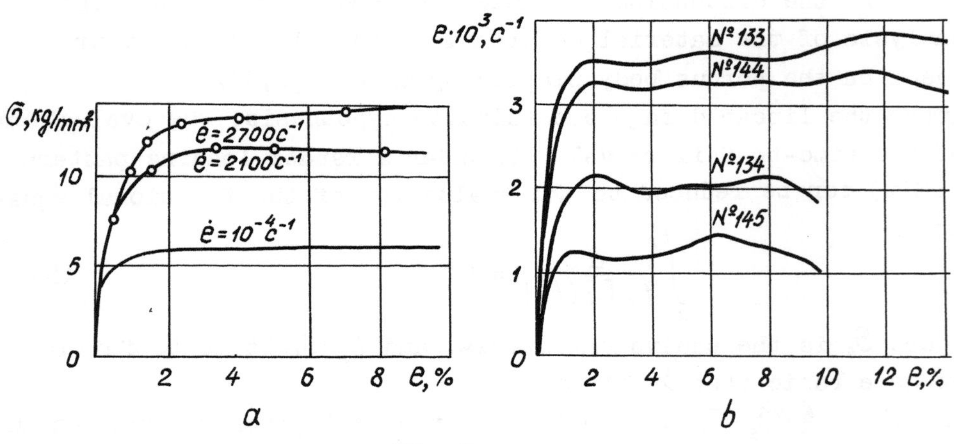

Fig.I.

The dynamic diagram construction according to Kolsky-HSPB method does not include such effects as the tested specimen radial inertia, effects of the specimen length finiteness, and face friction effects between the measure bars and the specimen.
A computer simulation was used to analyse these effects. The simulation was carried out in the axisymmetric formulation [II]. A large number of numerical experiments show, that Kolsky-HSPB method provides the dynamic diagrams, accurate enough for the design calculations.

<u>Practical Analysis of the Axisymmetric Processes in Continuous Bodies</u>

In the following, the analysis of several problems is outlined, illustrating characteristic features of the elastoplastic body dynamic deformation and failure. The impact test results for the circular 606I-I6-aluminium plates are reported in [I2]. The target thickness was four times the projectile thickness in all the tests. At a constant target (or projectile) thickness the tests were performed at various velocities, and a minimum velocity was determined, at which spalling initiated. The computational and the experimental data are compared in [6]. The x- and y-coordinates of Fig.2 represent, respectively, a total target plus projectile thickness, and the impact velocity. The triangles designate the computational results, the circles - the experimental data, and the solid triangles show the velocities, at which spalling occured. It is evident, that the computational results agree quite well with the experimental data of [I2].

Fig.2.

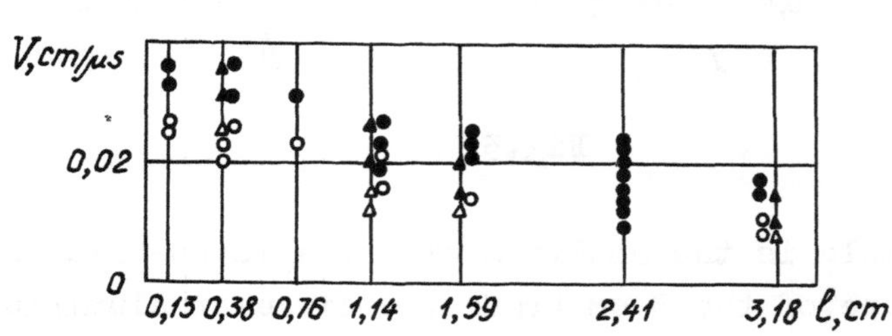

The work [9] gives the analytical treatment of the stress wave propagation under the thermal loads. The same problem was analysed using the above described method, assuming that the heating proceeded with a constant intensity, and the temperature decreased exponentially with the distance from the surface.
The effects of the heating rate t_0 , temperature level T_0 , and the heated layer thickness on the plastic fracture processes under the pulse thermal loads were investigated.
The time dependence of the damage level at various heating rates for $T=600^0$ and $k=I/6$ are shown at the heated layer midplane (Fig.3a), and at the interaction of the wave system with the free opposite surface (Fig.3b) (figure I shows the curves for $t_0=0,I$ mcsec, $k=I/5$).
Fig.4 shows the damage level distribution through the plate thickness for various heating conditions (the curves I-5, Fig. 4a, depict the values at $T=600^0$, $k=I/6$ and $t_0=0,6$, 0,4, 0,3, 0,2, 0,I mcsec, respectively, and the curve 6 - the values at $T=600^0$, $k=I/5$, $t_0=0,I$ mcsec; figure I, Fig.4b, shows the value for $T=700^0$, $t_0=0,2$ mcsec, $k=I/6$; figure 2 - the value for $T=600^0$, $t_0=0,I$ mcsec, $k=I/6$, and figure 3 - the value for $T=700^0$, $t_0=$ $=0,3$ mcsec, $k=I/5$).

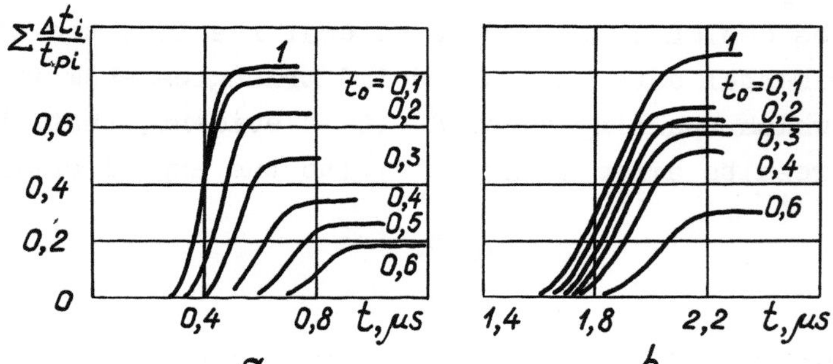

Fig.3.

The next example is the analysis of the axisymmetric deformation and failure of the 8 cm truncated circular aluminium cone with the major and minor radii of 6 and 2 cm, respectively, loaded by the triangle pressure pulse, applied to the major radi-

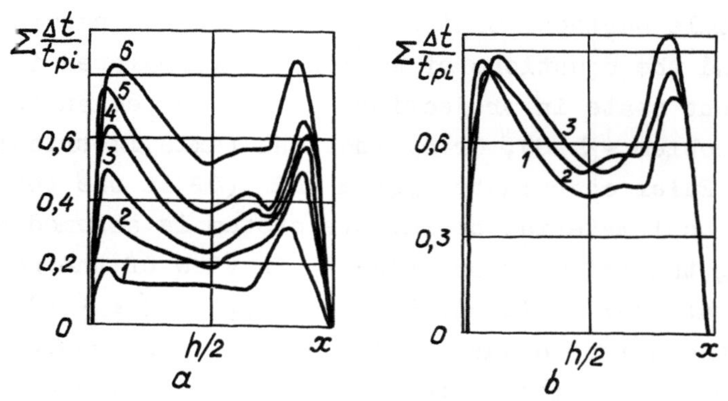

Fig.4.

us. Initially, significant compressing stresses appear (Fig.5, axial stresses along the cone length at various times (figures stand for mcsec), neglecting fracture), resulting in large plastic deformations and distortions (Fig.6). Fig.6 is the fracture situation at the time t=IO mcsec.

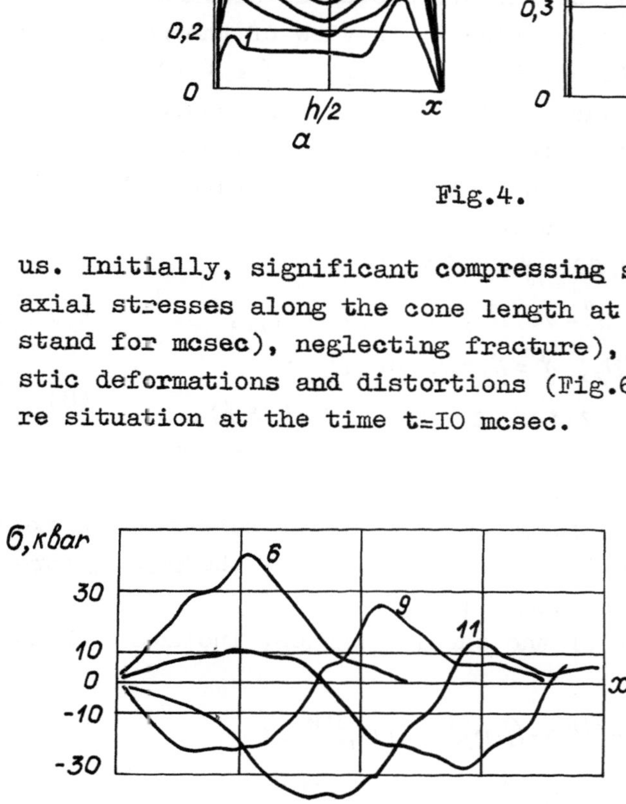

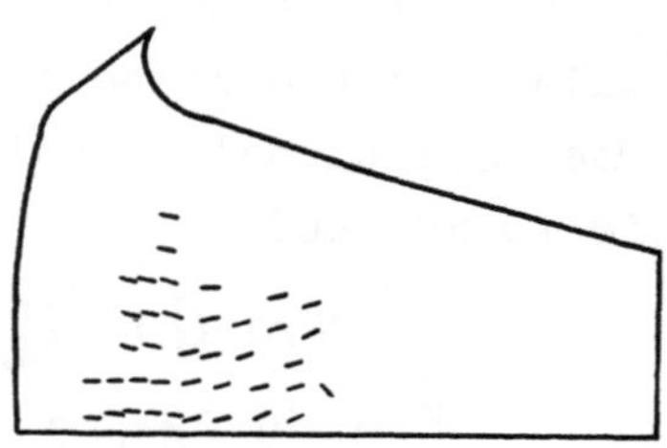

Fig.5. Fig.6.

<u>The Wave Problem Analysis Method for the Elastoplastic Shell-of-Revolution Dynamics</u>

A method is developed to analyse two-dimensional and axisymmetrical wave problems in the arbitrary thin-walled elastoplastic shell dynamics, providing for large distortions of the shells under small bending and shear strains. In the frame of the method the middle surface strains are described as natural (logarythmic), and the equations of motion are formulated in terms of the current state in projections on the Cartesian (two-dimensional problem $-\nu=0$), or cylindrical (axisymmetrical problem $-\nu=I$) Euler coordinate system rOz (Oz is the rotation axis). The local material coordinate system is assumed orthogonal during the deformation process, in view of the negligible shear deformations; S=s(t) is the arc length; $\xi=\xi(t)$ - distance from the point to the middle surface. Designating the middle surface displacement rates along the Or, Oz axes as $v_r(s,t)$; $v_z(s,t)$, and the cross-section rotation rate as $v_\varphi(s,t)$ gives the expressions for the deformation rates as

$$\dot e_{ii}=\dot e^0_{ii}+\xi\dot\varkappa_{ii}\ (i=1,2);\quad \dot\gamma=\dot\gamma^0[1-(2\xi/h)^2];$$
$$\dot e^0_{11}=v_{z,s}m+v_{r,s}n;\quad \dot e^0_2=\nu r^{-1}v_r;\quad m=z_{,s};\quad n=r_{,s}; \tag{7}$$
$$\dot\varkappa_{11}=v_{\varphi,s};\quad \dot\varkappa_2=\nu n r^{-1}v_\varphi;\quad \dot\gamma^0=v_\varphi-v_{r,s}m+v_{z,s}n;$$
$$dS=\sqrt{(dr)^2+(dz)^2};\quad r=r(s,t),\ z=z(s,t);\ h=h(s,t), \tag{8}$$

where the index after the comma shows differentiation about a corresponding variable. Total deformations are found by the time integration of the deformation rates. The stresses are expressed in terms of elastic deformations, using the Hooke's formulae for the plane stress state [I3].

The variational equation of the shell dynamics, including (7-8) is formulated as

$$\int_0^L\Big[(N_1m+Qn)\delta v_{z,s}+(N_1n-Qm)\delta v_{r,s}+M_1\delta v_{\varphi,s}+$$
$$+\nu r^{-1}N_2\delta v_r+(Q+\nu r^{-1}M_2n)\delta v_\varphi+\rho h(\dot v_z\delta v_z+\dot v_r\delta v_r+ \tag{9}$$
$$+(h^2/12)\dot v_\varphi\delta v_\varphi)-p_z\delta v_z-p_r\delta v_r\Big]r^\nu dS-\big[r^\nu(P_r\delta v_r+P_z\delta v_z+M\delta v_\varphi\big]_{s=0,L}=0.$$

where $p_r(s,t)$, $p_z(s,t)$ are the load surface components, $P_r(t)$, $P_z(t)$, M(t) - external force components and the moment, given on the shell boundaries (s=0,L) in the general basis.

The application of the explicit finite-difference scheme reduces the analysis of the problem to the recurrent count according to the formulae

$$v_j^{k+1/2}=v_j^{k-1/2}+(\tau^{k+1/2}+\tau^{k-1/2})\Phi_j^k/2A_i^k;\ (j=z,r,\varphi;\ i=M,J);$$
$$z^{k+1}=z^k+v_z^{k+1/2}\tau^{k+1/2};\quad r^{k+1}=r^k+v_r^{k+1/2}\tau^{k+1/2}; \tag{IO}$$

$$h^{\kappa+1} = h^{\kappa}(1+\Delta\mathcal{E}_3^{o})^{\kappa+1/2}; \quad (\Delta\mathcal{E}_3^{o})^{\kappa+1/2} = (\dot{\mathcal{E}}_1^{o}+\dot{\mathcal{E}}_2^{o})\tau^{\kappa+1/2},$$

where Φ_j are the node-reducted generalized forces, A_i - mass generalization, and $\tau^{\kappa}(\kappa=1,2..)$ - the integration time step.

Using the non-classical shell theory, including the through-thickness wave processes, the application range of the Timoshenko-type theory under the stress and thermal pulse loads, rapidly changing in time and space, was investigated in [I4-I5]. It is found, that in plastic deformation the local (in stresses) and integral (in displacements) criteria of the theory applicability practically coincide, whereas in perfectly elastic deformation the application range of the shell theory in the integral sense is much greater, as the stresses σ_{33} do not affect the stresses σ_{11}, σ_{22} through the yield condition. Generally, the theory is applicable, if the pulse has duration in time (multiplied by the sound velocity) and the application region in space greater than 2-3 times the shell thickness.

The method is readily generalized for the wave problems of the composite and branching shell dynamics, reinforced by circular stiffeners [I3]. The efficiency of the method is confirmed by a wide range of the analytical results and their comparison with the experimental data. Thus, the analysis of the elastoplastic plate deformation processes with finite deflections is presented in [I6]; large viscoplastic deformations of the cylindrical shells, characteristic of the magnet-pulse loads on the metals, are analysed in [I7] ; impact buckling of the smooth and circular stiffener reinforced viscoplastic cylindrical shells - in [I8] ; and the shell structure interaction processes with the liquid and gaseous media - in [I9].

The Analysis Method for the Problems of the Elastoplastic Shell Structure - Shock Wave Interaction in Perfectly Compressible Media

The two-dimensional and axisymmetric problems of the elastoplastic shell structure interaction with the shock waves in perfectly compressible media and the problems of the impact on a compressible liquid are formulated in [I9], allowing for the nonlinear properties of the media, hydro-gaso-dynamic parameter

field discontinuities, large distortions of the contact and free surfaces, and the cavitational effects in liquids.
The analysed region is assumed finite, generally, consisting of subregions with different medium physical properties, but identical equations of state. The structural element deformation processes are described using the Lagrange approach, and a combined Euler-Lagrange approach for media. The transient hydro-gaso-dynamics equations have, in Euler coordinates r , z , the form

$$
\begin{aligned}
&(\rho r)_{,t} + (\rho v_z r)_{,z} + (\rho v_r r)_{,r} = 0 \\
&(\rho v_z r)_{,t} + [(\rho v_z^2 + p) r]_{,z} + (\rho v_r v_z r)_{,r} = 0 \\
&(\rho v_r r)_{,t} + (\rho v_r v_z r)_{,z} + [(\rho v_r^2 + p) r]_{,r} = 0 \qquad \text{(I1)} \\
&(\mathcal{E} r)_{,t} + [(\mathcal{E} + p) v_z r]_{,z} + [(\mathcal{E} + p) v_r r]_{,r} = 0
\end{aligned}
$$

where p is the hydrostatic pressure, ρ -density, and $\mathcal{E} = \rho[e + (v_r^2 + v_z^2)/2]$; v_z, v_r -velocity components. The state equation is taken in the form

$$E = \frac{p + \alpha\beta}{\rho(\alpha - 1)} + \frac{c_0^2}{\alpha - 1}, \quad \beta = \rho_0 c_0^2 / \alpha ,$$

which allows, fixing corresponding parameter values, to apply it both to the gaseous and liquid perfect compressible media.
The pressure p^* and the velocity v^+, normal to the discontinuity surface, are found by evaluating the one-dimensional problem of the discontinuity break-up, formulated as a system of two transcendental equations [20]

$$
\begin{aligned}
&v^* - v^- + (p^* - p^-) a^- = 0 \\
&v^* - v^+ - (p^* - p^+) a^+ = 0 , \quad a^{\pm} = a^{\pm}(p^*) , \qquad \text{(I2)}
\end{aligned}
$$

where indexes (+, -) show the hydrodynamic parameter values on the both sides of the discontinuity surface.
The thin -wall assumption for the deforming elements makes it possible to neglect the through-thickness wave effects and take their middle surfaces to weightable contact surfaces, confining the analysed area, or being the interface surfaces. The equations, relating the contact pressures of the shell, follow from (I2), assuming $v^* = v_\xi$

$$p_-^* = p^- - \alpha^-(v_\xi - v^-);\quad p_\xi = p_-^* - p_+^*;\quad p_s = 0$$
$$p_+^* = p^+ - \alpha^+(v^+ - v_\xi);\quad p_r = -p_\xi m;\quad p_z = p_\xi n \qquad (13)$$

The transient hydro-gaso-dynamic problem is treated using the S.K.Godunov count-through exponential scheme [20] with mobile rectangular meshes, which make it possible to describe the mobile contact and the free surfaces. The contact pressures between the deforming structural elements and the media, interacting with them, are found by evaluating the combined (10), (13) equations iteratively. A role of the nonlinear effects was investigated on a number of applied problems. It is shown [19], that the geometric linearization of the boundary conditions in the problems of the shock wave-deforming structure interaction in liquid and gaseous media is relevant up to the 10-15%-dimensional change of the body in the cross-section, normal to the wave propagation direction. The cavitational effects are negligibly small at the pulse duration greater than the wave propagation time across a characteristic dimension of a plate or a shell (e.g., a diameter).

As an example, the axial impact and the submersion into the water of a cylindrical shell, attached to a deformable (Fig.7a), or perfectly rigid (Fig.7b) spherical bottom, is described. The top end of the shell is covered, in both cases, by a rigid disc. The main structural characteristics are: $R_1/R_0=1,45$, $h/R_0=0,05$; $L/R_0=2$; $m_2/m_1=5$; $\varphi_0=0,76$; and $v^0/c_0=0,3$, where L, R_0, R_1 is the length and the radii of the cylindrical and spherical shells (case "b" - the spherical bottom), respectively; φ_0 is the bottom half-angle. The material of the deforming structural elements is the Д16Т alloy. The dimensions of the analysed liquid flow region and the initial structural geometry (dashed line) are shown in Fig.8.

Figs.7,8 illustrate the submersion and deformation processes of the analysed structures, as well as the isobars in the liquid $(\bar{p} = p/(\rho_0 c_0^2), \tau = c_0 t/R_0)$.

At the time $\tau=1,75$ negative pressure zones appear in the vicinity of the polar point. The zones have a local (both in space and time) nature. The negative pressure contribution into the total hydrodynamic force is less than 5%. At the time $\tau=2,92$

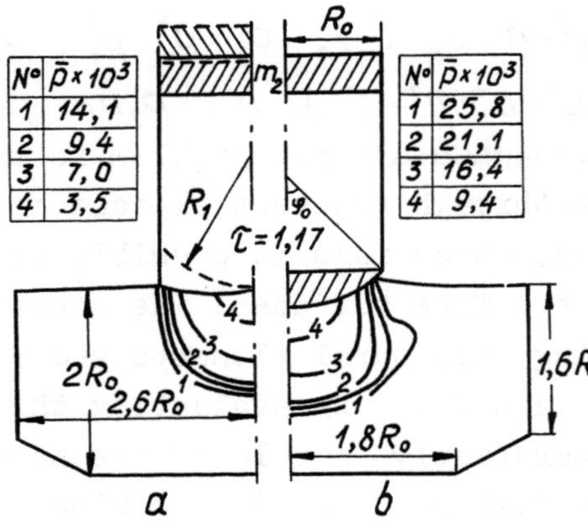

Fig.7.

(Fig.8), the deformed portion of the cylindrical shell comes in contact with the liquid. The results of the analysis show, that in case of the deformable bottom the hydrodynamic drag force is more smooth, it's maximum value being attained much later. At the time τ =I,2 the plastic deformation zone in case "a" spreads over the whole of the spherical shell. The intensive deflection growth results in the turn-out of the shell, which leads to the hydrodynamic force drop. In case "b", a similar drop occures at the time τ =I,6, corresponding to the beginning of the spray jet formation.

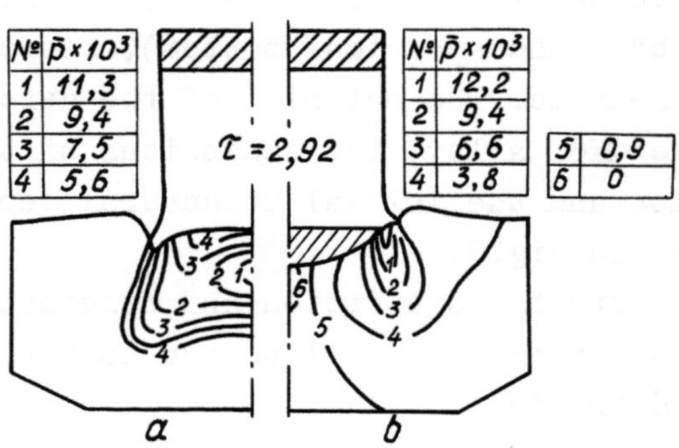

Fig.8.

References

I. Коротких Ю.Г., Угодчиков А.Г. Уравнения теории термовязко-пластичности с комбинированным упрочнением. Глава 6 в кн. "Уравнения состояния при малоцикловом нагружении", М., Наука, 1981.

2. Рузанов А.И. Численное моделирование процессов разрушения твердых тел при импульсных нагрузках. - Прикладные проблемы прочности и пластичности. Статика и динамика деформируемых систем. Всесоюз. межвуз. сб. /Горьк.ун-т, 1980, с.38-53.

3. Рузанов А.И. Моделирование разрушения твердых тел при динамических нагрузках как процесса образования и роста дискообразных микротрещин. - Прикладные проблемы прочности и пластичности. Статика и динамика деформируемых систем. Всесоюз. межвуз. сб. /Горьк. ун-т, 1981, с.23-31.

4. Seaman L., Curran D.R., Shockey D.A. Computational models for ductile and brittle fracture.- J. Appl. Phys., 1976, v.47, N II.

5. Рыбин В.В., Лихачев В.А. Статистика микротрещины на вязких (чашечных) изломах. - ФММ, 1977, т.44, вып.5.

6. Рузанов А.И. Численное исследование разрушения в упруго-пластических телах при динамических воздействиях. В сб.: Труды Ш Всесоюз. конф. по численным методам решения задач теории упругости и пластичности. Новосибирск, СО АН СССР, 1974.

7. Коротких Ю.Г., Рузанов А.И. Исследование динамического разрушения упругопластических тел при силовых и тепловых воздействиях. Прикл. мех., т.XIУ, № 7, 1978.

8. Коротких Ю.Г., Рузанов А.И., Садырин А.И. Анализ методом конечного элемента задач динамики сплошных сред. В сб.: Метод конечных элементов в строительной механике. Горький, 1975.

9. Рауш. Влияние времени нагрева на волны напряжений, вызванные тепловым воздействием. - Тр. амер. о-ва инж. - мех. Прикл. механика, 1969, 36, № 2.

10. Брагов А.М., Ломунов А.К., Русин Е.Е. Методика исследования динамических свойств материалов с использованием составных стержней Гопкинсона. - Прикладные проблемы прочности и пластичности. Всесоюз. межвуз. сб. /Горьк. ун-т, 1980, с.138-144.

II. Коротких Ю.Г., Пуртов Ю.Б. Численное исследование процесса динамического деформирования разрезного стержня Гопкинсона. Механика деформируемого тела. СО АН СССР, 1979, в.41, с.60-64.

I2. Blinsow D.W., Keller D.V. Experiments on the mechanism of spall.- Sympos. on Dynam. Behavior of Materials.- Albuque, N., I962, Philadelphia, Pa, Amer. Soc. Test. Mater., I963, p. 252-263.

I3. Баженов В.Г. Нелинейные задачи динамики тонкостенных конструкций при импульсных воздействиях. - Прикладные проблемы прочности и пластичности. Статика и динамика деформируемых систем. Всесоюз. межвуз. сб. /Горьк. ун-т, I98I, с.57-65.

I4. Абросимов Н.А., Баженов В.Г. Исследование динамического деформирования упругопластических сферических оболочек при тепловом ударе. Изв. АН СССР, МТТ, 1978, № I, с.I39-I43.

I5. Абросимов Н.А., Баженов В.Г. О применимости модели Тимошенко при локальном импульсном нагружении упругопластических цилиндрических оболочек. - Прикладные проблемы прочности и пластичности. Автоматизация решения задач упругости и пластичности. Всесоюз. межвуз. сб. /Горьк. ун-т, I980, с.50-56.

I6. Баженов В.Г., Батанин М.А. Исследование упругопластических процессов деформации круглых пластин при импульсном нагружении с учетом больших прогибов. - Прикладная механика, I978, № 3, с.74-78.

I7. Баженов В.Г., Ломунов В.К., Петров М.В. Упругопластическое деформирование цилиндрических оболочек при магнитно-импульсном нагружении. - Прикладные проблемы прочности и пластичности. Всесоюз. межвуз. сб. /Горьк. ун-т, с.73-78.

I8. Ломунов В.К. Исследование выпучивания дискретно-подкрепленных цилиндрических и конических упругопластических оболочек при осевом ударе. - Прикладные проблемы прочности и пластичности. Всесоюз. межвуз. сб. /Горьк. ун-т, I977, с.64-7I, вып.7.

I9. Баженов В.Г., Кочетков А.В., Михайлов Г.С., Угодчиков А.Г. Взаимодействие упругопластических тонкостенных элементов конструкций с ударными волнами в идеальных сжимаемых средах. - Изв. АН СССР, МТТ, I979, № 2, с.I4I-I49.

20. Численное решение многомерных задач газовой динамики (под редакцией С.К.Годунова), М., Наука, I979, с.400.

Nonlinear Inelastic Deformation Waves in Polymers and Metals

G.I.BARENBLATT

Institute of Oceanology,
USSR Academy of Sciences,
Moscow

The localization of inelastic deformation attracts now wide attention. In the main part of the existing literature the localization is analysed from the viewpoint of the classical continuum mechanics. From the viewpoint developed during some years in the present author's papers, the localization is a peculiar mechanical problem that can be stated and solved only by taking the transformation of the material explicitly into account. Such an alternative statement of the problem is presented here. It is shown that the procedure of determining a relation between the velocity of the wave propagation and the nominal acting stress leads naturally to a nonlinear eigenvalue problem similarly to an analogous problem in the flame propagation theory.

1. A typical localization phenomenon is appearence and uniform propagation of geometrically invariant inelastic deformation waves in stretched cylindrical bars of such polymeric materials as polycaproamide (Fig. 1)

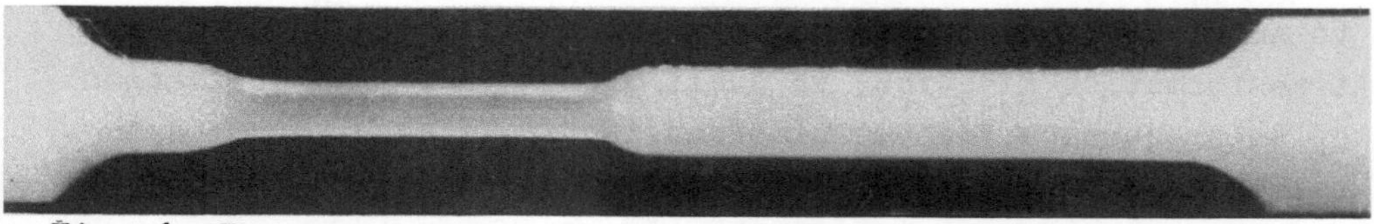

Fig. 1. The propagation of an inelastic deformation wave in a stretched polycaproamide bar

The experiments show that the acting load remains constant during the whole wave propagation along the bar. Furthermore, the strain rate distribution along the specimen is highly nonhomogeneous: the strain rate in those parts of the bar where the wave either did not yet come or already passed is negligibly small. Thus the strain is proceeding in the wave only whose length is small comparing to the bar length.

Nonlinear Deformation Waves
IUTAM Symposium Tallinn 1982
Editors U Nigul, J Engelbrecht

The inelastic orientation straining and strain-hardening of the material takes place in the wave. Microscopic and X-ray investigations showed that supramolecular structural elements of the polymer undergo a transformation under the local micro-stress action: they transfer from a rather isotropic state to one oriented in the drawing direction. Thus, the structural elements are elongated and their cross-sections are reduced. An average relative element's cross-section reduction will be denoted by α ; $0 < \alpha < 1$. Furthermore, the experiments show that the density of the material is held constant under orientation transformation.

It is essential that due to microinhomogeneity of the stress field the local stress reaches a critical value necessary for a microelement transformation not simultaneously through the whole bar cross-section. So, in a section within the wave some part of the material is in the transformed oriented state, meanwhile the remaining part is in the non-transformed state.

Introduce the concentration - relative volume - of transformed oriented structural elements of the material n(x,t), x is the coordinate along the specimen axis, t is the time. The explicit taking into account of the material transformation orders to supplement the mechanical equations by an equation of the transformation kinetics. We shall outline here the derivation of this equation for the simplest isotermic case. For a spatially homogeneous transformation process the equation of transformed material balance has the form:

$$\partial_t n = q(n, \sigma, T), \tag{1}$$

where $q(n, \sigma, T)$ is a certain function of the concentration, stress σ and temperature T. This function may be taken in the Arrhenius form known from physical chemistry

$$q = A(1-n)\exp[-(U-\gamma\sigma)/kT], \tag{2}$$

with the linearly stress depending activation energy. Here U is zero stress activation energy; A, γ - other kinetic constants, k - the Boltzmann constant. The value of q depends strongly on the stress and temperature; this is connected with large quantities U/kT, $\gamma\sigma$ /kT in the exponent.

Due to the spatial inhomogeneity of the wave over the specimen length the relation (1) does not hold. In fact the deformation of structural elements is correlated with neighbouring sections. Due to symmetry a compensation occurs in a spatially homogeneous process but not in the wave. Thus the resulting kinetical equation should be written in the form

$$\partial_t Sn + \partial_x Snu = S\int_{-\infty}^{\infty} q[n,\sigma(\xi,t),T]\rho(x-\xi)d\xi, \quad (3)$$

where S is cross-section area, u(x,t) is the displacement velocity, and ρ is a correlation function governing the influence of the transformation in neighbouring sections on the transformation rate in given section. The following relations hold

$$\int_{-\infty}^{\infty}\rho(x-\xi)d\xi = 1, \int_{-\infty}^{\infty}(x-\xi)\rho(x-\xi)d\xi = 0. \quad (4)$$

The second of these relations is valid because the function ρ is supposed to be even. The correlation length scale L may be introduced

$$L^2 = \frac{1}{2}\int_{-\infty}^{\infty}(x-\xi)^2\rho(x-\xi)d\xi. \quad (5)$$

The order of magnitude of this length scale L may vary from the structural element size to the bar width. Expanding the integrand in (3) in a series over (x - ξ) and using the rapid decreasing of the correlation function and the relations (4) and (5) we reduce the kinetic equation (3) to the form

$$S\partial_t n + Su\partial_x n = qS + S\partial_x[(L^2\partial_\sigma q)\partial_x\sigma]. \tag{6}$$

Thus the transformation non-locality leads to the appearence of a quasi-tenso-diffusion flow of transformed material under a strongly stress (and temperature) dependent tensodiffusion coefficient $L^2\partial_\sigma q$ which may reach rather large values under tension. The kinetic equations (6) closes the system of mechanical equations: the material balance equation

$$\partial_t S + \partial_x Su = 0, \tag{7}$$

together with quasistatic equilibrium equation and an obvious equation relating concentration and cross-section area:

$$\sigma S = \sigma_0 S_0 = Const, \; S/S_0 = 1-(1-\alpha)n. \tag{8}$$

Here σ_0, S_0 are correspondingly the stress and cross-section area in the non-deformed specimen part.

2. The uniform propagation of an inelastic deformation wave is described by a steady wave type solution of the obtained system for which all unknown functions depend on a single variable $\xi = x + Vt$. Note that the grip extension rate V_1 and wave velocity V are connected by an obvious relation $V_1 = (1-\alpha)V/\alpha$.

In the isotermic case the system of the equations for a steady wave is reduced to a single equation for the stress

$$V\frac{d\sigma}{d\xi} = q\sigma(1-\alpha) + \sigma(1-\alpha)\frac{d}{d\xi}(L^2\partial_\sigma q)\frac{d\sigma}{d\xi}, \tag{9}$$

under the boundary conditions

$$\sigma=\sigma_0, \frac{d\sigma}{d\xi}=0, (\xi\to-\infty); \sigma=\frac{\sigma_0}{\alpha}, \frac{d\sigma}{d\xi}=0, (\xi\to\infty).$$

This problem is similar to that of determination of the flame velocity in the combustion theory. Indeed, putting $z=(L_{to}^2\partial_\sigma q)\,d\sigma/d\xi$, we reduce the equation (9) to an equation of the f i r s t order

$$\frac{dz}{d\sigma}=\frac{V}{\sigma(1-\alpha)}-\frac{L^2 q\,\partial_\sigma q}{z} \tag{10}$$

under t w o boundary conditions:

$$z=0, \sigma=\sigma_0 \;;\; z=0, \sigma=\sigma_0/\alpha. \tag{11}$$

The solution of this problem does not exist for an arbitrary V however does exist for a selected one. Therefore, the velocity of the wave propagation V is obtained as an e i g e n-v a l u e. An asymptotic integration gives a relation between V and critical stress σ_0 in the linear form:

$$\ln\frac{V}{V_*}=-\frac{U}{kT}+\frac{\gamma\sigma_0}{\alpha kT}, \quad V_*=\frac{\alpha AL}{\sqrt{2}\ln(1/\alpha)}. \tag{12}$$

The experimental data confirm a linear relation of the stress and the logarithm of the wave propagation velocity and give some reasonable estimates for the kinetic constants agreeing with the values obtained from different independent experiments.

3. A most peculiar effect of non-isotermic processes in inelastic wave propagation is the appearence under certain conditions of the non-steady, oscillatory wave propagation (Fig. 2)

Fig. 2. The oscillatory propagation of an inelastic deformation wave. In the white regions the temperature increased sharply and the stress decreased.

Under the oscillatory propagation of the inelastic deformation wave the load acting on the specimen under a constant grip extension rate is also oscillating. A theory of this process is proposed. It is shown that the oscillatory propagation of the wave is connected with the loss of stability of the uniform steady-state wave propagation. Moreover, under further deformation the wave propagation becomes aperiodic, stochastic one. In the proposed theoretical models it corresponds to the appearence of "strange" attractors in phase spaces of suitable dynamic systems.

Analogous phenomena which can be described within the frames of the same theoretical scheme are also found for the metals: plastic deformation localization in tensed cylindrical specimens (Lüders lines), transition to a non-steady propagation of the inelastic deformation wave, stochastization of the wave propagation.

The participation in the experimental part of this work of Mr.D.Ya.Pavlov, Mr.B.S.Soloviev, Mr.V.A.Volodchenkov and Dr.I.M.Kershtein (Institute of Mechanics, Moscow University) is gratefully acknowledged.

Attenuation of Strong Shock Waves in Laminate Materials

V.F. NESTERENKO, V.M. FOMIN, P.A.CHESKIDOV

Special Design Office of Hydroimpulsive Engineering
USSR Academy of Sciences, Novosibirsk 630090
Institute of Theoretical and Applied Mechanics
USSR Academy of Sciences, Novosibirsk 630090

Summary

In the present work the process of attenuation of strong shock waves in periodic laminate materials by impulsive loading is examined by numerical and experimental methods. It is shown that the effect of loading waves overtaking a bow shock can play a determining role on the character of the bow shock amplitude change. The influence of the number of cells, amplitude and wave length on the bow shock attenuation character is also studied.

Formulation of the Problem. The Method of Solution

Let us consider a problem on the loading of a periodic laminate material $\mathfrak{A}_1 \ldots \mathfrak{A}_n$ by detonation products $\mathfrak{A}_0\mathfrak{A}_1$ of a charge. The thicknesses of layers are assumed to be much less than diameter of a laminate material (LM) that enables one to solve the problem in one-dimensional approximation. We shall describe the behavior of a solid by using an elastic-plastic model in Eulerian coordinates (t, x)

$$\frac{d\rho}{dt}+\rho\frac{\partial u}{\partial x}=0, \quad \rho\frac{du}{dt}+\frac{\partial \sigma}{\partial x}=0, \quad \frac{de}{dt}+\sigma\frac{dv}{dt}=0, \qquad (1)$$

$$p=(\gamma-1)\rho e + a_0^2(\rho-\rho_0), \qquad \frac{d\tau}{dt}=\begin{cases} -\mu\dfrac{\partial u}{\partial x}, & |\tau|<Y_0, \\ 0, & |\tau|\geqslant Y_0, \end{cases}$$

$$\sigma = p + \frac{4}{3}\tau,$$

where $v=1/\rho$, $\frac{d}{dt}=\frac{\partial}{\partial t}+u\frac{\partial}{\partial x}$, the rest of the notations is standard. In a particular case $\tau \equiv 0$ ($\mu=0$, $Y_0=0$) we obtain a hydrodynamic model. In such a way the mathematical problem is reduced to the following: find functions u, p, ρ, e, τ in the domain $Z=\{\mathfrak{A}_0 \leq x \leq \mathfrak{A}_n, 0 \leq t < \infty\}$ satisfying at $\mathfrak{A}_{i-1} < x < \mathfrak{A}_i$, $i=1,\ldots,n$ a system of equations (1) with i n i t i a l c o n d i t i o n s: the distribution of pa-

Nonlinear Deformation Waves
IUTAM Symposium Tallinn 1982
Editors U Nigul, J Engelbrecht

rameters of detonation products behind the detonation wave front is set from a self-similar solution which describes the Chapman-Jouguet detonation wave [1] . In LM at t = 0 we assume $p=0$, $\tau=0$, $u=0$, $\rho=\rho_i$ at $\mathcal{A}_{i-1}<x<\mathcal{A}_i$, $i=2,\dots,n$ and boundary conditions: at t=0 on free surfaces $x=\mathcal{A}_0$, $x=\mathcal{A}_n$ the stress are equal to zero, on interfaces $x=\mathcal{A}_i$, $i=1,\dots,n-1$ the continuity conditions of normal stresses and velocities should be satisfied. The problem posed was solved numerically with the help of an explicit difference scheme of Wilkins type [2].
Calculations were carried out using both elastic-plastic and hydrodynamic models.

Experiments Technique

The measurement of mass velocity in LM fluoroplastic-paraffin was performed using electromagnetic technique [3] . A ratio of layer thicknesses was $\varepsilon = \frac{3}{7}$. Characteristic oscillographs of mass velocity measured for LM with a cell size δ = 5 mm between 2,3 cells that corresponds at t = 0 to the distance ℓ = 10 mm from the contact boundary with detonation products (curve a), as well between 5,6 cells (curve b) and for LM with δ = 2.5 mm between 10,11 cells (curve c) are represented in Fig.I. The frequency of time marks was I MHz . The amplitude scale u = I mm/μs is referred to I b,c and for I a it is twice as much.

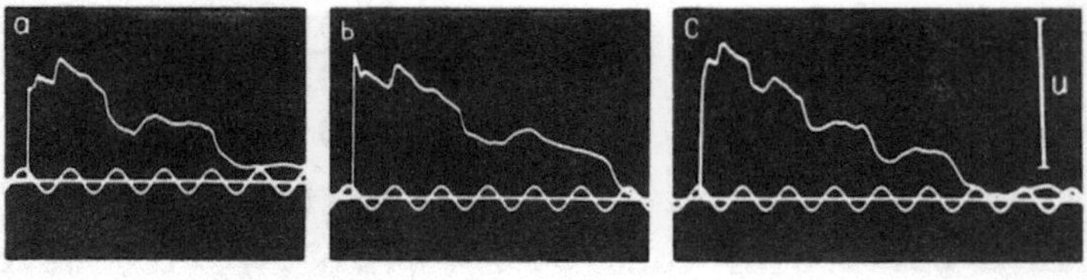

Fig.I. Oscillographs of mass velocity in LM fluoroplastic-paraffin

The measurement of free surface velocity of Al-Cu laminate material was carried out with the use of the technique described in [4] .

Analysis of Computational and Experimental Results

A bow shock behind which a complex wave picture is generated propagates along the laminate material when it is loaded by detonation products. Therein the bow shock interacts, firstly, with interfaces of different materials in LM that results in generating waves moving both in forward and backward direction, secondly, it interacts with waves which overtake a bow shock moving with a more high velocity along a preliminarily loaded material. The effect of the first factor is displayed in a considerable decrease in the bow shock amplitude at an equal depth with increasing the number of cells in a periodic LM. The influence of overtaking loading waves results in increasing its amplitude.

An estimate of the influence of both factors on the character of the bow shock amplitude change was carried out at the example of LM fluoroplastic-paraffin and aluminium-copper. TNT-RDX 50/50 with $D = 7.6$ mm/μs, $\rho = 1.68$ mg/mm^3; $\gamma = 3$, $\mu = 0$, $Y_0 = 0$, $a_0^2 = 0$ was employed as a charge. The thickness of a charge was 16.8 mm.

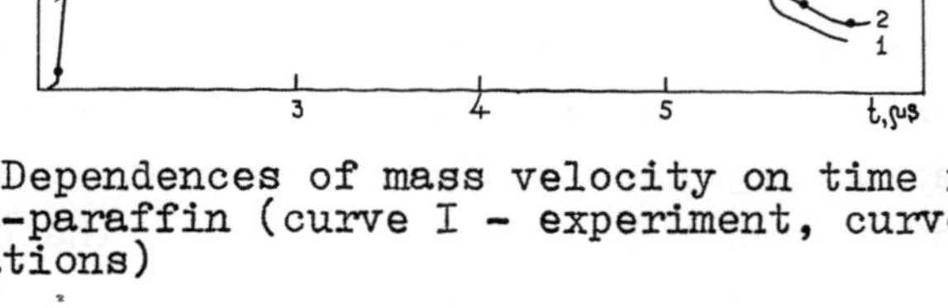

Fig.2. Dependences of mass velocity on time in fluoroplastic-paraffin (curve I - experiment, curves 2,3 - calculations)

A comparison of dependences of mass velocity on time obtained numerically and experimentally in fluoroplastic-paraffin (Fig.2) shows their satisfactory qualitative and quantitative agreement in detail of complex wave processes. Some deviation of computational dependence from experimental one at times $> 4.5\,\mu s$ is, apparently, connected with the **influence of lateral** unloading. The amplitudes of the bow shock mass velocity in LM with $\delta = 5$ mm and $\delta = 2.5$ mm coincide within the limits of computations accuracy, although in the second case the bow shock met in its way twice as many interfaces. The independence of the bow shock amplitude on the number of cells in LM fluoroplastic-paraffin was also observed in experiments at $\ell = 25$ mm as well in numerical calculations at $\ell = 25$ and 40 mm.

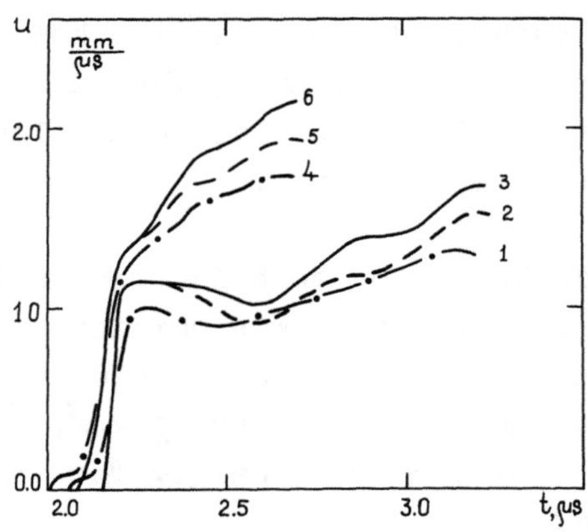

Fig.3. Velocity profiles of free surface in LM Al-Cu (I,4 - calculations carried out with the help of elasticoplastic model, 2,5 - calculations carried out with the help of hydrodynamic model; 3,6 - experiments)

The laminate material Al-Cu consisted of aluminium and copper layers of equal thickness. The aluminium layer was disposed the first near the explosive charge. Calculations carried out with the help of hydrodynamic model and experimental dependences of free surface velocity in LM Al-Cu which are represented in Fig.3 are sufficiently near to each other. A comparison of curves 1,4 and 2,5 shows that the existence of the stress deviator influences essentially the flow picture in LM even at triangular loading pulse. That fact that the bow shock amplitude in LM with a smaller cell is higher than in LM with a large one is unexpected. Analysis of the $p-u$ graphs without regard for the influence of the overtaking waves on the bow shock shows that the amplitude of the mass velocity in LM Al-Cu with δ = 2 mm should be nearly half the size than in LM with δ = 4 mm. This result points to the fact that the effects of the overtaking waves associated with the nonlinearity of the material behavior are of paramount importance by bow shock generation.

Figure 4 clearly represents the wave generation dynamics and the influence of different factors on the character of the bow shock amplitude change. Thus, at ℓ = 4 mm and ℓ = 8 mm (the overtaking effects are not essential at such depths) we can observe a considerable decrease in the bow shock amplitude with increasing the number of interfaces. However, already when the depth is 8 mm in LM with δ = 2 mm it is clearly seen that the second loading wave propagates following the bow shock, then it overtakes the first one increasing essentially the velocity and pressure behind it. When the depth is 12 mm the bow shock amplitude in LM with a smaller cell is already considerably bigger than in LM with a large cell. Moreover, the comparison of maximum pressure values shows that in LM with δ = 2 mm it is by 40% higher than in LM with δ= 4 mm. This result is also indicative of the fact that the possibilities of employing a laminate model to composite materials are limited.

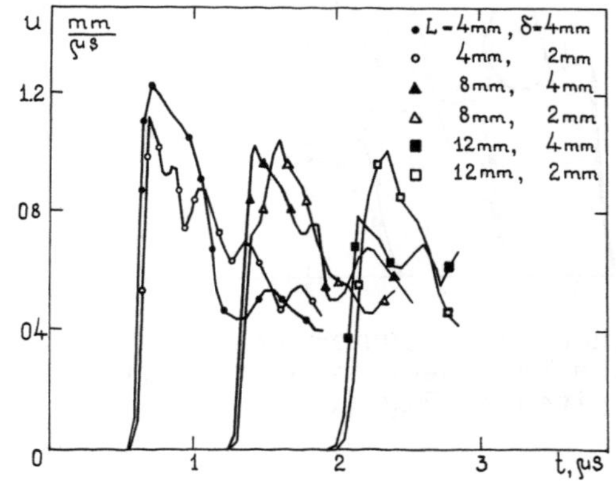

Fig.4. Dependences of mass velocity on time in LM Al-Cu at different depths (calculations carried out with the use of hydrodynamic model)

The study of the influence of the wave amplitude on the character of its attenuation is of interest as a role of overtaking effects for weak shock waves at the same distances should be to a considerable extent decreased because of the decrease in the nonlinearity of the material behavior. Figure 5 represents the pressure-time histories in LM Al-Cu when it is loaded by a weaker triangular pulse. Thereat one can note, firstly, a strong attenuation of the bow shock amplitude with increasing the number of interfaces, secondly, the absence of anomalously high pressure values in LM with a smaller cell.

Thus, the nonlinear effects of overtaking waves leading to qualitative peculiarities different from a linear approximation are essentially displayed for strong waves of finite duration. One can expect that the effects considered will be displayed weaker as decreasing the duration of the loading waves even at its large amplitude. Calculations carried out for a shock wave generated in LM Al-Cu by detonation of the same charge, but having the thickness 4.5 mm confirmed fully this conclusion.

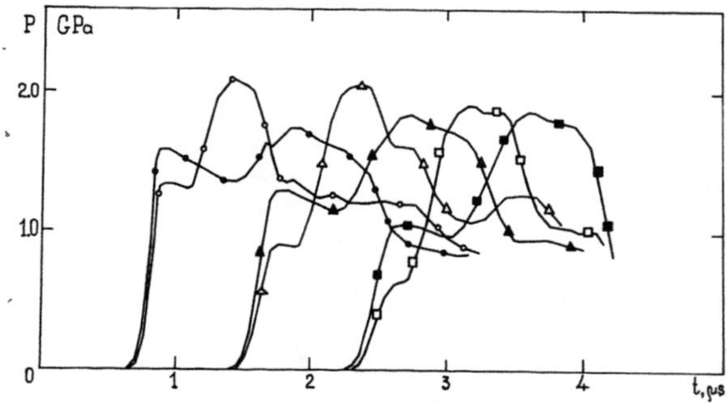

Fig.5. Dependences of pressure on time in LM Al-Cu at different depths by a weak triangular pulse loading. Notations are given in Fig.4.

References

1. Baum, F.A.; Orlenko, L.P.: In: Fizika vzryva. Moscow, Nauka 1975.

2. Wilkins M.L.: Calculation of elastic-plastic flows. In: Computational methods in hydrodynamics. Moscow, Mir, 1967.

3. Dremin A.N.; Shvedov,K.K.: Determination of Chapman-Jouguet pressure and reaction time in detonation wave. Journ. of Appl. Math. and Techn. Phys.2 (1964)

4. Deribas,A.A.; Nesterenko,V.F.; Sapozhnikov,G.A.; Teslenko T.S.; Fomin,F.M.: The study of shock wave attenuation process in metals by contact charge loading. The Physics of Explosion and Combustion. 2 (1979)

Application of a Multiparameter Group of Transformations to an Impact Problem of a Nonlinear Viscoelastic Rod

WACLAW FRYDRYCHOWICZ (1), MANSA C.SINGH (2)

(1) University of Warsaw
Institute of Mechanics
Warsaw, Poland

(2) University of Calgary
Department of Mechanical Engineering
Calgary, Alberta, Canada

The central feature of dimensional-similarity analysis is the use of a continuous r-parameter group of transformations

$$\overline{Z}_i = f_i\,(Z_1, Z_2, \ldots, Z_n;\ A_1, A_2, \ldots, A_r),\quad (i = 1, \ldots n)$$

and the absolute invariants associated with the group-functions π which satisfy, $\pi(Z_1, Z_2, \ldots, Z_n) = \pi(\overline{Z}_1, \overline{Z}_2, \ldots, \overline{Z}_n)$. The A's are a set of essential parameters. Z_i correspond to the variables of a set of differential equations and its associated boundary and/or initial conditions.

The first step in the application of the group theoretic approach to dimensional-similarity analysis is the establishment of a group under the transformations of which the set of governing equations is invariant in form. Next, the absolute invariants of the group are used to express the governing equations in terms of a less number of variables.

Experience reveals that for a wide range of problems in engineering and sciences a sufficiently general group of transformations is provided by a subclass of r-parameter groups in the form [1]

$$\overline{Z}_i = A_1^{\gamma_{i1}} \ldots A_r^{\gamma_{ir}} Z_i,\ (i = 1, \ldots, n).$$

Many of the manipulative difficulties inherent in the foregoing approach are eliminated by initiating an analysis with a group of transformations in the above form.

Applications of a multiparameter dimensional group of transformations to a nonlinear viscoelastic semi-infinite rod subjected to a time dependent velocity impact is considered. The solution of this problem has been made possible by seeking a group of transformations under which the similarity trans-

Nonlinear Deformation Waves
IUTAM Symposium Tallinn 1982
Editors U Nigul, J Engelbrecht

formations can be obtained. The most obvious advantage is the reduction of number of independent variables so that the problem is reduced to the integration of a system of ordinary differential equations. In addition, closed form solutions are often recognizable when the differential equations are expressed in similarity variables. Also, for hyperbolic problems (wave propagation) the similarity characteristic relationship on the moving front can be made use of to obtain a boundary condition. The system of equations for a nonlinear viscoelastic rod is: [2,3]

$$\frac{\partial \sigma}{\partial x} = \rho \frac{\partial v}{\partial t} , \quad \text{(equation of motion)} \tag{1a}$$

$$\frac{\partial v}{\partial x} = \frac{\partial e}{\partial t} , \quad \text{(compatibility relation)} \tag{1b}$$

$$\frac{\partial e}{\partial t} = \frac{\partial}{\partial t} \left[\frac{\sigma}{\mu}\right]^q + \left[\frac{\sigma}{\lambda}\right]^n . \quad \text{(constitutive law for a non-linear viscoelastic material)} \tag{1c}$$

In the above equations the coordinate x is measured along the axis of the rod, t denotes the time, σ is the normal tensile stress, v is the component of velocity in the x direction, u is the displacement, e is strain, ρ is the density, q, n, λ, μ are material constants. By allowing μ to become infinite with λ remaining finite we obtain a non-linear viscous material, and with λ becoming infinite for finite μ, a non-linear elastic model results. For n = q = 1, equation (1c) reduces to the constitutive law for a classical linear viscoleastic material.

The boundary and initial conditions for the velocity impact are:

$$V\ (x = o,t) = V_c t^{\delta} , \tag{2a}$$

$$u\ (x = x_w(t),t) = 0, \text{or } u\ (x \to \infty,t) = 0, \tag{2b}$$

$$\sigma\ (x \to \infty,t) = 0, \tag{2c}$$

$$u\ (x,t = 0) = 0, \quad x > 0 \tag{2d}$$

$$\frac{\partial u}{\partial t}\ (x,t = 0) = 0, \quad x > 0 \tag{2e}$$

where $x_w(t)$ gives the location of the wave front for hyperbolic problems, δ, and V_c are the parameters of the velocity impact. The restrictive relation among δ, q and n is obtained on the basis of the existance of the similarity transformations. The

4 - parameter group of transformations G_4 under which the system of equations (1) and auxiliary conditions (2) are invariant has the form:

$$G_4 \begin{cases} S_4 : \begin{cases} \bar{x} = A_x\, x,\ \bar{t} = A_t\, t, \\ \bar{\mu} = A_\mu\, \mu,\ \bar{\lambda} = A_\lambda\, \lambda, \\ \bar{\rho} = A_x^{-2}\, A_t^{\frac{1+q-2n}{q-n}}\, A_\mu^{\frac{q(1-n)}{q-n}}\, A_\lambda^{-\frac{n(1-q)}{q-n}}\, \rho, \\ \bar{V}_c = A_x\, A_t^{\frac{1}{q-n}-\delta}\, .A_\mu^{\frac{q}{q-n}}\, A_\lambda^{-\frac{qn}{q-n}}\, V_c, \end{cases} \\ \bar{u} = A_x\, A_t^{\frac{q}{q-n}}\, A_\mu^{\frac{qn}{q-n}}\, A_\lambda^{-\frac{qn}{q-n}}\, u \\ \bar{v} = A_x\, A_t^{\frac{q}{q-n}}\, A_\mu^{\frac{qn}{q-n}}\, A_\lambda^{-\frac{qn}{q-n}}\, v \\ \bar{\sigma} = A_t^{\frac{1}{q-n}}\, A_\mu^{\frac{q}{q-n}}\, A_\lambda^{-\frac{n}{q-n}}\, \sigma \end{cases} \tag{3}$$

In the above relations S_4 is 4 - parameter subgroup of G_4 and $q \neq n$.

G_4 group of transformations leads to the following similarity transformations: [1,4]

$$\eta = K_1 \frac{x}{t^{\frac{1+q-2n}{2(q-n)}}} \tag{4a}$$

$$u(x,t) = K_2\, t^{\frac{3q+1-2n}{2(q-n)}}\, F_1(\eta) \tag{4b}$$

$$v(x,t) = K_2 t^{\frac{1+q}{2(q-n)}}\, F_2(\eta), \tag{4c}$$

$$\sigma(x,t) = \frac{\mu^{\frac{q}{q-n}}}{\lambda^{\frac{n}{q-n}}}\, t^{\frac{1}{q-n}}\, F_3(\eta) \tag{4d}$$

$$\tilde{\pi}_4 = \left[\frac{\lambda^{\frac{n(1+q)}{q-n}} \rho}{\mu^{\frac{q(1+n)}{q-n}}} \right]^{\frac{1}{2}} V_c \, , \tag{4e}$$

where

$$K_1 = \left[\frac{\lambda^{\frac{n(1-q)}{q-n}} \rho}{\mu^{\frac{q(1-n)}{q-n}}} \right]^{\frac{1}{2}} , \tag{5a}$$

$$K_2 = \left[\frac{\mu^{\frac{q(1+n)}{q-n}}}{\lambda^{\frac{n(1+q)}{q-n}} \rho} \right]^{\frac{1}{2}} . \tag{5b}$$

The similarity representation of equations (1) assumes the form:

$$F_3'(\eta) = \frac{1+q}{2(q-n)} F_2(\eta) - \frac{1+q-2n}{2(q-n)} \eta F_2'(\eta) , \tag{6a}$$

$$F_2(\eta) = \frac{3q+1-2n}{2(q-n)} F_1(\eta) - \frac{1+q-2n}{2(q-n)} \eta F_1'(\eta) , \tag{6b}$$

$$F_2'(\eta) = \frac{q}{q-n} F_3^q(\eta) + F_3^n(\eta) - \frac{q(1+q-2n)}{2(q-n)} \eta F_3'(\eta) F_3^{q-1}(\eta) . \tag{6c}$$

The boundary conditions (2), for velocity impact, transform to

$$F_1(o) = \frac{2(q-n)}{3q+1-2n} \tilde{\pi}_4 , \tag{7a}$$

$$F_2(o) = \tilde{\pi}_4 , \tag{7b}$$

$$F_3'(o) = \frac{1+q}{2(q-n)} \tilde{\pi}_4 , \tag{7c}$$

$$F_1(\infty) = F_2(\infty) = F_3(\infty) = 0 , \tag{7d}$$

where

$$\tilde{\pi} = K_2^{-1} V_c . \tag{7e}$$

Combining equations (6a) and (6c) we obtain the following two ordinary differential equations:

$$F_2' \left[1 + \frac{q(1+q-2n)^2}{4(q-n)^2} \eta^2 F_3{}^{q-1}\right] = \frac{q}{q-n} F_3{}^q +$$

(Equation contd. on the next page)

$$F_3^{\,n} + \frac{q(1+q)\ (1+q - 2n)}{4\ (q-n)^2}\ \eta\ F_2 F_3^{\,q-1}, \tag{8a}$$

$$F_3'\ [1 - \frac{q(1+q - 2n)^2}{4\ (q-n)^2}\ \eta^2\ F_3^{\,q-1}] = \frac{1+q}{2(q-n)}\ F_2$$

$$- \frac{q\ (1+q - 2n)}{2\ (q-n)^2}\ \eta\ F_3^{\,q} - \frac{1+q - 2n}{2(q-n)}\ \eta\ F_3^{n}\ . \tag{8b}$$

Making use of the boundary conditions (7) equations (8) can be solved numerically. Then, from equations (6b), F_1 can be evaluated. Use of equations (4) gives the values of the original functions namely, u, v and σ.

References

1. Frydrychowicz, W.; Singh, M.C., Group Theoretic and Similarity Analysis of Hyperbolic Partial Differential Equations, Report #233, 1982, Department of Mechanical Engineering, The University of Calgary.

2. Seshadri, R.; Singh, M.C., Similarity Analysis for Impact of Rods of Nonlinear Rate sensitive Strain Hardening materials Archives of Mechanics, Warsaw, Vol. 28, 1976.

3. Taulbee, B.D.; Cozzarelli, F.A.; and Dym, C.L. Similarity Solution to some Nonlinear Impact Problems, Int. J. Nonlinear Mechanics, Vol. 6, 1971.

4. Singh, M.C.; Frydrychowicz, W., Wave Propagation in Nonhomogeneous Thin Elastic Rods Subjected to Time Dependant Velocity Impact. Journal of the Acoustical Society of America, Vol. 71, 1982.

Shock Wave Propagation in a Viscoelastic Rod

N. SUGIMOTO, Y. YAMANE, and T. KAKUTANI

Department of Mechanical Engineering
Faculty of Engineering Science
Osaka University, Toyonaka, Osaka 560
JAPAN

Summary

The propagation of torsional and longitudinal shock waves in a thin circular viscoelastic rod is investigated theoretically. Assuming a "nearly elastic" rod, new approximate governing equations are derived for the respective mode. Two typical relaxation functions are compared. One is the exponential function type well known as Maxwell-Voigt model. The other is the power function type newly proposed here as a model having a continuous relaxation spectrum. Based on the approximate equations, the steady shock conditions and profiles are shown for both types. Also the unsteady unidirectional wave propagation is briefly discussed.

Introduction

Recently nonlinear viscoelastic wave propagation has attracted much attention with the diverse applications of polymers [1]. The propagation of shock waves is one of the main features in contrast to the linear wave propagation. It results from a balance between the nonlinearity and the dissipation. The viscoelastic materials are not only dissipative but dispersive. In particular, if we consider a body with a boundary, there appears another dispersion which is attributed to the geometrical configuration of the body and is distinguished from the material dispersion. Most previous works are concerned with a wave propagation in an infinite body without geometrical dispersion and little has been made for a body with a boundary. One of the main purposes of this paper is to consider, as a fundamental example, the propagation of shock waves in a thin circular viscoelastic rod. The other is concerned with the relaxation character of viscoelastic materials. The viscoelastic behavior is often modeled by a combination of several springs and dashpots [2]. Such models have succeeded in explaining intuitively the inner relaxation mechanism. But they are inadequate in describing the actual viscoelastic behavior[3].

Nonlinear Deformation Waves
IUTAM Symposium Tallinn 1982
Editors U Nigul, J Engelbrecht

It is usually characterized by infinitely many relaxation times, whereas the spring-dashpot model has only finite ones. In other words, the continuous relaxation spectrum is required in place of the discrete one characterizing the spring-dashpot model. Here we propose, as a plausible model, a new relaxation function of power function type, whose relaxation spectrum has a continuous distribution approximated by a power law.

In what follows, we consider the torsional and longitudinal shock waves, respectively. We remark that since the torsional waves concerned here are the lowest mode, they are geometrically non-dispersive in spite of the presence of the boundary. For comparison, Maxwell-Voigt model is considered in addition to this power function type. The shock conditions and the explicit shock profiles are obtained for both types. Finally for the unsteady unidirectional wave propagation, the reduction of the approximate equations is briefly discussed.

Approximate Governing Equations

The approximate governing equations for the torsional and the longitudinal waves are presented. Their derivation should be referred to [4,5], but is briefly outlined here. Based on the three dimensional nonlinear theory of viscoelasticity in Lagrangian formulation, the boundary value problem is posed at a free lateral surface of an infinite rod. With the thermal effect ignored, a "nearly elastic" rod is assumed in which the dependence of stress on a history of strain is weak. In the constitutive equations, the elastic response is taken up to the third order in strain, while the viscoelastic one is taken in a form of the linear hereditary integral. In derivation, we make use of three small parameters ε, δ, and γ, $(0<\varepsilon,\delta,\gamma<<1)$ designating, respectively, the thinness of the rod, the magnitude of displacement, and the weakness of viscoelasticity, a lengthscale being normalized by a wavelength. By the assumption of thin rod, the displacement is sought in a power series of the radial coordinate. The approximate governing equations are derived so that the boundary conditions may be fulfilled.

For the torsional waves, the finite torsional deformation couples with the induced longitudinal one. Taking account of both geometrical and material nonlinearity, we have the coupled equations:

$$\phi_{tt} - G\phi_{zz} - \gamma\int_{-\infty}^{t} K_T(t-t')\phi_{zzt'}dt' = -2\varepsilon^2[(\psi\phi_t)_t - G(\psi\phi_z)_z]$$
$$+ \varepsilon^2[(a w_z + b_1\phi_t^2 + b_2\phi_z^2)\phi_z]_z , \quad (1)$$

and $$w_{tt} - Ew_{zz} = (c_1\phi_t^2 + c_2\phi_z^2)_z , \quad (2)$$

with $$\psi = -\sigma w_z + d_1\phi_t^2 + d_2\phi_z^2 , \quad (3)$$

where ϕ and w represent the finite angle of torsion and the longitudinal displacement, respectively, and the suffixes t and z imply partial differentiation with respect to t and z; t being time and z the axial coordinate. Here G, E, and σ denote the modulus of rigidity, Young's modulus, and Poisson's ratio, respectively, and K_T denotes the shear stress relaxation function. The constants a, b_i, c_i, and d_i (i=1,2) are determined by G, E, and the higher order (equilibrium) elastic constants.
For the longitudinal waves, on the other hand, account is taken of not only the finite deformation but also the lateral contraction of the rod. Then we have

$$w_{tt} - Ew_{zz} - (\varepsilon\sigma)^2 w_{ttzz}/2 + \gamma\int_{-\infty}^{t} K_L(t-t')w_{zzt'}dt' = \delta\beta(w_z)_z^2, \quad (4)$$

where K_L denotes the tensile stress relaxation function and β is a constant given by the elastic constants.

Relaxation Functions and Linear Dispersion Relations

The viscoelastic behavior of rod is specified by the relaxation functions K_T and K_L. For the spring-dashpot model, they can be expressed by a sum of several exponential functions including the delta or step function as a limiting case:

$$K_{T,L}(t) = \sum_i K_{T,L}^{(i)}\exp(-t/T_i), \quad (5)$$

where T_i's imply the relaxation times and $K_{T,L}^{(i)}$'s are constants. But in order to describe actual viscoelastic materials that have infinitely many relaxation times, the relation (5) should be extended as

$$K_{T,L}(t) = \int_{-\infty}^{\infty} H_{T,L}(T')\exp(-t/T')d(\ln T'), \quad (6)$$

where $H_{T,L}(T')$ are relaxation spectra. Here we consider two simple but typical types for K_T and K_L. One is the exponential function type (Type I) and the other the power function type (Type II)

$$\text{Type I: } K_{T,L}(t) = \exp(-\kappa t), \quad (\kappa>0), \quad (7)$$

$$\text{Type II: } K_{T,L}(t) = t^{-\nu}, \ (0<\nu<1), \tag{8}$$

where Type I has a single relaxation time κ^{-1} in (5) and Type II is a model in which $H_{T,L}(T')$ is given by $H_{T,L}(T')=T'^{-\nu}/\Gamma(\nu)$, $\Gamma(\nu)$ being the gamma function.

For both types, we note the linear dispersion relations. For the torsional waves, the phase velocity c is given by $c^2=G+\gamma/(1+i\kappa/\omega)$ for Type I and $c^2=G+\gamma\Gamma(1-\nu)(-i\omega)^\nu$ for Type II, ω being a frequency. For Type I, we remark two characteristic sound speeds, the instantaneous sound speed $c_i=(G+\gamma)^{\frac{1}{2}}$ as $\omega\to\infty$ and the equilibrium one $c_e=G^{\frac{1}{2}}$ as $\omega\to 0$. But for Type II, c_i becomes infinite because of the infinite stiffness at t=0, though $c_e=G^{\frac{1}{2}}$. For the longitudinal waves, c is given by $c^2=E-(\varepsilon\sigma\omega)^2/2+\gamma/(1+i\kappa/\omega)$ for Type I and $c^2=E-(\varepsilon\sigma\omega)^2/2+\gamma\Gamma(1-\nu)(-i\omega)^\nu$ for Type II. The geometrical dispersion becomes important as $\omega\to\infty$.

<u>Steady Shock Waves</u>

We investigate the steady shock wave propagation into the unstrained state far ahead ($z\to+\infty$) with an equilibrium strain state far behind ($z\to-\infty$). Putting $\eta=t-z/\lambda$, $\lambda(>0)$ being a shock velocity and retaining only the lowest order terms in ε^2, δ, and γ, eqs. (1)-(3) and eq.(4) are reduced, respectively, to

$$U\Phi - \mathrm{sgn}\alpha\,\Phi^3 = \int_{-\infty}^{\eta} K_T(\eta-\eta')\Phi_{\eta'}d\eta' , \tag{9}$$

with $U=(\lambda^2-G)/\gamma$ and $\Phi=(\varepsilon^2|\alpha|/\gamma)^{\frac{1}{2}}d\phi/d\eta$, where $\alpha=[a(c_1G+c_2)/(G-E)+b_1G+b_2]/G$, and

$$VW - W^2 - \mu W_{\eta\eta} = \int_{-\infty}^{\eta} K_L(\eta-\eta')W_{\eta'}d\eta', \tag{10}$$

where $V=(\lambda^2-E)/\gamma$, $W=-\delta\beta/(\gamma\lambda)dw/d\eta$, and $\mu=(\varepsilon\sigma)^2/(2\gamma)$. We note that for the same W, the difference in sign of β produces physically compression or expansion.

By examining the necessary conditions for the shock waves to exist, it is found for the torsional waves that α and U should be positive and that the equilibrium value at $\eta=\infty$ is given by $\Phi_\infty^2=U$. For the longitudinal waves, it is found that V should be positive and that the equilibrium value is given by $W_\infty=V$. Thus from U,V>0, each shock velocity is always greater than the respective equilibrium sound speed $G^{\frac{1}{2}}$ and $E^{\frac{1}{2}}$. The velocity becomes faster as the shock strength Φ_∞ and W_∞ increase. Although these conditions are derived on assuming smooth solutions, we remark the existence of a discontinuous solution in Φ. For Type I and $\alpha>0$, eq.(9) allows a discontinuity with a jump $\Phi_0^2=U-1$. For Type II, no discontinuity is allowed. In eq.(10), on the other hand, $\mu W_{\eta\eta}$ prevents an occurrence of a discontinuity, which should be compared with eq.(9). For Type I, eqs.(9) and (10) can be rewritten in the nonlinear differential equations, which are solved analytically and numerically. For Type II, however, they remain the nonlinear and singular integral equations, which are solved by the numerical quadrature.

Torsional Shock Waves

For Types I and II, the typical shock profiles are displayed in Figs.1 and 2, respectively. For Type I, there is the critical speed $(G+\gamma)^{\frac{1}{2}}$ (U=1) beyond which the shock profiles change remarkably. For a velocity below it (0<U<1), a smooth and monotonous transition appears, while beyond it (U≥1), a discontinuity followed by a monotonous profile appears. The discontinuity is introduced according to the jump condition $\Phi_0^2=U-1$.

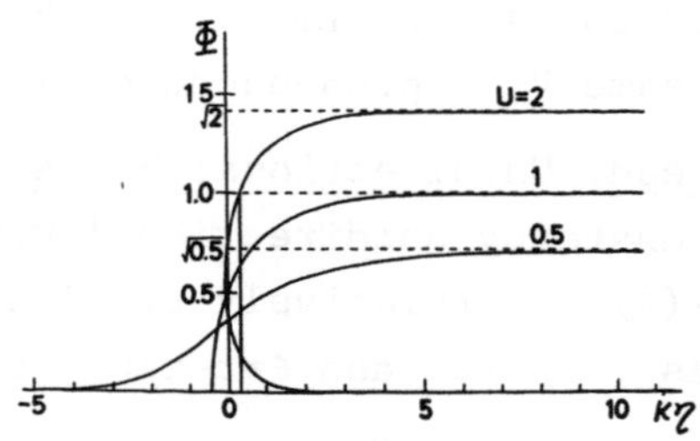

Fig.1.

For Type II, the profiles are always smooth and monotonous. The step-up is exponential but there appears a very slow relaxation region in a trail. This tendency becomes prominent as ν decreases. Thus no sharp shock layer can be seen in Type II, which should be compared with the result in Type I.

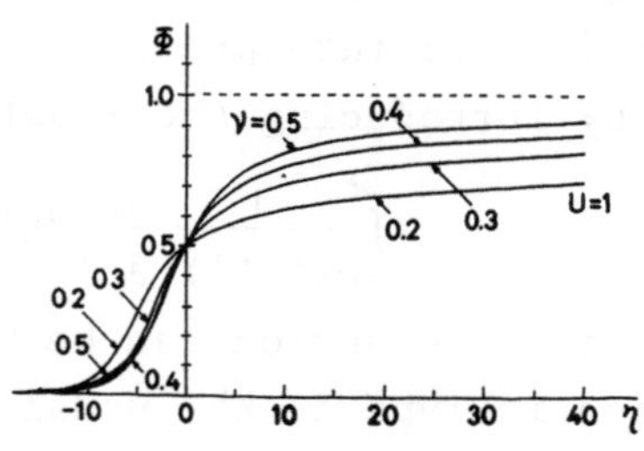

Fig.2.

Longitudinal Shock Waves

For Types I and II, the typical profiles are shown, respectively, in Figs.3 and 4. The geometrical dispersion gives rise to oscillatory profiles. This is the essential difference from the torsional waves. For Type I, the shock waves step up exponentially, then overshoot the equilibrium value, and approach it exponen-

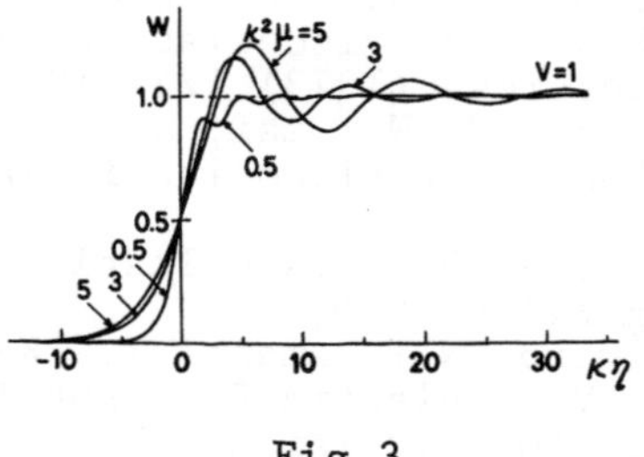

Fig.3.

tially with an oscillation. As $\kappa^2\mu$ decreases, the step-up becomes steep while as it increases, a strong oscillation appears. For a weak shock wave, the profiles are monotonous for $\kappa^2\mu \gtrsim 4$ and oscillatory for $0<\kappa^2\mu \lesssim 4$. For Type II, there also appears a very wide oscillatory relaxation region. Except for large V, the profiles do not overshoot the equilibrium value but undulate below it. The tendency with respect to μ is similar to that in Type I. We only note that for the same V, a pronounced oscillation appears as ν decreases.

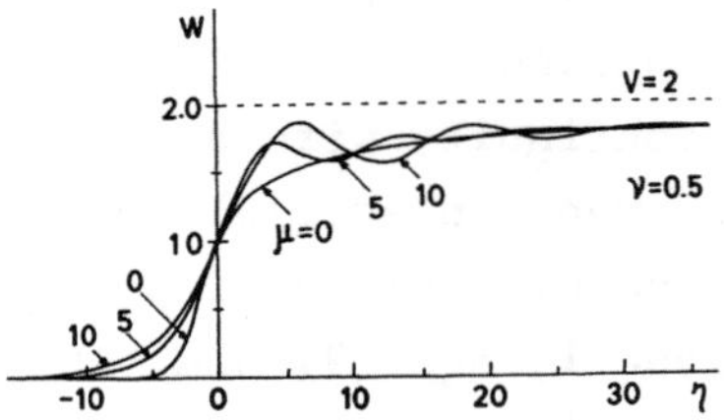

Fig.4.

Unsteady Unidirectional Waves

For unsteady unidirectional waves into the unstrained state, eqs. (1)-(4) are relatively simplified. Introducing the new coordinates $\tau=\varepsilon^2 z/G^{\frac{1}{2}}$ and $\xi=t-z/G^{\frac{1}{2}}$, eqs.(1)-(3) are reduced to

$$f_\tau - 3\alpha f^2 f_\xi/2 = \gamma/(2\varepsilon^2 G)\int_{-\infty}^{\xi} K_T(\xi-\xi')f_{\xi'\xi'}d\xi', \quad (11)$$

where $f=\phi_z$. For Type I, eq.(11) can be rewritten in the differential equation. If, in particular, the rapid relaxation is assumed, $\kappa^{-1}\sim\varepsilon^2/\gamma \ll 1$, then "cubic Burgers' equation" is reduced with the right hand side of (11) replaced by $\gamma f_{\xi\xi}/(2\varepsilon^2 G\kappa)$. For Type II, eq.(11) is interpreted as "generalized cubic Burgers' equation" by introducing the fractional derivative defined by

$$\int_{-\infty}^{\xi}\frac{1}{(\xi-\xi')^\nu}\frac{\partial f}{\partial \xi'}d\xi' \equiv \Gamma(1-\nu)\frac{\partial^\nu f}{\partial \xi^\nu}, \quad (0<\nu<1). \quad (12)$$

The similar reduction can be made for eq.(4). For the rapid relaxation in Type I, the usual K-dV-Burgers' equation is reduced. For Type II, we have the "generalized K-dV-Burgers' equation".

References

1. Schuler, K. W.; Nunziato, J. W.; Walsh, E. K.: Recent Results in Nonlinear Viscoelastic Wave Propagation. Int. J. Solids Structures. (1973) 1237-1281.
2. Findley, W. N.; Lai, J. S.; Onaran, K.: Creep and Relaxation of Nonlinear Viscoelastic Materials. Amsterdam: North-Holland 1977.
3. Ferry, J. D.: Viscoelastic Properties of Polymers. New York: John Wiley 1970.
4. Sugimoto, N.; Yamane, Y.; Kakutani, K.: Torsional Shock Waves in a Viscoelastic Rod. submitted to TRANS. ASME J. Appl. Mech.
5. Sugimoto, N.; Yamane, Y.; Kakutani, K.: Oscillatory Shock Waves in a Viscoelastic Rod. submitted to TRANS. ASME J. Appl. Mech.

Nonlinear Wave Propagation in Plastic Fracturing Materials – A Constitutive Modelling and Finite Element Analysis

Larsgunnar NILSSON and Mats OLDENBURG
Division of Structural Engineering
University of Luleå
S-951 87 Luleå, Sweden

Summary

In wave propagation analysis of materials like concrete and rock, the result is strongly affected by the constitutive model of tensile fracturing being used. Traditionally, tensile cracking has been modelled as an ideal brittle failure, although numerous tests have shown a continous strain-softening type of failure.

In the present analysis the material is treated as plastic fracturing. A continuum approach to tensile fracturing is adopted, and related to fracture mechanical concepts. The rate of strain-softening is obtained from the rate of fracture energy being released. Material anisotropy, due to the formation of crack-planes is accounted for.

The proposed strain-softening tensile fracturing model is implemented in a finite element program for wave propagation analysis. The modelling of the tensile fracturing is objective in the sense that it is independent of the element mesh size. The present strain-softening model is compared with an ideal brittle model in analyses of wave propagation in a concrete rod.

Introduction

Wave propagation in materials with a small tensile to compressive strength ratio is to a large extent affected by the formation of cracks. Apparent illustrations of this fact are the phenomena of spalling and scabbing. Thus, each tentative constitutive model for this kind of material must include a tensile fracture mechanism.

In recent years, the finite element method (FEM) has been extensively used in continuum analysis of tensile fracture, cf Chen [1]. In the previous FEM analyses, the tensile fracture has been modelled either as a perfect brittle failure or as a continously (or stepwise continously) softening failure. The perfect brittle failure model yields results which strongly depend on the mesh size being used. Thus by decreasing the element size at the crack tip, it is always possible for the crack to propagate also into that element, cf Bazant and Oh [2]. Previously used strain-softening models of tensile

Nonlinear Deformation Waves
IUTAM Symposium Tallinn 1982
Editors U. Nigul, J. Engelbrecht

fracture generally suffer from mesh dependence too. Since the crack is a localized deformation, the strain across the crack will depend on the size of the element penetrated by this crack.

Heterogeneous materials like concrete and rock behave nonlinearly in a relatively large zone ahead of the fracture front. Considerable microcracking takes place in this zone. In the fictitious crack model of Hilleborg et al [3], the stress σ across the crack opening is assumed to be a function of the crack width w, such that

$$G_f = \int_0^\infty \sigma(w)dw \tag{1}$$

where G_f is the fracture energy. G_f has been found to be a function of the rate of crack width increase $\dot{w}$, cf Nilsson [4], with a constant value for small and medium $\dot{w}$.

In previous analyses of wave propagation in heterogeneous materials, Nilsson [5], [6] and Nilsson and Glemberg [7], we emphasized studies of nonlinear constitutive modelling and rate effects. A combined elastic-viscoplastic-plastic-brittle constitutive model was used. In the present analysis we replace the brittle failure model by an extended fictitious crack model. Unlike previous finite element analyses using the fictitious crack and similar models, we allow cracks to appear within elements and to propagate in any direction.

Constitutive theory

The actual material behaviour is governed by the location of the stress point relative to the viscoplastic ($F^v = 0$) and plastic ($F^p = 0$) loading surfaces and to the brittle or strain-softening ($F^c = 0$) failure surfaces in stress space, see Figure 1.

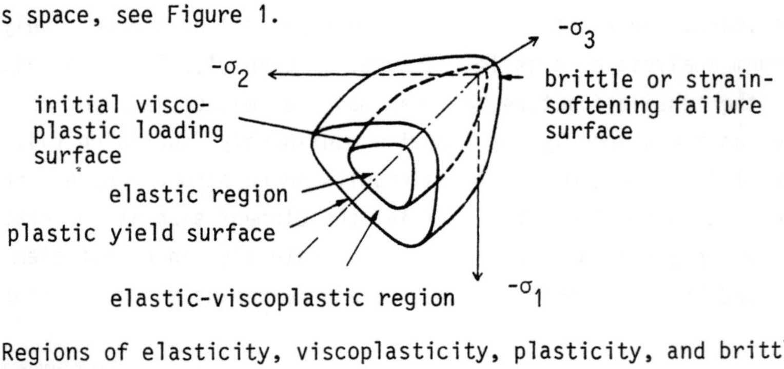

Fig.1. Regions of elasticity, viscoplasticity, plasticity, and brittle or strain-softening failure

The viscoplastic loading surface and the plastic loading surface represent elastic and crack initiation limits, respectively. From true triaxial tests it is found that these surfaces are closed. A simple geometric surface which can be fitted to experimental data fairly well is the generalized ellipsoid

$$F = \left[\left(\frac{2\sigma_o/H_r - \xi_u - \xi_\ell}{\xi_u - \xi_\ell}\right)^2 + \left(\frac{\tau_o/H_r}{b(\Theta)}\right)^2\right]^{1/2} - 1 \qquad (2)$$

where σ_o, τ_o and Θ are the octahedral normal and shear stresses and the angle of similarity, respectively. The material parameters ξ_ℓ, ξ_u, and $b(\Theta)$ must be obtained from experimental data. The rate-hardening parameter H_r is given by

$$H_r = f_{cu}[c_1 + c_2 \ln(\dot{\varepsilon}^{ef}) + c_3 \ln^2(\dot{\varepsilon}^{ef})] \qquad (3)$$

where f_{cu} is the uniaxial compressive strength and c_1, c_2, and c_3 are obtained by fitting Eq. 3 to experimental data, cf Figure 2. In Eq. (3), $\dot{\varepsilon}^{ef}$ is a scalar measure of the strain rate tensor $\dot{\underset{\sim}{\varepsilon}}$, usually $\dot{\varepsilon}^{ef} = (\dot{\underset{\sim}{\varepsilon}}:\dot{\underset{\sim}{\varepsilon}})^{1/2}$. Further details on the fit of the loading surfaces and appropriate hardening and softening functions can be found in Nilsson [6].

Within the elastic-viscoplastic region ($F^p < 0$, $F^c < 0$) the stress-strain relation is

$$\dot{\underset{\sim}{\sigma}} = \underset{\sim}{S} : (\dot{\underset{\sim}{\varepsilon}} - \dot{\underset{\sim}{\varepsilon}}^{vp}) \qquad (4a)$$

where $\underset{\sim}{S}$ is the elastic stiffness tensor and $\dot{\underset{\sim}{\sigma}}$ is the stress rate tensor. To account for stiffness degradation effects, we have assumed nonlinear bulk and shear moduli dependent on the plastic dilatancy. The viscoplastic strain rate $\dot{\underset{\sim}{\varepsilon}}^{vp}$ is governed by a modified Perzyna theory. Thus

$$\dot{\underset{\sim}{\varepsilon}}^{vp} = \gamma(\underset{\sim}{\dot{\varepsilon}}) \frac{1}{2} [1 + \mathrm{sgn}(F^v/F^{v1})] \, |F^v/F^{v1}|^\alpha \, \underset{\sim}{n}_F \qquad (4b)$$

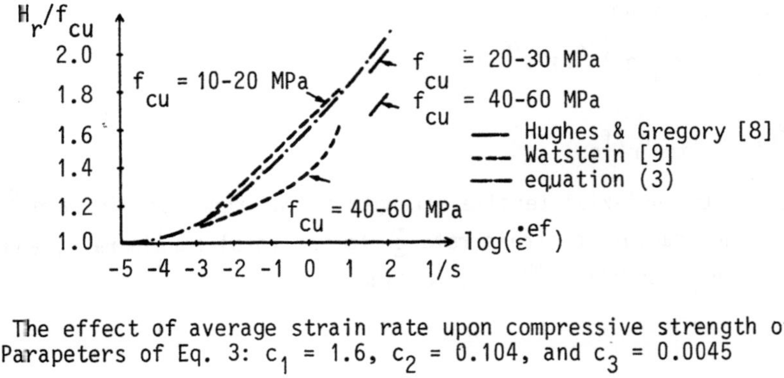

Fig.2. The effect of average strain rate upon compressive strength of concrete. Parapeters of Eq. 3: $c_1 = 1.6$, $c_2 = 0.104$, and $c_3 = 0.0045$

where $\underset{\sim}{n}_F$ is the normalized gradient of the viscoplastic loading surface $F^v =$ = const., and

$$\gamma(\dot{\underset{\sim}{\varepsilon}}) = \dot{\varepsilon}^{ef} \, |\dot{\varepsilon}^{ef}/{}^{o}\dot{\varepsilon}^{ef}|^{-r} \tag{4c}$$

In Eqs. (4b) and (4c) F^{v1} and ${}^{o}\dot{\varepsilon}^{ef}$ denote reference values, and r and α are material constants.

During plastic loading ($F^p = 0$, $F^v \geq 0$, $F^c < 0$)

$$\dot{\underset{\sim}{\sigma}} = \underset{\sim}{S}^{ep}:(\dot{\underset{\sim}{\varepsilon}} - \dot{\underset{\sim}{\varepsilon}}^{vp}) \tag{5a}$$

where the elastic-plastic stiffness tensor $\underset{\sim}{S}^{ep}$ is derived under the assumption of an associated flow rule

$$\underset{\sim}{S}^{ep} = \underset{\sim}{S} - \frac{1}{D}\left(\underset{\sim}{S}: \frac{\partial F^p}{\partial \underset{\sim}{\sigma}}\right)\left(\frac{\partial F^p}{\partial \underset{\sim}{\sigma}} : \underset{\sim}{S}\right) \tag{5b}$$

$$D = \frac{\partial F^p}{\partial \underset{\sim}{\sigma}} : \underset{\sim}{S} : \frac{\partial F^p}{\partial \underset{\sim}{\sigma}} - \frac{\partial F^p}{\partial \underset{\sim}{\varepsilon}^p} : \frac{\partial F^p}{\partial \underset{\sim}{\sigma}} \tag{5c}$$

Strain-softening tensile fracture is assumed to occur whenever the fracture criterion $F^c = 0$ is met. The tensile fracture creates a crack plane, which is memorized for all future time. The tensile strength across the crack plane depends on the crack width w and the rate of straining $\dot{\varepsilon}^{ef}$. A compressive stress, normal to the crack plane, can exist first after crack closure (w = 0). Secondary cracks may develop across preexisting crack planes as long as a tensile strength persists in any principal direction. A maximal stress function governs the tensile fracturing. Thus,

$$F^c(\underset{\sim}{\sigma}, H_r) = \sigma_i - \sigma_{fi}(w_i, H_r) \qquad i = 1, 2, 3 \tag{6a}$$

where σ_i is the principal stress and σ_{fi} is the strength in principal direction i. We have assumed

$$\sigma_{fi} = {}^{o}\sigma_{fi} \, e^{-({}^{o}\sigma_{fi} w_i / G_f)} \tag{6b}$$

$${}^{o}\sigma_{fi} = f_{tu} \, H_r / f_{cu} \tag{6c}$$

where f_{tu} is the uniaxial tensile strength at reference strain rate ${}^{o}\dot{\varepsilon}^{ef}$. By adding the fracturing strain rate $\dot{\underset{\sim}{\varepsilon}}^c$ to the total strain rate, and using the consistency condition $\dot{F}^c = 0$, we find

$$\dot{\underset{\sim}{\sigma}} = \underset{\sim}{S}^{c} : (\dot{\underset{\sim}{\varepsilon}} - \dot{\underset{\sim}{\varepsilon}}^{vp}) \tag{7a}$$

where the fracture stiffness tensor for the material with one crack plane is

$$\underset{\sim}{S}^{c} = \underset{\sim}{S} - \frac{1}{D^{c}} \left(\underset{\sim}{S} : \frac{\partial F^{c}}{\partial \underset{\sim}{\sigma}} \right) \left(\frac{\partial F^{c}}{\partial \underset{\sim}{\sigma}} : \underset{\sim}{S} \right) \tag{7b}$$

$$D^{c} = \frac{\partial F^{c}}{\partial \underset{\sim}{\sigma}} : \underset{\sim}{S} : \frac{\partial F^{c}}{\partial \underset{\sim}{\sigma}} - \frac{\partial F^{c}}{\partial \underset{\sim}{\varepsilon}^{c}} : \frac{\partial F^{c}}{\partial \underset{\sim}{\sigma}} \tag{7c}$$

The derivation of the fracture stiffness tensors $\underset{\sim}{S}^{cc}$ or $\underset{\sim}{S}^{ccc}$ when two or three crack planes have been formed is analogous. From Eqs. (6a) and (7c) we obtain

$$D^{c} = S_{ii} + \frac{\partial \sigma_{fi}}{\partial w_i} \frac{\partial w_i}{\partial \varepsilon^{c}_{i}} \tag{7d}$$

(no summation), where $\partial \sigma_{fi} / \partial w_i$ is found from Eq. (6b). In order to relate the crack width w to the fracture strain tensor $\underset{\sim}{\varepsilon}^{c}$, we consider a control volume V_e enclosing a crack plane of area A_c. For this volume

$$\int_{A_c} \sigma_i \dot{w}_i \, dA = \int_{V_e} \underset{\sim}{\sigma} : \dot{\underset{\sim}{\varepsilon}}^{c} \, dV \tag{8}$$

if only mode I cracks are considered. Mode II and III cracks can be included by adding appropriate terms to the left hand side of Eq. (8). By using Eq. (6a) and assuming constant conditions in the control volume, we find

$$\partial w_i / \partial \varepsilon^{c}_{i} = V_e / A_c \tag{9}$$

which can be interpreted as a characteristic length. In the finite element analysis, we have taken V_e equal to the element volume and used the rough estimate $V_e/A_c \simeq V_e^{1/3}$. Figure 3 shows the influence of V_e/A_c on the uniaxial stress-strain relation.

By memorizing strength and maximal crackwidth, Eq. (6b) can be made to account for cyclic loading effects, cf Gylltoft [10].

Numerical procedure

The equation of motion of the finite element system

$${}^{t}\ddot{\underset{\sim}{u}} = M^{-1} ({}^{t}\underset{\sim}{F}^{e} - {}^{t}\underset{\sim}{F}^{i}) \tag{10a}$$

is integrated in time by the central differences

$${}^{t+\Delta t/2}\dot{\underset{\sim}{u}} = {}^{t-\Delta t/2}\dot{\underset{\sim}{u}} + \Delta t \, {}^{t}\ddot{\underset{\sim}{u}} \tag{10b}$$

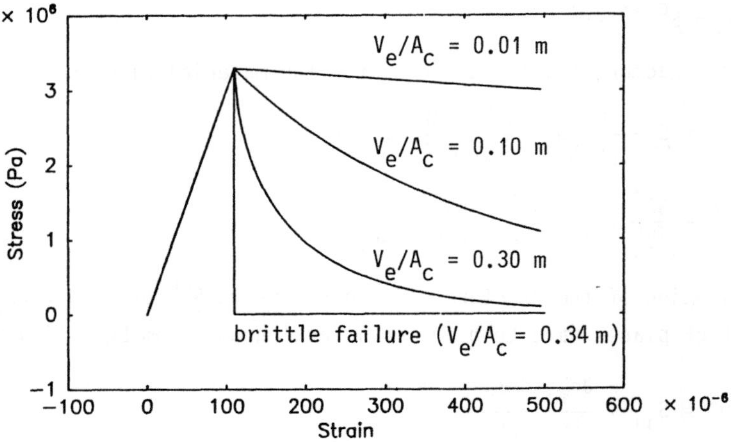

Fig.3. Influence of characteristic length V_e/A_c on the uniaxial tensile stress-strain curve. (f_{tu} = 3.3 MPa, E = 30 GPa, ν = 0.2, G_f = 127 N/m). Critical characteristic length is 0.34 m.

$$^{t+\Delta t}\tilde{u} = {}^{t}\tilde{u} + \Delta t\,{}^{\Delta t+\Delta t/2}\dot{\tilde{u}} \tag{10c}$$

where $\tilde{u}$, $\dot{\tilde{u}}$, and $\ddot{\tilde{u}}$ are the nodal displacement, velocity, and acceleration vectors, respectively, M is the diagonalized structural mass matrix, and $\tilde{F}^e$ is the external nodal force vector. Left superscripts denote actual time and Δt is the time increment. The stress divergence vector at time $t+\Delta t$ is

$$^{t+\Delta t}\tilde{F}^i = \sum_{k=1}^{N_e} L_k^T \int_{V_k} (\tilde{\nabla}\phi_k)^T ({}^{t}\underset{\sim}{\sigma} + \Delta\underset{\sim}{\sigma})\, dV_k \tag{11a}$$

$$\Delta\underset{\sim}{\sigma} = {}^{t}\dot{\underset{\sim}{\sigma}}\,\Delta t \tag{11b}$$

where ϕ_K is the spatial interpolation matrix, $\tilde{\nabla}$ is a gradient operator, L_K is the nodal indices vector, N_e is the number of finite elements, and T is the transpose operator.

In order to minimize the computation time, Eq. (11a) is spatially integrated by a one point Gaussian quadrature. Possible hourglass deformations are controlled by adding the hourglass resisting vector $\tilde{F}^h$ to the stress divergence vector $\tilde{F}^i$. Thus,

$$^{t}\tilde{F}^h = \sum_{k=1}^{N_e} L_k^T\, {}^{t}\tilde{F}_k^h \tag{12a}$$

$$^{t}\tilde{F}_k^h = -Q_h \rho c\, V^{2/3} \sum_{j=1}^{N_h} ({}^{t}\dot{\tilde{u}}_k \tilde{\Gamma}_j)\tilde{\Gamma}_j \tag{12b}$$

where for element k: ρ, c, and V are the mass density, speed of dilational waves, and volume, respectively, N_h is the number of possible zero energy modes, $\tilde{\Gamma}_j$ is zero energy mode j, and Q_h is a constant (usually $0 < Q_h < 1.5$), cf Hallquist [11]. Spurious oscillations due to the discretization are damped by the use of standard linear and quadratic artificial viscosities, cf Zukas et al [12].

Further details on the numerical procedure including contact-impact relations are found in Nilsson [6].

Numerical example

The sphere-rod impact problem, illustrated in figure 4a, was experimentally investigated by Goldsmith et al [13]. The sphere was made of steel with Young's modulus 170 GPa, Poisson's ratio 0.3, and mass density 7800 kg/m^3. The rod consisted of 7 days old concrete with Young's modulus 27 GPa (obtained from sound speed measurements), Poisson's ratio 0.16, mass density 2170 kg/m^3, uniaxial compressive strength (f_{cu}) 25 MPa, and uniaxial tensile strength (f_{tu}) 2 MPa. Material parameters, which were not reported in [13], have been chosen in accordance with standard concrete data. Thus, we have assumed the fracture energy G_f = 150 N/m, the hydrostatic elastic limit 2.4 f_{cu}, the biaxial strength 1.3 f_{cu}, and reference effective strain rate $\dot{\varepsilon}^{ef}_{o} = 2 \cdot 10^{-6}$ 1/s. In the finite element analysis axisymmetric conditions have been assumed. Figure 4b shows the mesh consisting of 818 isoparametric 4 node bilinear elements resulting in a total of 1858 active degrees of freedom. The time increment was 0.1 μs.

Figure 5 shows the calculated axial surface strain records at two stations along the rod compared to experimentally found strain records. It is found that the strain-softening tensile fracture mechanism yields axial strains fairly close to measured strains. The brittle failure mechanism, on the other hand, yields axial strains with too large a magnitude and duration. Figure 6 shows the crack pattern obtained from the finie element analysis with strain softening tensile fracture. The first spalling crack appears 168 μs after impact time and at 20 mm distance from the rod end. Goldsmith et al [13] found spalling cracks at distances between 44 and 57 mm from the rod end. The first calculated crack may, however, not be the one creating the complete failure.

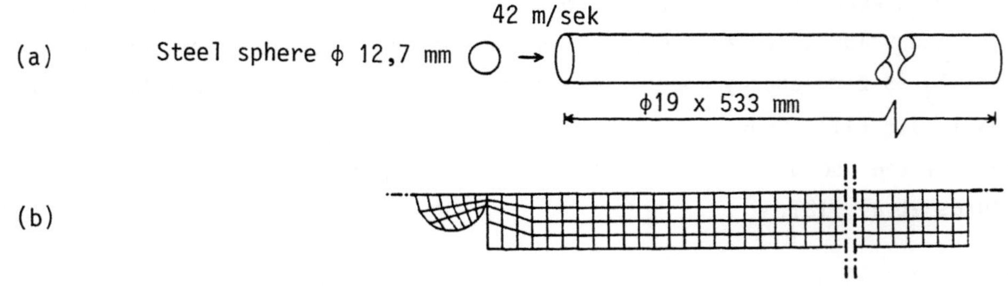

Fig.4. Sphere-rod impact problem. (a) Geometry and initial condition. (b) finite element axisymmetric model

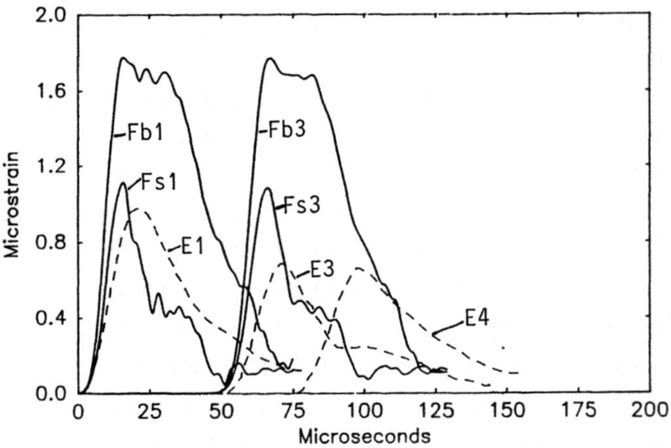

Fig.5. Axial strain in the concrete rod 89 mm (1), 267 mm (3), and 356 mm (4) from the contact face. Notation: E = Experiment Goldsmith et al [13], Fb = FEM with brittle failure, and Fs = FEM with strain-softening tensile fracture. Time measured from first arrival at station No. 1.

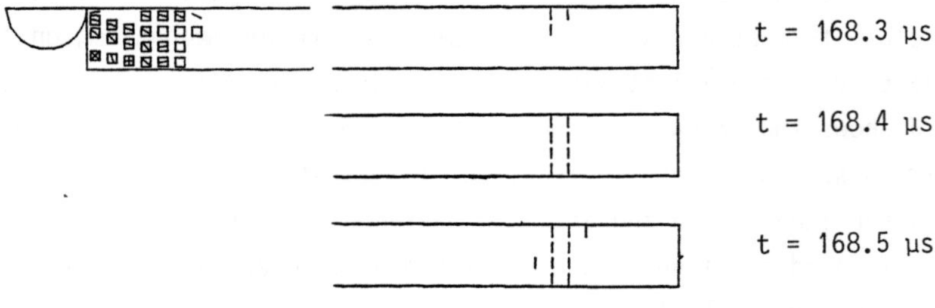

Fig.6. Crack patterns from FEM analysis with strain-softening tensile fracture. Cracks shown have width w > 1 μm. Notation / = ring crack (out of the plane of the paper) ▫ = radial crack (in the plane of the paper)

Concluding remarks

The present tensile fracture model relates strain-softening to the opening of cracks. It is related to the fictitious crack model used in nonlinear fracture mechanics. The new model, however, can be used in continuum analysis and allows cracks to develop in any spatial region and to propagate in any direction. Unlike previously used finite element models of strain-softening tensile failure, the present model does not suffer from mesh dependence. In the analysis of the sphere-rod impact problem, the strain-softening fracture model yields results which are superior to those obtained by a brittle failure model. These results indicate that energy dissipation due to cracking must be accounted for in order to obtain realistic results.

References

1. Chen, W.F.: Plasticity in reinforced concrete. New York: McGraw-Hill 1982
2. Bazant, Z.P.; Oh, B.H.: Concrete fracture via stress-strain relations. Rep. 81-101665c, Evanstone: Northwestern University, Technological Institute 1981
3. Hilleborg, A.; Modéer, M.; Petersson, P.-E.: Analysis of crack formation and crack growth in concrete by means of fracture mechanics and finite elements. Cement and Concrete Research 6 (1976) 773-782
4. Nilsson, F.: On crack propagation. Rep. 8, Stockholm: The Royal Institute of Technology, Department of Strength of Materials 1975
5. Nilsson, L.: Finite element analysis of impact on concrete structures. In Finite elements in nonlinear mechanics (Ed. Bergan, P.G.). Trondheim: Tapir 1978
6. Nilsson, L.: Impact loading on concrete structures. Publ. 79:1, Göteborg: Chalmers University of Technology, Department of Structural Mechanics 1979
7. Nilsson, L.; Glemberg, R.: A constitutive model for concrete in high rate of loading conditions. Delft: IABSE Colloquium 1981
8. Hughes, B.P.; Gregory, R.: Concrete subjected to high rates of loading in compression, Magazine of Concrete Research 24 (1972) 25-36
9. Watstein, D.: Properties of concrete at high rates of loading, ASTM, STP176 (1976) 156-169
10. Gylltoft, K.: A fracture mechanics model for fatigue of reinforced concrete. To be presented at the ACI Convention, Detroit Sept. 19-24 (1982)
11. Hallquist, J.O.: Theoretical manual for DYNA3D. Rep. UCID-19401, Livermore: Lawrence Livermore Laboratory 1982
12. Zukas, J.A.; Nicholas, T.; Swift, H.F.; Greszczuk, L.B.; Curran, D.R.: Impact Dynamics, New York: John Wiley 1982
13. Goldsmith, W.; Polivka, M.; Yang, T.: Dynamic behaviour of concrete. Experimental Mechanics 6 (1966) 65-79

Adiabatic Elastic/Viscoplastic Plane Waves in a Rod at Finite Strain

M. MALATYNSKI , W.K. NOWACKI

Institute of Fundamental Technological
Research, Polish Academy of Sciences, Warsaw

Summary

It is discussed how the temperature field depends on the viscoplastic deformation in the dynamically deformed bar under the assumption that the process is locally adiabatic. Experimental investigation is performed in the system of a modified Hopkinson bar. The initial-boundary value problem is solved by employing the method of characteristics. The results of numerical computations are compared with experimental data.

Description of Experiments

The first papers on temperature variation in solids due to their deformation were published a few years ago in connection with the appearance of infrared cameras on the market. Thermovision apparatus is capable of recording the distribution of radiation emitted from the surface. The measured signal of the distribution is processed to obtain the relevant temperature distribution. No inertial effect is virtually involved in the temperature measurement based on the detection of infrared radiation. Moreover, the measuring system has no contact with the sample and this is an important advantage, particularly in dynamic tests.

The test stand is based on a modified Hopkinson pressure bar system and AGA 680 thermovision set. It is designed for dynamic compression of cylindrical samples and for thermovision recording of temperature distribution. The Hopkinson pressure bar system consists of a compressed air-gun and a set of bars. The tested sample leaves the gun barrel at a high velocity, strikes against the face of the bar and becomes considerably deformed.

Nonlinear Deformation Waves
IUTAM Symposium Tallinn 1982
Editors U Nigul, J Engelbrecht

The sample velocity is measured by means of a laser, the beam of which is split by a set of prisms into two parallel beams directed towards the phototransistors. Once the first laser beam is cut off by the striker sample, the phototransistor **gives** an electric signal actuating a time recorder which is stopped when the striker sample cuts off the second beam.

As a result of the impact an elastic wave of compression is produced in the measuring bar which after reflection travels back in this bar as an elastic wave of extension. A potentiometric strain gauge system is mounted at the mid-length of the bar. Its signal is processed in an amplifier and recorded in one of the channels of a digital storage oscilloscope. This signal represents the initial and reflected elastic waves as functions of time. The oscilloscope time is triggered by an other potentiometric strain gauge system fixed at the front of the bar. Such an arrangement makes it possible to obtain a "waiting" time base, to record the full shapes of the initial and reflected waves, and to establish a fixed reference point on the time axis.

The impact point of the sample at the instant of striking against the active measuring bar is observed by a thermovision camera. Temperature variation is determined by analysing the amplified detector signal of the infrared radiation emitted by the sample surface points lying along the line parallel to the axis of the sample, that is, by analysing the signal of the thermovision image line. This signal is observed on the screen of an oscilloscope attachement and recorded on the screen of the oscilloscope.

The samples were made of a soft aluminium alloy PA1. The cylindrical samples were $d_o \approx 9.80$ mm in diameter and $l_o \approx 95$ mm in length. The carrying part of the sample was limited by two teflone rings in order to ensure coaxial motion of the sample inside the gun barrel. The sample had no side contact in the motion inside the gun what made it possible to avoid friction effects and misinterpretation of the measurement results.

The power of the radiation observed by the camera varies both in the time and space. The principle of operation of the thermovision camera used in the present experiments consists in observation of the succesive points of a selected sample line by means of a rotating prism. In the diagram θ (X_1),a complete observation cycle,is **constrained** between two points at which the quantity observed changes its **sign**. In the same diagram prepared for a striker velocity of $v \approx 70$ m/s - it is possible to distinguish three maxima - cf.Fig.1. The first maximum /rather weak/ located

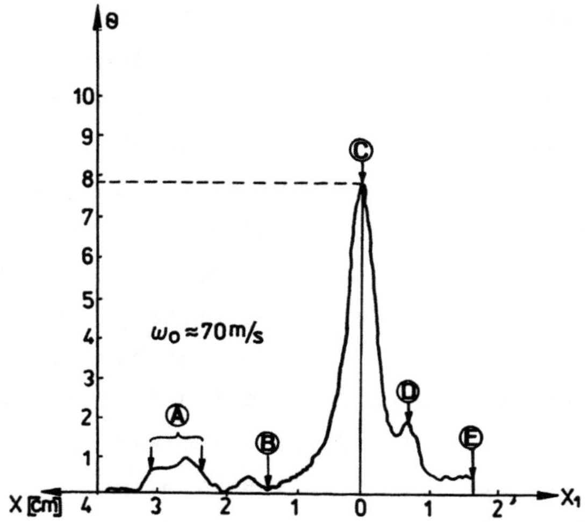

Fig.1. Typical curve of sample radiation power versus X_1

at the left is caused by the heat radiated from the indicator heater wires /point A/;the second maximum /most distinct at point C/, and the third maximum /at point D/ correspond to the end of the striker bar and the end of the incident bar,respectively,both ends being heated up during impact.
From the records of deformation $\varepsilon(t)$ in the incident bar it is possible to determine the strain and stress at the contact point of the sample and the bar /under the assumption that the incident bar is longitudinally elastic/. Permanent longitudinal and transversal deformations Δ_{11} and Δ_{22} ,respectively,were measured.

Formulation of the Thermomechanical Problem

Solution is found to the problem of wave propagation in an elastic/viscoplastic bar which strikes against an undeformable obstacle. The theory of finite deformation of thermoviscoplasticity [1] is used. The effect of transverse inertia is disregarded, but that of transverse deformation is taken into account because it may be important in case of finite deformation.

In the rectangular coordinate system X_i the motion is described by $x = x(X,t)$. The gradients of elastic and plastic deformation assume the following form:

$$\underline{E} = \begin{vmatrix} E_{11} & & \\ & E_{22} & \\ & & E_{22} \end{vmatrix} \qquad \underline{P} = \begin{vmatrix} P_{11} & & \\ & P_{22} & \\ & & P_{22} \end{vmatrix} \qquad /1/$$

The gradient of the total transformation is $\underline{F} = \underline{E}\ \underline{P}$. The Cauchy tensor has only one component σ_{11} in the coordinate system X_i. The velocity of elastic deformation can be written in the form [2] :

$$D^e_{11} = \left[\dot{\sigma}_{11} - \sigma_{11}\left(\frac{\partial v}{\partial x} - 2\,\dot{E}_{11}\,E^{-1}_{11}\right)\right] \varrho_o\,E^{-4}_{11} / \varrho\,E_o - \alpha\,\dot{T}\,E^{-2}_{11} \qquad /2/$$

where $v = \frac{\partial x}{\partial t}$, E_o is the Young's modulus, α is the coefficient of linear thermal expansion, and ϱ is the density: $\varrho = \varrho_o / F_{11}\,F^2_{22}$. The velocity of viscoplastic deformation is defined by /cf. [2]/:

$$D^{vp}_{11} = E_{11}\dot{P}_{11}\,P^{-1}_{11}\,E^{-1}_{11} = E_{11} < \Phi(\sigma_{11}, \alpha_n, T) > E^{-1}_{11} \qquad /3/$$

with the equation of evolution for the internal parameters α_n:

$$\dot{\alpha}_n = \Pi_n\ (\sigma_{11}, \alpha_k, T) \quad , \quad n,k, = 1,\ldots N. \qquad /4/$$

The functions Φ and Π_n are determined on the basis of experimental investigations.

Equation for the temperature in the case of adiabatic processes takes the form:

$$\dot{T} = \frac{1}{c_\varepsilon}\left[\sigma_{11}\,\dot{P}_{11}\,P^{-1}_{11} / \varrho - T\,\dot{E}_{11}\,E_o / \varrho_o - \frac{\partial U}{\partial \alpha_n}\ \dot{\alpha}_n\right] \qquad /5/$$

where U - stored internal energy, c_ε - specific heat at constant deformation.

The system of equations /2/-/5/ together with the equation of motion:

$$\frac{\partial \sigma_{11}}{\partial X} F^2_{22} = \varrho_o\,\dot{v}_1 \qquad /6/$$

and relation $D_{11} = D^e_{11} - D^{vp}_{11}$ enable an unique statement of the problem of adiabatic wave propagation in the elastic/viscoplastic bar under given boundary and initial conditions.

Analysis of Experimental Results. Comparison with Theoretical Calculations

On the basis of the method of characteristics, an initial-boundary value problem is solved for the same boundary and initial conditions as those used in the experiments described in the present work. Numerical data were close to those used for the samples tested in the experiments. The calculated permanent longitudinal and transversal deformations, Δ_{11} and Δ_{22}, /for a fixed time/ of the sample under test are shown in Fig.2a in comparison with the corresponding strains measured on the deformed samples.

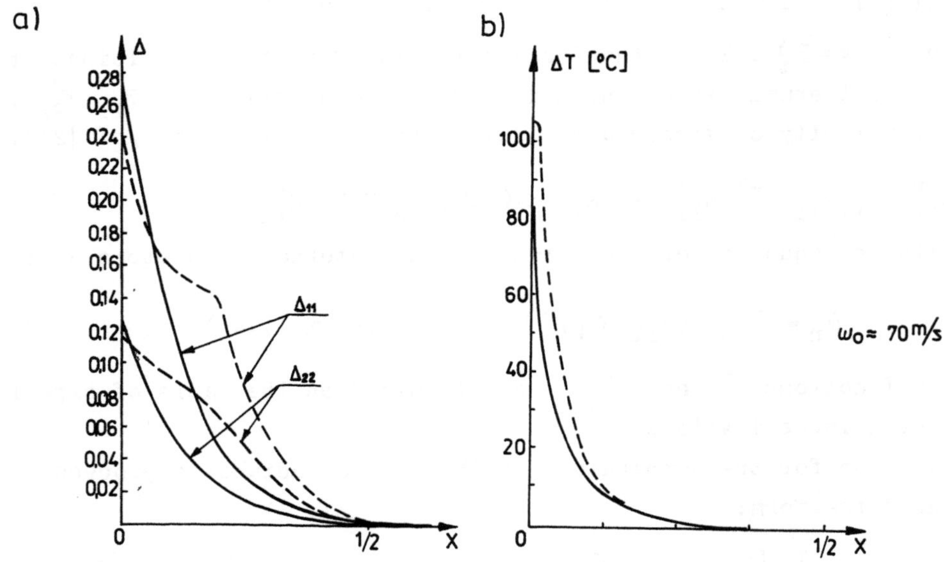

Fig.2. a/ Permanent deformations of sample: Δ_{11} longitudinal deformations, Δ_{22} transversal deformations,

b/ Temperature increase along the sample tested, on line $X_1 = v_{cam}/t-t_D/$, where v_{cam} is the "sweeping velocity" of the camera,

—— calculated, - - - measured.

The maximum deformations appear on the boundary of the sample under test. It is seen that the maxima calculated and measured coincide. The differences between the measured and calculated deformations over the entire range of permanent strains can be explained by disregarding the term of instantaneous deformation in the constitutive equation -cf.[2]. From the mathematical point of view it means that only disturbances propagating with the elastic wave velocities can move in the material under consideration, and plastic waves do not appear. It follows from the experiments described here that in the theoretical discussion on long dynamically deformed samples made of strain-rate-sensitive materials it is necessary to take into account not only the deformations of a relaxation type, but also the instantaneous deformations.

Dashed line in Fig.2b represents temperature variation ΔT along the sample /that is, along the straight line $X_1 = v_{cam}/t-t_D/$. The distribution of the power of infrared radiation along the sample is shown in Fig.1. Solid line in Fig.2b represents the temperature increase /also along the line $X_1 = v_{cam}/t-t_D/$ calculated from the numerical solution of the initial-boundary value problem. The impact velocity of the sample was almost identical in the both cases. The difference between the calculated and measured results can also be explained by disregarding the instantaneous plasticity effects in the theoretical discussion. The experimental temperature increase and the heating region of the sample face are larger that those predicted by the theory.

References

1. Mandel, J.: Plasticité classique et viscoplasticité. CISM Publications. Udine: Springer Verlag 1971.
2. Nowacki, W.K.; Kurcyk, T.: Sur le champ des températures en thermoélastoviscoplasticité finie. Journal de Mécanique Théor. et Appl. 2, 1/1982/.

Dynamics of Two-Dimensional Thermoviscoelastic Waves in Laminated Media

V. K. RYMSKY

Department of Energetical Cybernetics,
Academy of Sciences of the Moldavian SSR, Kishinev

P. F. SABODASH

Lomonosov Technological Institute of Food Industry, Odessa

Summary

On the basis of characteristic relations in the difference form and of the principle of the electrothermal similarity an effective approach is suggested to solve an axially symmetric problem of thermoviscoelasticity of polylaminated media with non-linear boundary conditions. Sufficient stability and convergence conditions for the first order finite difference method are obtained by making use of a priori estimation method. A transient axially symmetric wave process in a stack of plies generated by an impact with the flat end of a circular cylinder moving with a constant velocity has been investigated.

In this research on the basis of thermomechanics of continuous medium the non-linear theory of thermoviscoelasticity with a non-linear equation of heat conductivity is built by means of introduction of additional postulates. With the non-linear theory as a basis, the concrete mathematical models have been constructed assuming the first derivatives of displacements with respect to the material coordinates and the change of the temperature to be small.

Let the flat end of a continuous circular cylinder with radius R and height h_0 being strained and translated with the constant velocity $\bar{v}_0$, the vector of which is directed along the axis of the cylinder, encounter the surface of a polylami-

Nonlinear Deformation Waves
IUTAM Symposium Tallinn 1982
Editors U. Nigul, J. Engelbrecht

nated plate which in the cylinder coordinates $r z \varphi$ occupies a region $0<z<h$, $0<r<\infty$, $0\leq \varphi < 2\pi$ at initial moment $t=0$. The plies in the plate are numbered downwards, while index "0" is assigned to the striker which is being strained.

Three-dimensional axially symmetric thermoviscoelastic wave process in a laminated plate is described inside each ply by a following system of linear coupled equations.

$$\left[\frac{1}{\theta_1^{(k)}}+\frac{\partial}{\partial t}\right]\sigma^{(k)}=\left[3\lambda^{(k)}+2\mu^{(k)}\right]\frac{\partial}{\partial t}\left[\varepsilon^{(k)}-3\alpha^{(k)}\left(T^{(k)}-T_0\right)\right],$$

$$\left[\frac{1}{\theta_2^{(k)}}+\frac{\partial}{\partial t}\right]S_{ij}^{(k)}=2\mu^{(k)}\frac{\partial e_{ij}^{(k)}}{\partial t},$$

$$\rho^{(k)}c_v^{(k)}\left[\frac{\partial T^{(k)}}{\partial t}+\tau_0^{(k)}\frac{\partial^2 T^{(k)}}{\partial t^2}\right]+\left[3\lambda^{(k)}+2\mu^{(k)}\right]\alpha^{(k)}T_0\frac{\partial \varepsilon^{(k)}}{\partial t}= \qquad (1)$$

$$=\operatorname{div}\left[K^{(k)}\operatorname{grad}T^{(k)}\right],$$

$$\sigma^{(k)}=\sigma_{11}^{(k)}+\sigma_{22}^{(k)}+\sigma_{33}^{(k)},\quad \varepsilon^{(k)}=\varepsilon_{11}^{(k)}+\varepsilon_{22}^{(k)}+\varepsilon_{33}^{(k)},$$

$$S_{ij}^{(k)}=\sigma_{ij}^{(k)}-\delta_{ij}\sigma^{(k)}/3,\quad e_{ij}^{(k)}=\varepsilon_{ij}-\delta_{ij}\varepsilon^{(k)}/3,\quad k=\overline{1,N}.$$

Here $\sigma^{(k)}$ is average (hydrostatic) stress, $\varepsilon^{(k)}$ - cubic strain and $e_{ij}^{(k)}$ - stress and strain deviators, δ_{ij} - Kronecker's symbol, $\lambda^{(k)}$ and $\mu^{(k)}$ - Lamé constants, $\alpha^{(k)}$ coefficient of linear expansion, $\theta_1^{(k)}$ and $\theta_2^{(k)}$ - relaxation time of the spherical and deviation parts of the stress field, $\rho^{(k)}$ - material density of the plate plies, $c_v^{(k)}$ specific heat of the materials, $\tau_0^{(k)}$ - relaxation time of heat flows, $K^{(k)}$ - thermal conductivity, T_0 - initial temperature of the mechanical system, $[\lambda^{(k)}+2\mu^{(k)}]\alpha^{(k)}T_0$ - coupling coefficient of heat and mechanical fields which is responsible for thermoviscoelastic dissipation of a part of the mechanical energy into the heat, N - quantity of plies constituting the plate, t - running time. As far as we are concerned, the mechanical and thermophysical properties of the materials are constant and independent of the temperature T.

The relations (1) together with the two equations of motion

$$\rho^{(k)}\frac{\partial u^{(k)}}{\partial t}=\frac{\partial \sigma_r^{(k)}}{\partial r}+\frac{\partial \sigma_{rz}^{(k)}}{\partial z}+\frac{\sigma_r^{(k)}-\sigma_\varphi^{(k)}}{r},$$
$$\rho^{(k)}\frac{\partial v^{(k)}}{\partial t}=\frac{\partial \sigma_{rz}^{(k)}}{\partial r}+\frac{\partial \sigma_z^{(k)}}{\partial z}+\frac{\sigma_{rz}^{(k)}}{r}$$

form a closed system of equations in partial derivatives which depends upon the three variables r, z, t wherein $u^{(k)}$ and $v^{(k)}$ stand for the components of the velocity vector along the axis r and z respectively. The problem is solved in dimensionless variables $r=r^\circ/h^\circ$, $z=z^\circ/h^\circ$,

$$t=t^\circ a^\circ_m/h^\circ,\quad \sigma_{ij}=\sigma^\circ_{ij}/(\lambda^\circ_m+2\mu^\circ_m),\quad u^{(k)}=u_0^{(k)}/a^\circ_m,$$
$$v^{(k)}=v_0^{(k)}/a^\circ_m,\quad \theta^{(k)}=(T_0^{(k)}-T_0)/T_0.$$

The outer boundaries of the strained system are free of stress. Over the circular area of the cylinder's collision with the plate surface $z=0$, $0\le r<R$ the conditions of rigid adhesion exist or there are no friction forces (smooth impact).

Over the interface the conditions of ideal mechanical and heat integration of the solutions in continuous plies (or the conditions of their relative slip) are satisfied.

The heat exchange of the mechanical system with the environment the temperature of which remains constant is performed (takes place) according to Newton law with the coefficient of the connective heat transfer from the surface depending non-linearly upon the temperature

$$-\frac{k^\circ_k}{h}\frac{\partial\theta}{\partial n}=\alpha_{Tk}\left[T_c\theta_n+\frac{\tau^\circ_{0k}a^\circ_m}{n}T_c\frac{\partial\theta_n}{\partial t}\right]. \qquad (3)$$

Here α_{Tk} is the **summed** coefficient of the convective heat transfer from the surface of the material with the indices $k(k=0,1,M)$, which is determined by convective heat transfer,

emission and convection $\alpha_{TK} = \alpha_{1K} + \alpha_{2K}$,
where

$$\alpha_{1K} = 0{,}23 \cdot 10^{-6} \varepsilon_{1K} \left(\frac{T_n + T_c}{2}\right)^3, \quad \alpha_{2K} = A_2 \left(\frac{T_n - T_c}{h_0}\right)^{1/4}, \tag{4}$$

where ε_{1K} is emissivity factor, A_2 is a constant. In the case of laminar flow of the air at $T_n - T_c$ = 20 ° K the value of A_2 is 1,4.

The formulated non-linear problem of thermoviscoelasticity is solved numerically with the help of an iterative procedure based on the usage of net-characteristic relations and the principle of electrothermal analogy.

Some results of the numerical analysis of the dependence of the energy $f(t) = 1 - E_y(t)/E_0$, transferred by an aluminium striker to a one-ply plate made of the same material upon the time are represented in the following table

t	$Al_{0,25} \to Al$	$1{,}5\,Al_{0,25} \to 0{,}5\,Al$	$0{,}5\,Al_{0,25} \to 1{,}5\,Al$
0.5	0.22I	0.I46	0.438
I.0	0.4I2	0.278	0.783
I.5	0.529	0.393	0.784
2.0	0.630	0.52I	0.744
2.5	0.6I0	0.763	0.748
t_K	3.05	4.2	I.75

Here $E_0 = m_y \frac{v_0^2}{2}$ is initial (kinetic) energy of the striker, $E_y(t)$ - complete energy of the striker after the contact, t_K - time of the striker's contact with the plate, in the $\beta_1 Al_{\beta_2} \to \beta_3 Al$ the coefficient β_1 denotes the height of the striker, β_2 - radius of the base, β_3 - thickness of the plate.

The analysis of the numerical information shows that the quantity of energy, transferred by a cylinder being strained to a plate made of the same material depends on the geometrical dimensions of the cylinder and on time. Within the period of time $0 < t < 2$ a longer cylinder transfers less energy to the aluminium plate. There exist instants when the quantity of the energy transferred to the plate achieves the peak values. The increase in the length of an aluminium cylinder by 1,5 times leads to the increase in the time of the complete contact by 1,4 times. Hence the time of the complete contact is proportional to the length of cylinder.

References

1. Nigul, U.K.; Engelbrecht, J.K.: Non-linear and linear transient deformation waves in thermoelastic and elastic bodies. Tallinn: Academy of Sciences of the Estonian SSR, Institute of Cybernetics, 1972 .
2. Naval, I.K.; Sabodash P.F.: Numerical solution of dynamical coupled problem of thermoviscoelasticity for a ply with the final velocity of heat conduction. - Izvestiya Academii Nauk SSSR. Mekhanika tvyerdogo tela, 4 (1976), 108-114.

Representing Shear Band Damage at High Strain Rates

LYNN SEAMAN and JAMES L. DEIN

SRI International, Menlo Park, California, USA

Abstract

During construction of a microstructural model for shear band damage at high strain rates, means were developed for representing the bands with discrete size and orientation distributions and for computing the stresses while accounting for the damage. For each strain increment, the model computes the stress tensor and updates plastic strains and shear band populations. From comparison calculations we concluded that accurate representation of shear band distributions required discrete groupings both by size and orientation. With the damage limited to discrete orientations, it was convenient to choose a Tresca-like yield condition with Coulomb friction, with yielding allowed only on the damage planes. Plastic deformation and work hardening also were assigned to these planes. A stress-relaxation method was chosen for solving for stress and distributing the plastic strain and damage to the active orientations. The model treats naturally the complex anisotropy of plastic strain that develops in nonproportional loading problems and the anisotropy of shear band damage.

Introduction

Recently, we constructed a microstructural model that describes the microprocesses of shear band nucleation, growth, and damage [1]. The model treats arrays of shear bands in several size groups and discrete orient-tions. Under large plastic strains the bands are nucleated, grow, and coalesce to form fragments. Here the model is outlined briefly, with emphasis on the distributions of bands in orientation and size, and on the procedure for solving stress.

Model Description

The shear band model treats a statistical distribution of bands as a probability density at each point in the material. The bands are envisioned in the model as small, circular, planar regions that have lost their cohesive strength. In our experiments with fragmenting cylinders, projectile penetrations, and Taylor tests (blunt cylinder impacts on

Nonlinear Deformation Waves
IUTAM Symposium Tallinn 1982
Editors U Nigul, J Engelbrecht

unyielding targets), the shear bands occur in a range of sizes. As an example, Fig. 1 shows the size distributions of bands seen on the inside of a cylinder expanded by explosives detonated along the cylinder axis. The specimen cylinder was contained inside a large steel cylinder to control the amount of expansion, so that the shear bands formed but did not coalesce to produce fragments. In the model, the statistical size distributions are represented by a series of number and length coordinates, as indicated in Fig. 2. At nucleation the distribution is assumed to be exponential, with the number N_g per unit volume with a radius greater than R being

$$N_g = N_o \exp(-R/R_n) \tag{1}$$

where N_o is the total number of bands per unit volume and R_n is the shape parameter of the distribution. After growth and further nucleation the size distribution retains an exponential form between the designated coordinates, as shown in Fig. 2.

In our experiments the bands usually occur in discrete orientations, on the planes of maximum shear strain. In the model, we chose to represent the damage in nine orientations, as shown in Fig. 3. Each orientation can be considered to consist of bands on that specific plane, or to represent bands with orientations that range over a solid angle of $4\pi/9$. A size distribution such as that in Fig. 2 is assigned to each orientation. As deformation occurs the bands rotate with respect to each other, so they do not retain the angular relationship shown.

For three-dimensional motions, the nine band orientations represent all possible bands and all quadrants are equally well represented. In two-dimensional motions, some reduction in orientations can be made by using only planes that have unique normal and shear stresses. Also normal and shear strains on the plane must represent permitted strains on the material; that is, no ε_{xz} and ε_{yz}. We combined Orientations 5 and 8, and 6 and 9, so that only seven orientations are required. In two dimensions these pairs of orientations have identical stresses. Strains on these pairs of planes combine to provide none of the ε_{xz} and ε_{yz} strains.

Bands begin to appear in the model when the plastic shear strain on any plane exceeds a critical value at which the increase in strength caused by work-hardening is just balanced by the decrease in strength associated with thermal softening. After the threshold strain is reached, bands are

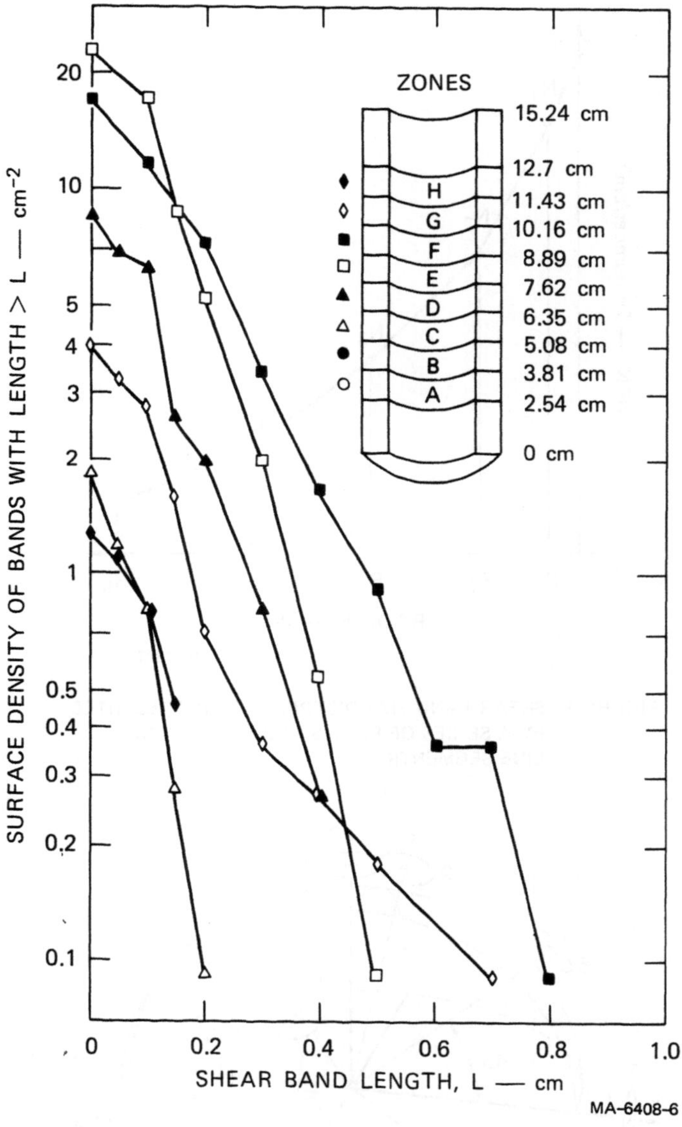

FIGURE 1 SHEAR BAND DISTRIBUTION FOR EXPERIMENT 2 IN 4340 STEEL, R_c40

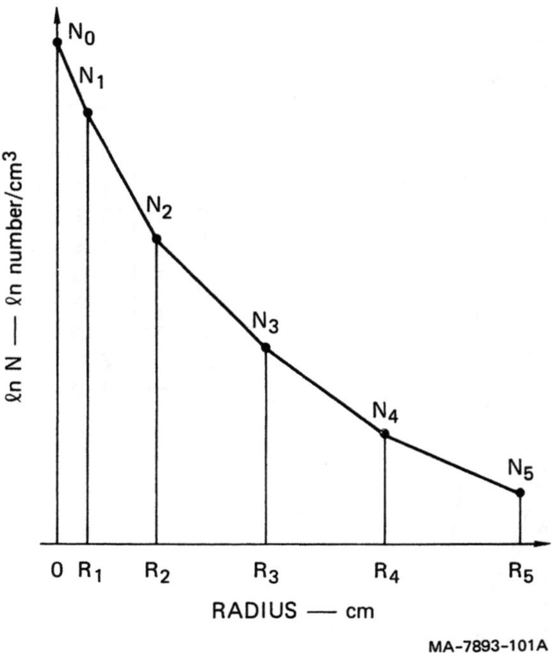

FIGURE 2 SHEAR BAND SIZE DISTRIBUTION REPRESENTED BY A SERIES OF POINTS AND EXPONENTIAL LINE SEGMENTS

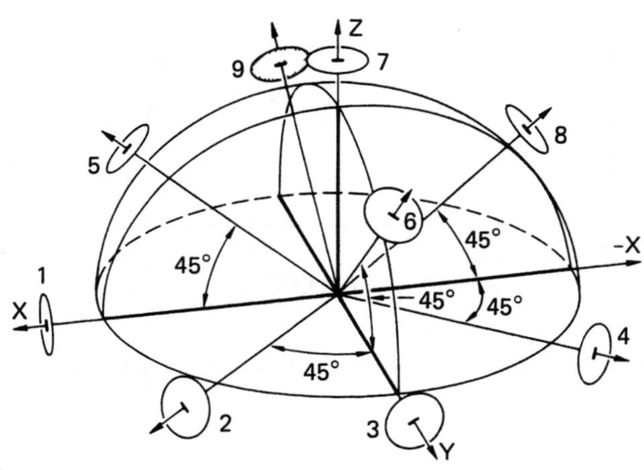

FIGURE 3 RELATIVE LOCATIONS OF THE COORDINATE DIRECTIONS AND INTIAL ORIENTATION OF THE SHEAR BAND DAMAGE GROUPS

nucleated according to a rate function containing plastic strain, strain rate, the critical strain, and a critical strain rate. The critical strain rate is that required to maintain approximately adiabatic conditions. The nucleation rate is

$$N = \alpha N_o H(\varepsilon_m^p - \varepsilon_o) \, \varepsilon_m^p \exp(-\varepsilon_o/\varepsilon_m^p) \qquad (2)$$

where $\alpha = \sigma_y/(\rho E_m)$, σ_y and ε_m^p are the yield stress and plastic strain rate, and N_o is the number of potential nucleation sites per unit volume. H is a Heaviside function, ε_m^p and ε_m^p are the continuum in-plane plastic strain and strain rate, ε_o is the plastic strain at the maximum shear stress (about 20% in 4340 steel), and ε_o is the strain rate required to attain adiabaticity at the nucleation sites.

The bands grow under the imposed strain rates according the following relation, subject to some restrictions:

$$\frac{dR}{dt} = T_1 \frac{d\varepsilon^{PS}}{dt} R \qquad (3)$$

where R is the band radius in plan and T_1 is the growth coefficient. The restrictions are that dR/dt cannot exceed the shear velocity V_s, and that the total strain taken by the band cannot exceed the total imposed strain. The strain taken by the band is

$$\Delta\varepsilon = \pi b \sum_i N_i (R_{2i}^3 - R_{1i}^3) \qquad (4)$$

where b = B/R, B is the average slip over the plane of the band, and R_1 and R_2 are the band sizes before and after growth, respectively.

Stress Calculation

To solve for the stress state resulting from an imposed strain increment, the plastic strains and developing shear band damage must be determined. For our model, these strains and damage are computed on each band orientation plane. Yielding on each plane is allowed to develop gradually, following the stress-relaxation process:

$$\frac{d\sigma_\ell'}{dt} = 2G \frac{d\varepsilon_\ell'}{dt} - \frac{\sigma_\ell' - Y_{\ell s}}{T} \qquad (5)$$

where σ_ℓ' is the shear stress on the planed ℓ, G is the shear modulus,

ε'_ℓ is the tensor shear strain acting on the plane, T is the time constant (either a material constant or a constant chosen for computational convenience), and $Y_{\ell s}$ is the shear yield strenght on the plane ℓ. Eq. 5 is integrated numerically over four to ten subcycles within each imposed time increment. During these subcycles, the overstress, $\sigma'_\ell - Y_{\ell s}$, will gradually decrease and plastic strain will increase on the yielding planes. Each time increment begins with a cell-centered elastic calculation of the stresses, based on the strain increments. From these stresses the shear strains ε'_ℓ on each plane are determined. On yielding planes, plastic strain is developed and used to modify the imposed strain for the next cell-centered calculation. During the subcycles, the overstress, $\sigma'_\ell - Y_{\ell s}$, gradually decreases and plastic strains increase on the yielding planes. The stress-relaxation process therefore provides a means for spreading the plastic strains over several orientation planes.

Preliminary Results

Initial tests have been made to compare the plastic strains computed with the model to those obtained from a standard Mises model. The equivalent plastic strain is equal to that from the Mises model, but the strains on each plane are not. Hence some alteration of the work-hardening relation is required.

Calculations are beginning now to simulate fragmenting cylinders. Later the model will be used on Taylor test impacts and other projectile problems.

References

1. Erlich, D. C., Seaman, L., Caligiuri, R. D. and Curran, D. R., Computational model for armor penetration. Annual Report to Ballistic Research Laboratory, Aberdeen Proving Ground, MD 21005; and to Army Materials and Mechanics Research Center, Watertown, MA; 02172 (November 1980).

PART 3
INVERSE PROBLEMS

Detection of Deformation Waves in Solids

K. Kawata

Institute of Interdisciplinary Research
Faculty of Engineering, University of Tokyo
4-6-1 Komaba, Meguro-ku, Tokyo 153, Japan

Summary

Methods for detection of deformation waves up to the range of nonlinear deformation waves are reviewed based upon the recent researches by the author and coauthors. The first is "One bar method" as simple and precise as possible for the material characterization in high velocity tension up to breaking. This method certifies the acquisition of precise tensile stress-strain relation of solids up to breaking at high strain rates, through the continuous detection of nonlinear deformation waves. New results on high-velocity brittleness and high-velocity ductility are found on metallic and composite materials including Aluminum alloys, steels for structural use, GFRP and CFRP. The second is high-speed photoelasticity of transmission type and of reflection type. Dynamic stress concentration behaviours around a circular hole in a strip under dynamic tension are studied specifically covering

1) elastic range of isotropic materials (by transmission type),
2) elastic range and fracture initiation of anisotropic(composite) materials, and
3) elasto-plastic range of metallic materials (by reflection type using photoelastic coating method for both 2) and 3)).

Several characteristic results are compared with the FEM results. The behaviour of dynamic stress concentration factors compared with static ones in elastic range for isotropic and anisotropic materials and the behaviour of plastic region propagation for various metals are clearly analyzed.

1. Introduction

Recently, advanced structures encounter high-velocity deformation more often than before. Therefore, methods for detection of deformation waves in solids have urgently been required. Recent progress in this field is reviewed based upon the authors' recent researches. The first is "One bar method" newly developed for obtaining high-velocity tensile stress-strain curves up to breaking, detecting nonlinear deformation waves. Distinctive individualities of various metallic or composite materals in high velocity tension are clarified by the method. The second is high-speed

Nonlinear Deformation Waves
IUTAM Symposium Tallinn 1982
Editors U. Nigul, J. Engelbrecht

photoelasticity of transmission type and of reflection type, as the means of detecting deformation waves in two dimensional fields. Dynamic stress, strain concentrations around a circular hole are investigated with three different high-speed photoelastic techniques, covering elastic to elasto-plastic range of transparent to opaque and isotropic to anisotropic materials.

2. Acquisition of High-Velocity Tensile Stress-Strain Curves up to Breaking of Solids

2.1 One Bar Method for Material Characterization in High Velocity Tension [1]

Material characterization in simple high-velocity tensile deformation up to breaking is most fundamental for precise designing of impact-resistant structures, but there has been little progress in its measuring techniques.

The behaviour of high velocity deformation is intimately related to crystal lattice systems, as clearly shown by the crystal lattice systems effect on the relation of breaking elongation ε_b vs. tensile velocity V_1 found by a systematic series of high velocity tension tests for pure FCC, BCC and HCP metals and alloys especially in the annealed state (Fig.1) [2]. From the consideration of the effect, the authors clarify that there are clear differences in the elasto-plastic wave propagation behaviour for FCC and BCC systems, based upon the respective constitutive equations, derived micromechanically, that is,

$$\dot{\varepsilon} - \frac{\dot{\sigma}}{E} = \frac{4}{3} b \left[N_0 + \frac{3}{4} M \left(\varepsilon - \frac{\sigma}{E} \right) \right] v^* \exp \left(- \frac{2\left[D + \frac{3}{4} H \left(\varepsilon - \frac{\sigma}{E} \right) \right]}{\sigma} \right) \tag{1}$$

for a material modelling a mild steel [3], and

$$\dot{\varepsilon} - \frac{\dot{\sigma}}{E} = C \exp\left\{ - \frac{U_0}{kT} \left(1 - \frac{\sigma - \sigma_g}{\sigma_0 \left(1 + H \left(\varepsilon - \frac{\sigma}{E} \right) \right)} \right) \right\} \tag{2}$$

for materials modelling high strength Aluminum alloys [4].

As a typical example, a sharp stress peak appears at the stress wave front in mild steel, and in contrast does not appear at the stress wave front in high strength Al alloy (Fig.2) [3-5]. Data acquisition in high velocity tension of solids should be performed using measuring systems enough to detect the differences of elasto-plastic stress wave front as theoretically predicted.

Through systematic investigations of impact responses of various candidate testing systems such as block to block, bar to block, and bar to bar systems, carried out by FEM, assuming specimen materials obeying the above-mentioned constitutive equations, one bar method of bar to block type is adopted, because of its accuracy and simplicity [1].

The testing system of one bar method consists of a hammer, an impact block, a specimen, and an output bar (Fig.3). When the impact block is impacted by the hammer, the specimen is deformed in high velocity tension and the strain $\varepsilon_g(t)$ caused by a travelling stress wave in the output bar at time t is measured from semiconductor gages at a distance a from the impact end. Dynamic tensile stress $\sigma(t)$ and strain $\varepsilon(t)$ of the specimen are calculated by the one-dimensional stress wave propagation theory as follows:

$$\begin{aligned} \sigma(t) &= (S_0/S)E_0\varepsilon_g(t+a/c) \\ \varepsilon(t) &= \frac{1}{l}\int_0^t [V(\tau) - c\varepsilon_g(\tau+a/c)]d\tau \end{aligned} \tag{3}$$

where l and S are the length and the cross-sectional area of the specimen, and S_0, E_0 and c are the cross-sectional area, Young's modulus and the elastic wave velocity of the output bar. V(t) is the velocity of the impact block and is measured by an electro-optical extensometer. Two types of testing machines developed are shown in Figs. 4 and 5. The hammer speed used is 10 m/s for the rotating disc type (obtainable up to 100 m/s) and 3.5 m/s for the pendulum type. Dimensions of straight specimens used are shown in Fig. 6(a). The strain rate in high velocity tension is, therefore, in the order of 10^3 s^{-1}, and about 10^6 times larger than 10^{-3} s^{-1} in conventional quasi-static tension. In this case the length of the output bar is 2,000 mm and a sufficient length for above-mentioned strain gages to be swept by a single sweep wave without a reflected wave from the other end of the output bar. The order of 10^4 s^{-1} is easily obtained.

2.2 Characteristic Behaviour of Materials in High Velocity Tension

Detecting of wave front: The existence and non-existence of dynamic upper yield point for a mild steel and an Al alloy are clearly obtained as stress wave front travelling in the output bar of one bar method, and the data are recorded satisfactorily up to breaking, as shown in Fig. 7 [1]. So, the one bar method is quite suitable for the detailed analysis of phenomena of high velocity brittleness and high velocity ductility found by Kawata et al. [2,4] for various materials.

Al alloys and steels: The data on some Al alloys are shown in Fig. 8. Peak stress σ_p, total strain ε_t and absorbed energy per unit volume up to breaking E_{ab} increase with the increase of the strain rate. The data on some steels for structural use are shown in Figs. 9 and 10.

Sample	Chemical Composition C	Si	Mn	P	S	Cr	Ni	Al	N
SUS 304	0.06	0.61	0.93	0.030	0.006	18.2	8.8	-----	------
YUS 304N	0.06	0.72	1.90	0.031	0.004	18.7	8.4	-----	0.21(Nb 0.11)
Dual Phase	0.12	1.36	1.60	0.002	0.005	----	---	0.017	0.0015

SUS 304	rolled and solution treated (1100°C x 30 min.).
YUS 304N	as above.
Dual Phase	hot rolled, heated (800°C x 10 min.), oil quenched to 50°C and then air cooled.

In contrast to the decreasing tendency of ε_t and E_{ab} for SUS 304 and YUS 304N with increasing strain rate from 10^{-3} s^{-1} to 10^3 s^{-1}, Dual Phase steel has the tendency of increasing with increasing strain rate. This is a remarkable feature of Dual Phase steel.

Composites [6-8]: Composites also show distinctive individualities in high velocity tension. Examples of GFRP and CFRP are shown in Figs. 11 and 12. The fracture appearances of GFRP and CFRP for static and dynamic tension are shown in Fig. 13. Both are nearly equal for CFRP, but quite different for GFRP. Anyway, the facts that have been unknown hitherto are being disclosed in the field of high velocity tension. Fracture-mechanical study (Fig. 6(b)) [8] and low-temperature high-velocity tension of FRP [9] have also been done.

3. Two Dimensional Analysis by Means of Photoelasticity of Transmission Type and Reflection Type---Dynamic Stress, Strain Concentration Problems

3.1 High-Speed Photoelasticity of Transmission Type

The relation of dynamic stress concentration factor vs. static stress concentration factor in the dynamic tension of a very long strip with a circular hole or notches, is a fundamental problem of importance [10]. An example of analysis by transmission type is stated [11,12]. Experimental arrangement and specimen shape are shown in Figs. 14 and 15. The high-speed camera is a multiple spark gap camera of 16 frames. The energy and spark duration are 14 J and 0.8 μs. Stress pulse as shown in Fig. 16 is applied to the specimen and successive isochromatic patterns (Fig.17) are obtained. The sharpness of the patterns is by no means inferior to the static ordinary patterns. As shown in Figs. 18 and 19, dynamic stress concentration factors increase with increasing time, and tend to the corresponding static stress concentration factors.

3.2 Photoelastic Coating Analysis of Elastic Wave Propagation in Composite Structures [13,14]

When the photoelastic analysis of elastic wave propagation in opaque composite materials is intended, photoelastic coating method is taken and a dynamic reflection-type camera is needed. A dynamic photoelastic system of reflection type is shown in Fig. 20. Specimen dimensions and some examples of dynamic isochromatic patterns are shown in Figs. 21 and 22. In this case also, the dynamic stress concentration factors tend to the corresponding static stress concentration factors (Fig.23). This method makes real materials, not model materials, objects of study. The initiation of fracture can also be detected by this method, as shown in Fig. 24.

3.3 Photoelastic Coating Analysis of Elasto-Plastic Wave Propagation and Plastic Region Propagation Problems in Metallic Structures [15]

Experimental setup is shown in Fig. 25. A high-speed framing camera is very useful for this purpose as well as a spark gap camera. In this study emphasis is laid on the analysis of the difference of plastic region propagation behaviour in a mild steel and an Al alloy. Dynamic isochromatic patterns for two metals under the same dynamic tension are shown in Fig. 26. The distributions of principal strain differences at the free boundary of the circular hole are shown in Fig. 27. The propagation behaviours of plastic region at the stress concentrated area, that is, at the circular hole side, are known by Fig. 28. It is observed that the plastic region propagation behaviours are clearly different for these two materials. In other words, the velocity of propagation of plastic region from the original stress concentrated area is quite larger in this mild steel than in this Al alloy.

4. Conclusions

Methods for detection of deformation waves up to the range of nonlinear deformation waves are treated in two areas. The first is "One bar method" for material characterization in high velocity tension up to breaking. This method is simple and precise enough for data acquisition in high velocity tension. New facts on high velocity brittleness and high velocity ductility of various metallic and composite materials are found. The second is high-speed photoelasticity of transmission type and of reflection type, as the means of detecting deformation waves in two dimensional fields. The relation of dynamic stress concentration factors vs. corresponding static stress concentration factors in simple tension of a very long strip with a circular hole etc., for elastic region of isotropic and anisotropic (composite) materials, and the plastic region propagation behaviour in stress concentrated area of metallic structures and its difference for a mild steel and an Al alloy, are analyzed clearly.

Acknowledgments

The author wishes to express his hearty thanks for the continued and earnest cooperation in my laboratory to Mr. S. Hashimoto, Mr. A. Hondo, Dr. S. Kuriyama, Dr. K. Kurokawa, Dr. N. Takeda, Mr. N. Kanayama, Mr. M. Kawase, Mr. T. Ide, Mr. T. Oda and other members. Support by the Ministry of Education through Grant-in-Aid for Scientific Research 00385065 and 00542027, the Light Metal Educational Foundation through continued grant, the Japan Society for Composite Materials, Asahi Fiber Glass Co., and Toray Co. are gratefully acknowledged.

References

1. Kawata, K.; Hashimoto, S.; Kurokawa, K.; Kanayama, N.: A new testing method for the characterization of materials in high velocity tension, Mechanical Properties at High Rates of Strain 1979(Conf.Ser. No. 47), Harding, J. ed., Inst. of Phys., Bristol and London, 1979, 71-80.

2. Kawata, K.; Fukui, S.; Seino, J.; Takada, N.: Some analytical and experimental investigations on high velocity elongation of sheet metals by tensile shock, Behaviour of Dense Media Under High Dynamic Pressure(Proc. of Symp. HDP, IUTAM, Paris, 1967), Dunod, Paris, 1968, 313-323.

3. Kuriyama, S.; Kawata, K: Propagation of stress wave with plastic deformation in metal obeying the constitutive equation of the Johnston-Gilman type, J. Appl. Phys., 44 (1973) 3445-3454.

4. Kawata, K.; Hashimoto, S.; Kurokawa, K.: Analysis of high velocity tension of bars of finite length of BCC and FCC metals with their own constitutive equations, High Velocity Deformation of Solids(Proc. of IUTAM Symp., Tokyo, 1977), Springer-Verlag, Berlin, 1979, 1-15.

5. Kawata, K.: Micromechanical study of high velocity deformation of solids, Theoretical and Applied Mechanics(Proc. of 15th Int. Congr. of Theo. and Appl. Mech., Toronto, 1980), North-Holland, 1980, 307-317.

6. Kawata, K.; Hondo, A.; Hashimoto, S.; Takeda, N.; Chung, H. L.: Dynamic behaviour analysis of composite materials, Composite Materials: Mechanics, Mechanical Properties and Fabrication (Proc. of Japan-US Conf. on Comp. Mater., Tokyo, 1981), Japan Soc. for Comp. Mater., Tokyo, 1981, 2-11.

7. Kawata, K.; Hashimoto, S.; Takeda, N.: On high velocity brittleness and high velocity ductility of compocites, Proc. of Solid Shock Technology Symp., University of Tokyo, 1981, 1-6. (in Japanese).

8. Kawata, K.; Hashimoto, S.; Takeda, N.: Mechanical behaviours in high-velocity tension of composites, Proc. of Int. Conf. on Comp. Mater. IV(ICCM-IV, Tokyo, 1982), in printing.

9. Kawata, K.; Hashimoto, S.; Takeda, N.: On mechanical properties in high velocity tension of FRP at low temperature, Proc. of Int. Cryogenic Mater. Conf.(ICMC, Kobe, 1982), in printing.

10. Kawata, K.: High speed photoelastic analysis of stress wave propagation, high velocity deformation, and fracture in impact of solids, Proc. of 13th Int. Congr. on High Speed Photography and Photonics(Tokyo, 1978), Japan Soc. of Precision Engr., Tokyo, 1979, 115-122.

11. Kawata, K.; Hashimoto, S.; Hondo, A.; Ide, T.: On the dynamic stress concentration of a very long strip with notches under high velocity tension, Bull. Inst. of Space and Aeronautical

Science, University of Tokyo, 17, 1(B), 1981, 427-447.

12. Kawata, K.: Three topics in the mechanics of high velocity deformation of solids, Proc. of 30th Japan Nat. Congr. for Appl. Mech. (Tokyo, 1981), University of Tokyo Press, 1981, 3-16.

13. Kawata, K.; Takeda, N.: Photoelastic-coating analysis of dynamic stress concentration for fiber-reinforced composites, Proc. 1982 Joint SESA-JSME Conf. on Exper. Mech. (Hawaii, 1982), SESA, Brookfield, CT, 1982, 405-410.

14. Kawata, K.; Hashimoto, S.; Takeda, N.: Dynamic stress concentrations in some composite strips with a circular hole under high-velocity tension, Proc. of Int. Conf. on Comp. Mater. IV (ICCM-IV, Tokyo, 1982), in printing.

15. Kawata, K.; Hashimoto, S.; Takeda, N.: Dynamic elasto-plastic behaviour of FCC and BCC metals in concentrated stress fields, unpublished.

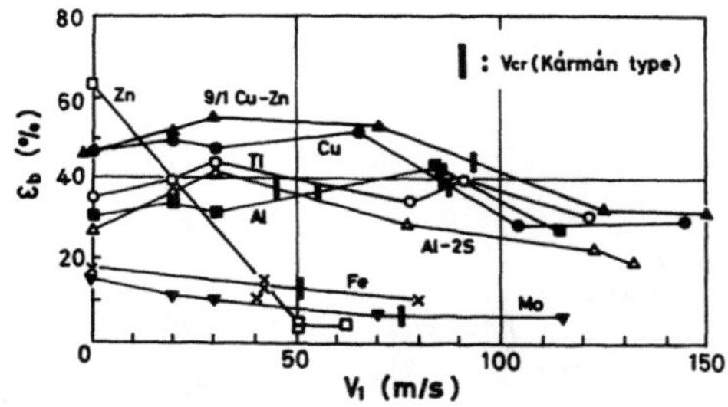

Fig.1. Breaking elongation ε_b vs. tensile velocity V_1 relation in high velocity tension of annealed pure metals and alloys [2]

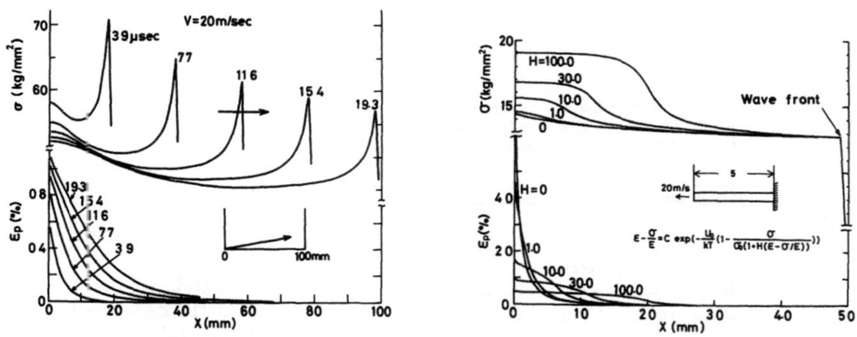

(a) mild steel (b) Aluminum alloy

Fig.2. Examples of theoretical prediction of stress wave form and plastic strain distribution in a bar fixed at the right end in high velocity tension, based upon eq.(1) for a mild steel and eq.(2) for an Aluminum alloy. X: distance from the tension end (left end), σ : stress, and ε_p: plastic strain

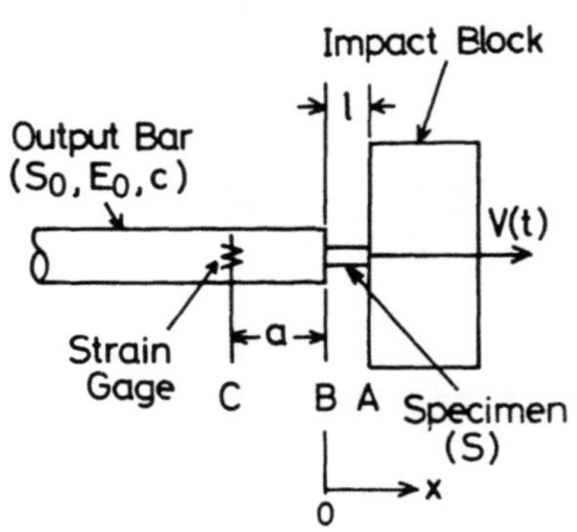

Fig.3. Fundamental system of "one bar method" for material characterization in high velocity tension

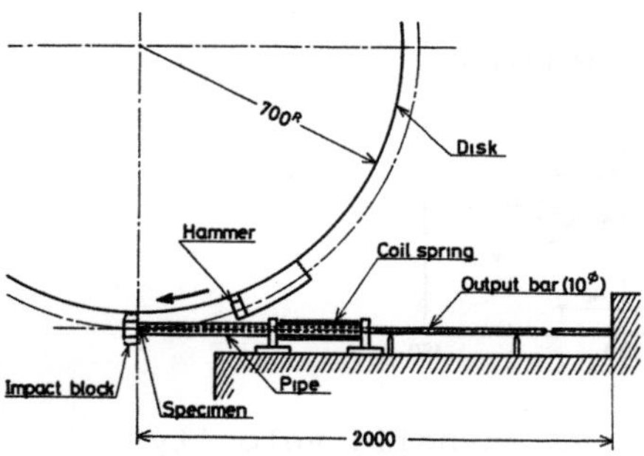

Fig.4. High velocity tensile testing machine A of a rotating disc type [1]

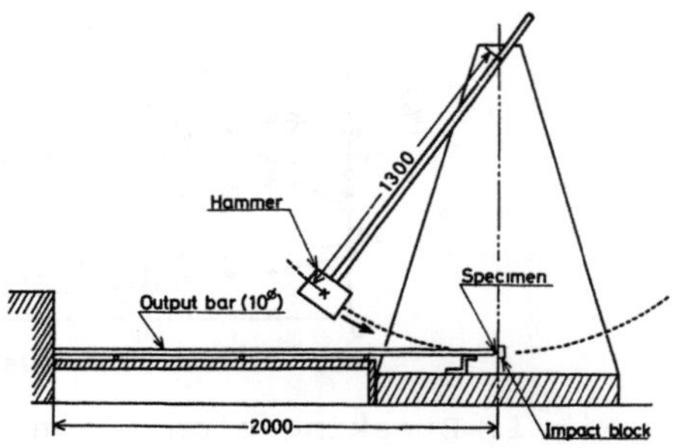

Fig.5. High velocity tensile testing machine B of a pendulum type [1]

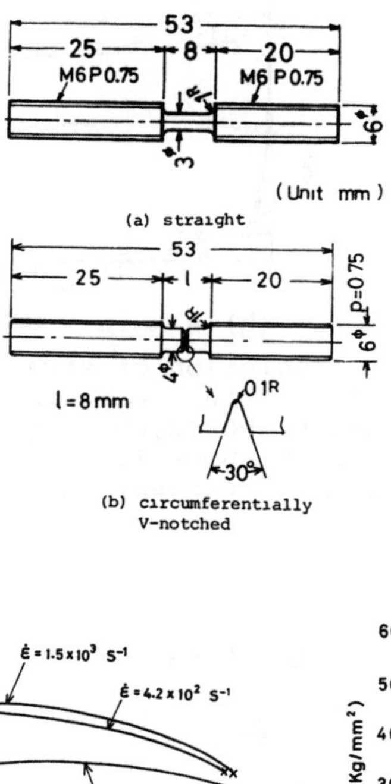

Fig.6. Specimen dimensions
(a) straight
(b) circumferentially V-notched

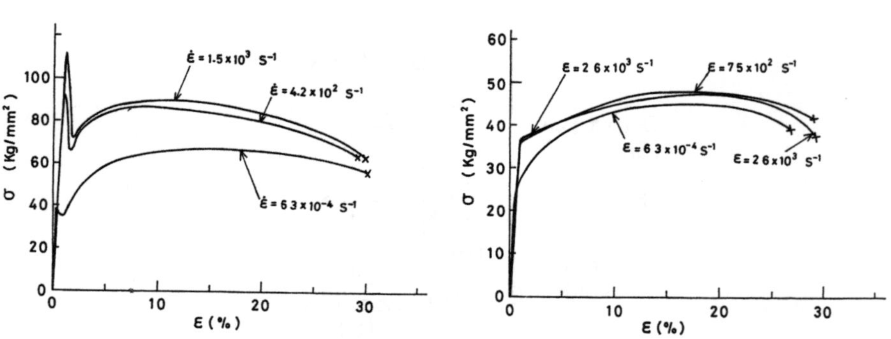

(a) a mild steel S 45C

(b) an Aluminum alloy 2017-T4

Fig.7. Typical experimental results obtained by machine A for (a) a mild steel S 45C and (b) an Aluminum alloy 2017-T4

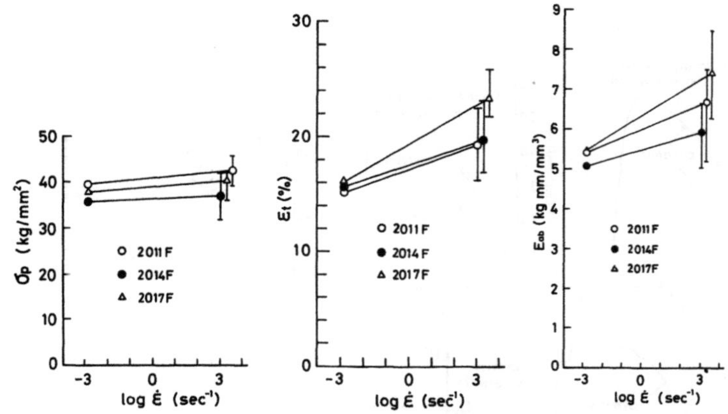

Fig.8. Characteristic behaviour of some Aluminum alloys in high velocity tension (revealing high velocity ductility etc.)

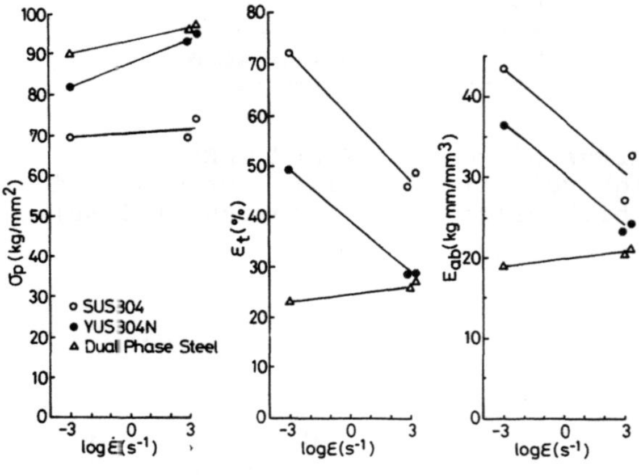

Fig.9. Characteristic behaviour of some steels for structural use in high velocity tension (revealing high velocity brittleness and ductility)

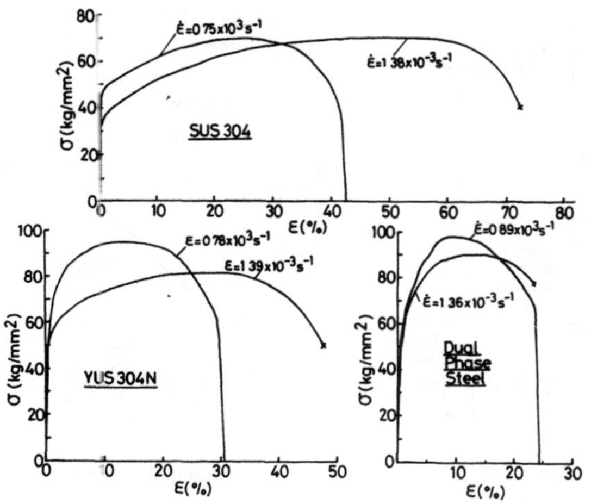

Fig.10. Stress-strain curves for SUS 304, YUS 304N and Dual Phase steel

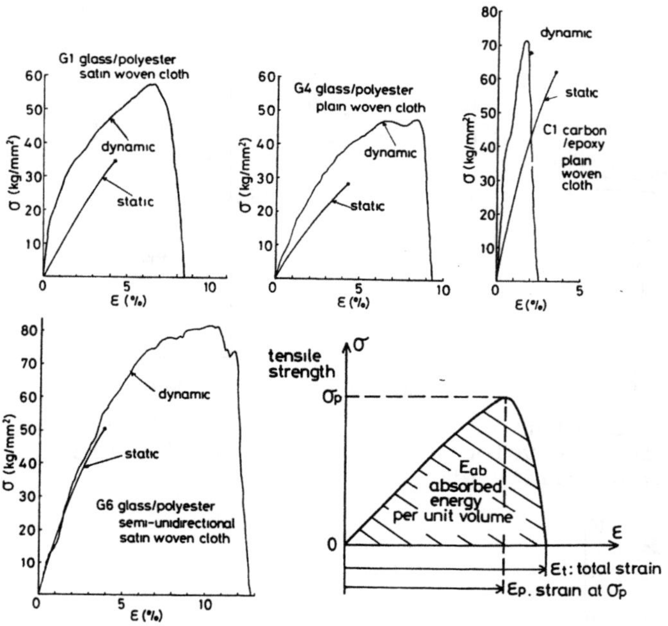

Fig.11. Stress-strain curves of some GFRP and CFRP (revealing remarkable high velocity ductility of GFRP and insensitivity or slight high velocity brittleness of CFRP)

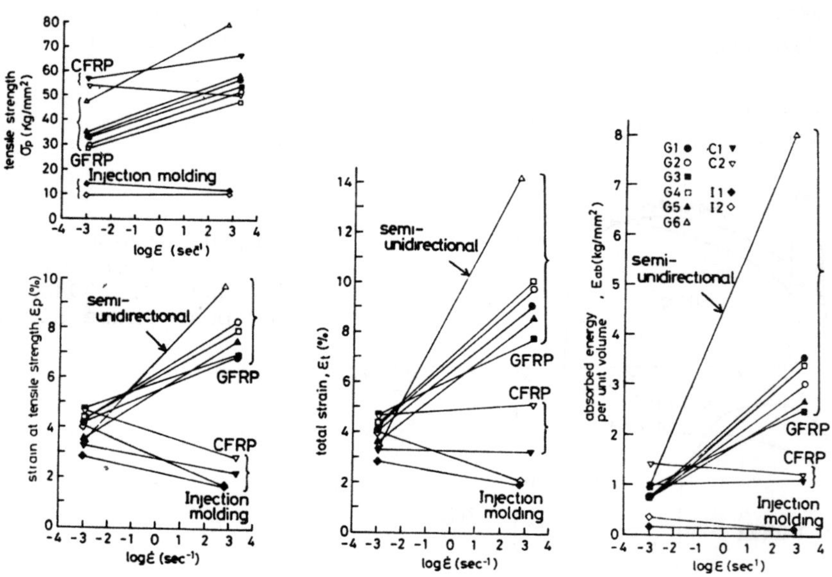

Fig.12. Characteristic behaviour of some composites in high velocity tension

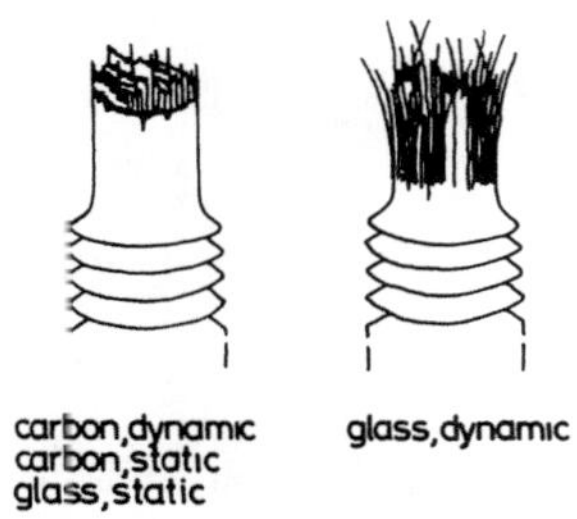

Fig.13. Sketches of fracture appearances of GFRP and CFRP in static and dynamic tension

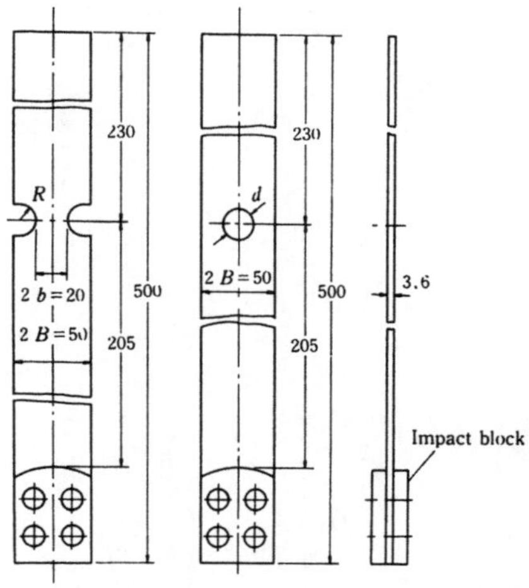

Fig.15. Specimen dimensions (epoxy resin)

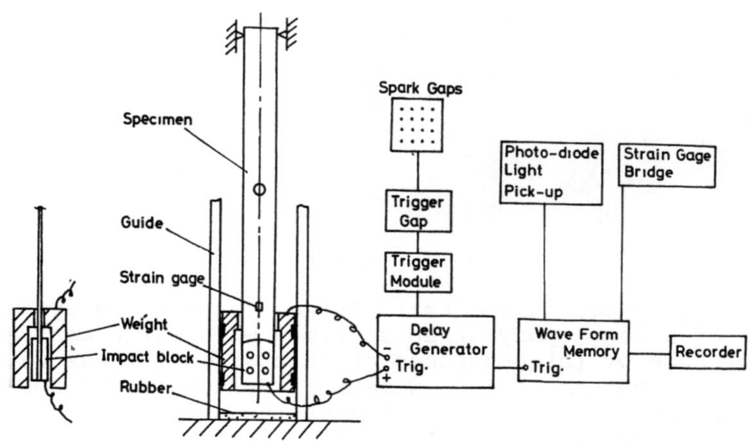

Fig.14. Experimental arrangement for dynamic stress concentration analysis in high velocity deformation of epoxy resin

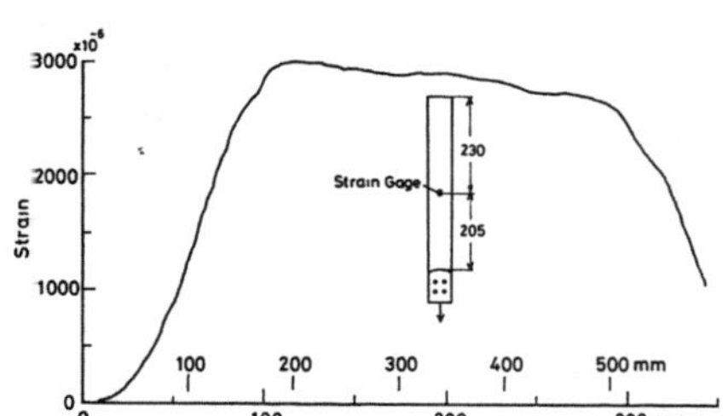

Fig.16. An example of input stress wave form

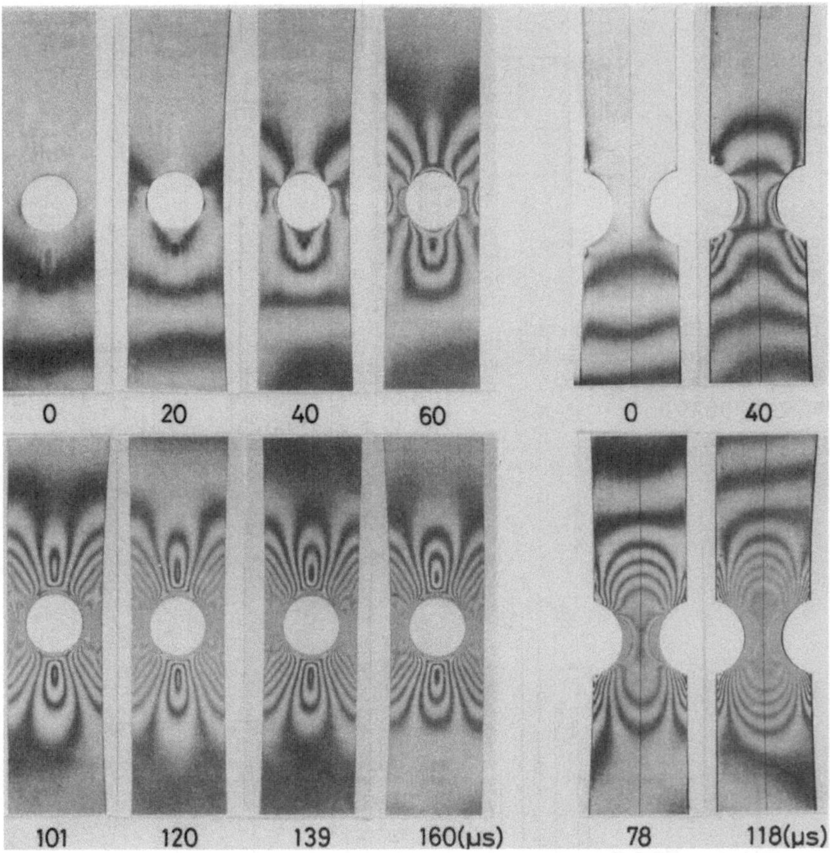

Fig.17. High speed photoelastic fringe patterns in dynamic tension of a long epoxy strip specimen with (a) a central circular hole or (b) U-notches at both sides (Stress wave travels from bottom to top. Numerical values show elapsed time. When wave front passes through the hole, stress concentrating point deviates from the narrowest section to up stresm side.)

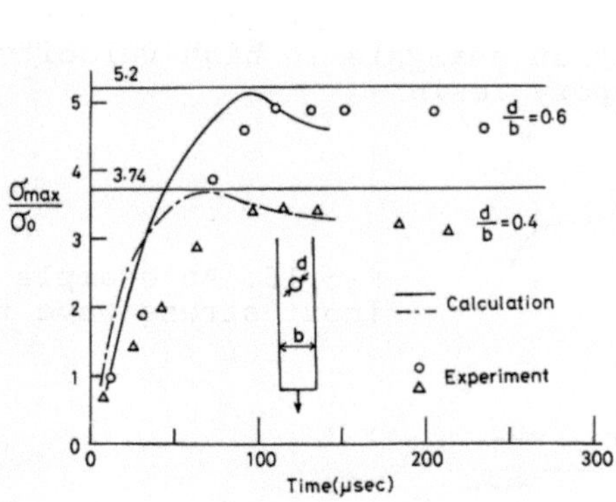

Fig.18. Dynamic stress concentration factor vs. static stress concentration factor relation in dynamic tension of a long strip with a central circular hole

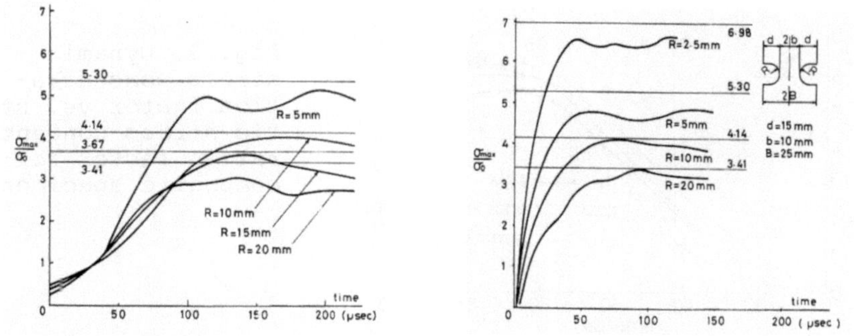

(a) photoelasticity (b) FEM

Fig.19. Dynamic stress concentration factor vs. static stress concentration factor for a long strip with U-notches at both sides under dynamic tension

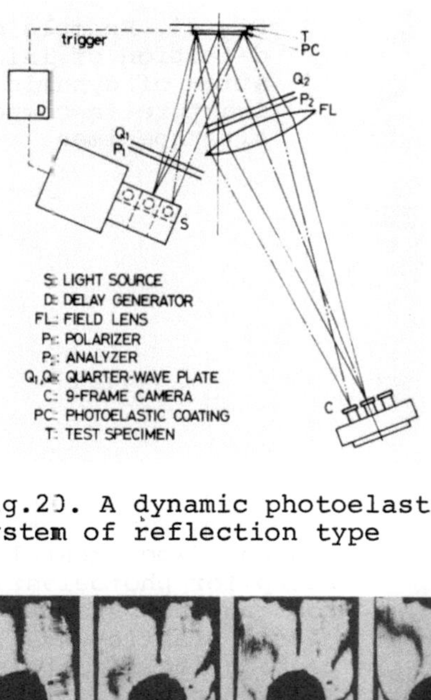

Fig.20. A dynamic photoelastic system of reflection type

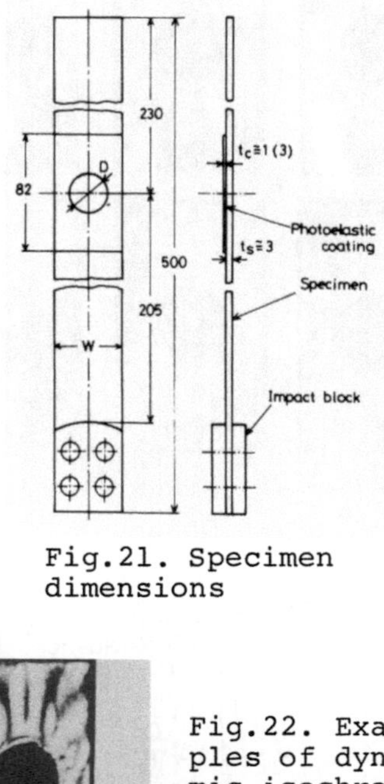

Fig.21. Specimen dimensions

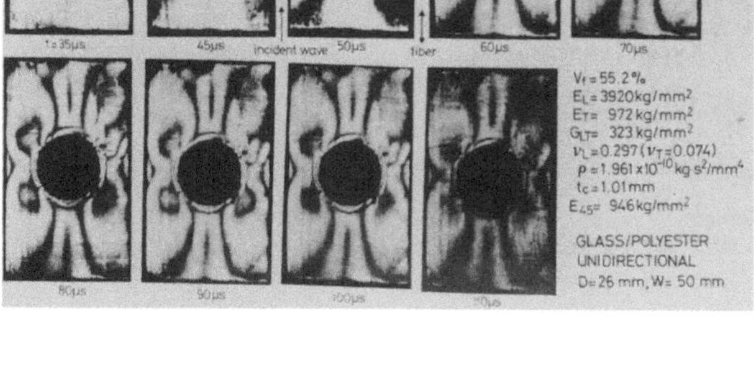

Fig.22. Examples of dynamic isochromatic pattern for a GFRP specimen

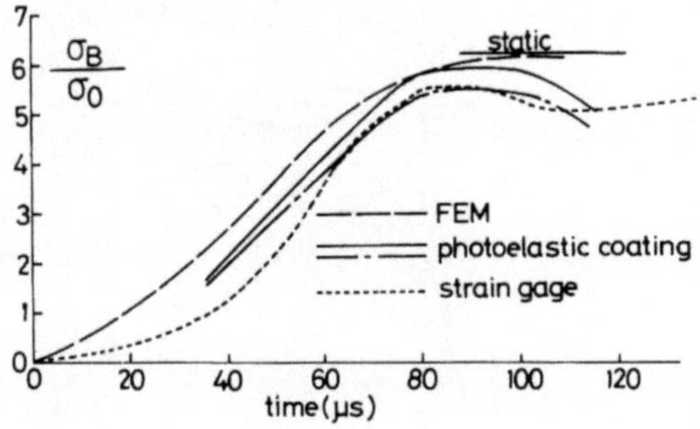

Fig.23. Dynamic stress concentration factor vs. static stress concentration factor for composite specimen

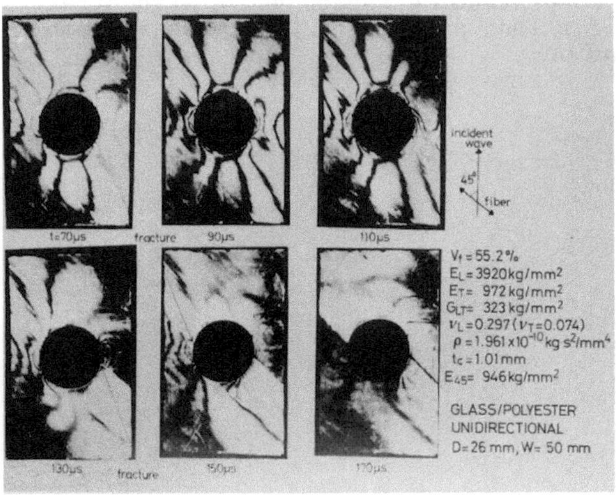

Fig.24. Example of detection of initiation of dynamic fracture in composite specimen

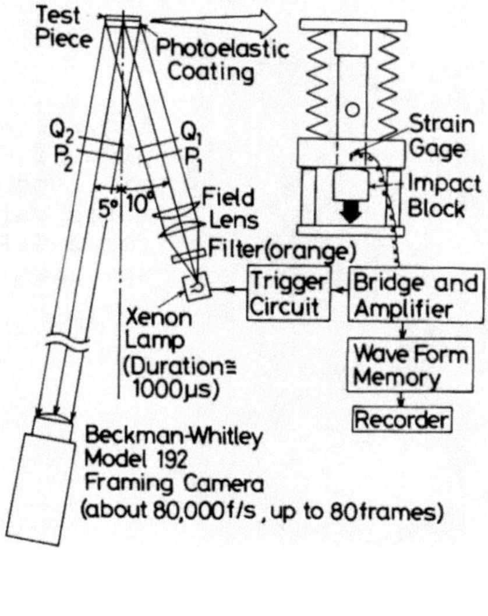

Fig.25. Experimental setup for photoelastic coating analysis of elasto-plastic wave propagation in metallic structures

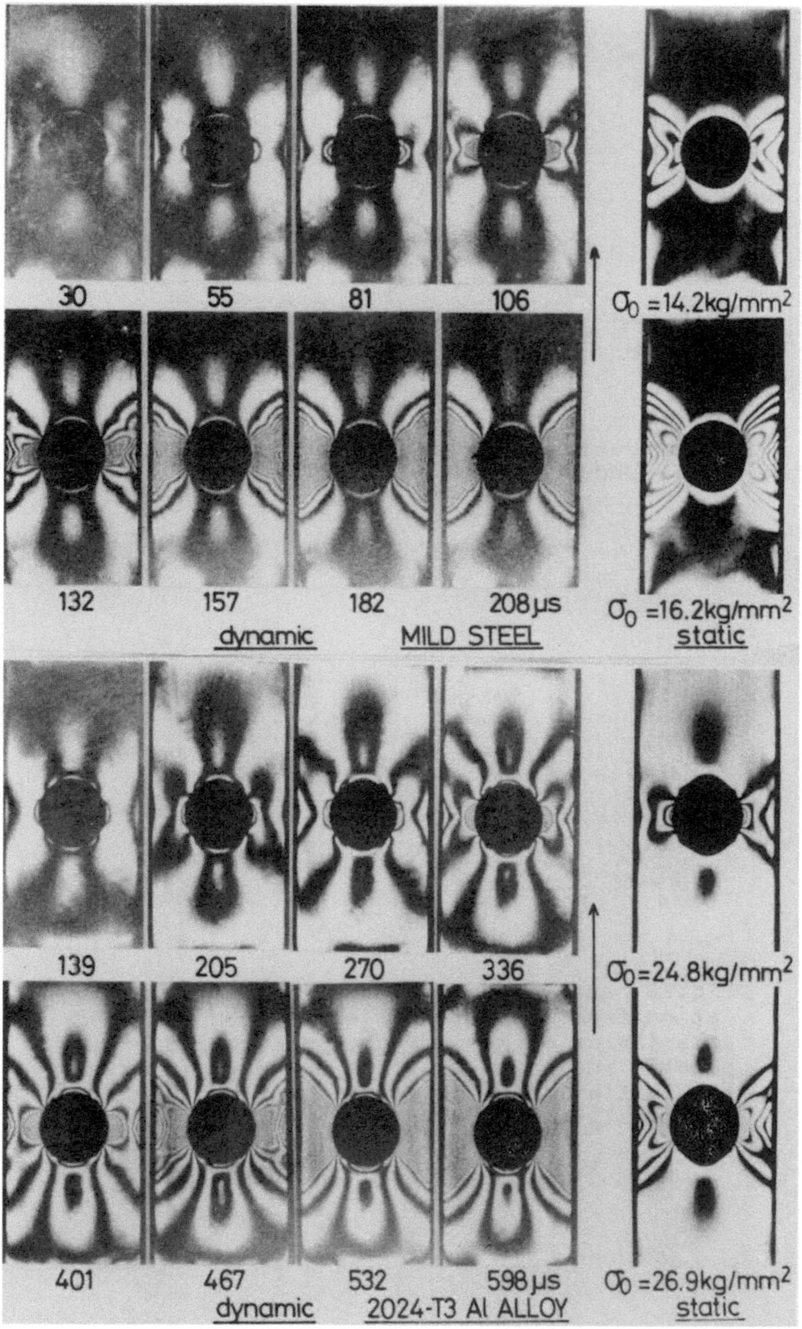

Fig.26. Dynamic photoelastic coating fringe patterns for metallic strip specimens with a central circular hole, under dynamic tension, showing elasto-plastic wave propagation and plastic region propagation

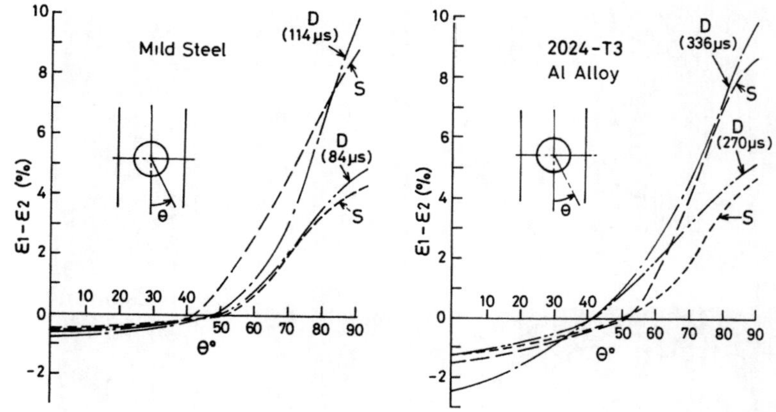

Fig.27. Distributions of principal strain differences at the free boundary of the circular hole

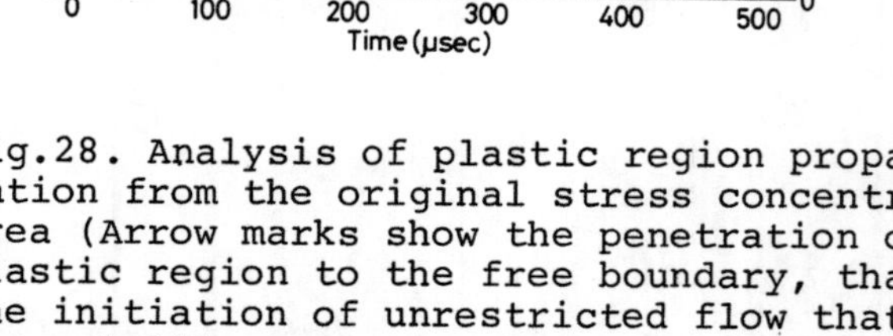

Fig.28. Analysis of plastic region propagation from the original stress concentrated area (Arrow marks show the penetration of the plastic region to the free boundary, that is, the initiation of unrestricted flow that makes the metallic structure unstable.)

Asymptotic Analyses of the Pulse Shape Evolution and of the Inverse Problem of Acoustic Evaluation in Case of the Nonlinear Hereditary Medium

U.NIGUL

Institute of Cybernetics, Academy of Sciences of the Estonian SSR, Akadeemia tee 21, 200 026 Tallinn, USSR

Summary

The propagation of the one-dimensional longitudinal deformation wave type pulses in the nonlinear hereditary medium is considered. The initial shape of the pulse is arbitrary, but the maximum value of its amplitude is supposed to be a small quantity ε. The governing equation is a quasi-linear integro-differential equation, the kernel function of which is supposed to be bounded positive and monotonically decreasing. The integral over the kernel function from zero to infinity is assumed to be another small quantity ε_0. The asymptotic $\varepsilon \to 0$, $\varepsilon_0 \to 0$ solutions are presented for description of a moderate distortion of the pulse shape during its propagation. The approximate evolution equations are derived. The propagation of a pulse, the initial shape of which is nearly sinusoidal, is analysed by presenting the solution as a sum of a stationary and of two transient components. Some aspects of the inverse problem of acoustic evaluation of the properties of the hereditary media are briefly discussed.

1. Statement of the Problem

We shall consider the one-dimensional transient longitudinal deformation waves in the rectangular Lagrangian coordinates. Let us denote by t the time, by X the coordinate, by ρ =const>0 the density of the medium in the natural state at t=0, by U(X,t) the longitudinal displacement, and by σ(X,t) the longitudinal stress. Let the dot denote a derivative with respect to t, and the prime a derivative with respect to X. Then the equation of motion has the form

$$\rho \ddot{U}(X,t) - \sigma'(X,t) = 0 \tag{1}$$

We shall use the initial conditions

$$U(X,0) = 0 \ , \ \dot{U}(X,0) = 0 \tag{2}$$

and define the nonlinear hereditary medium with the aid of the

Nonlinear Deformation Waves
IUTAM Symposium Tallinn 1982
Editors U Nigul, J Engelbrecht

constitutive equation [1,2]

$$Q(U') = E^{-1}[\sigma(X,t) + \varepsilon_0 \int_0^t R(\tau)\sigma(X,t-\tau)d\tau] \qquad (3)$$

where E=const > 0, ε_0=const > 0, $R(t)$ denotes the normalized kernel defined in the interval $0 < t < \infty$,and $Q(U')$ denotes the function of nonlinearity. Supposing the fading memory we assume $R(t)$ to be a bounded, positive and continuously decreasing function which has been normalized (by introducing ε_0) so that

$$\int_0^\infty R(t)dt = 1 \qquad (4) \quad (4)$$

We assume that $Q(U')$ and its derivative

$$q(U') = dQ(U')/dU' \qquad (5)$$

are continuous functions of $U'(X,t)$, and $Q(0)=0$. We also assume that at small values of $|U'(X,t)|$ is valid the expansion

$$q(U') = 1 + k_1U'(X,t) + k_2[U'(X,t)]^2 + \cdots \qquad (6)$$

where $k_1, k_2, \ldots$ denote the coefficients of nonlinearity of the medium. In the limiting case of the linear hereditary medium $q(U') = 1$.

It follows from (2), (3), and from condition $Q(0)=0$ that $\sigma(X,0)=0$.By eliminating $\sigma(X,t)$ from the system of the equations (1) and (3) we obtain the governing equation [2]

$$U^{\cdot\cdot}(X,t) - c^2q(U')U''(X,t) + \varepsilon_0\int_0^t R(\tau)U^{\cdot\cdot}(X,t-\tau)d\tau=0 \qquad (7)$$

where $c^2=E/\rho$.

By making use of the equation (7) and the initial conditions (2) we shall analyse waves which are caused by an input at $X=0$ and propagate in the X-axis positive direction. Of our primary interest will be the finite pulse type inputs

$$U'(0,t) = \varepsilon[H(t) - H(t-t_0)]\psi(t) \qquad (8)$$

$$U^{\cdot}(0,t) = -\varepsilon c[H(t) - H(t-t_0)]\psi(t) \qquad (9)$$

Here ε=const, t_0=const>0, $H(t)$ and $H(t-t_0)$ denote the Heaviside step-functions, and $\psi(t)$ is an arbitrary continuous function which satisfies the conditions

$$\psi(0) = 0, \qquad \psi(t_0) = 0, \qquad \max|\psi(t)| = 1 \qquad (10)$$

and has continuous derivatives.
In some discussions we make use of the semi-infinite pulse type input defined as the limiting case (9) by the formulae

$$U^{\bullet}(0,t) = -\varepsilon c H(t)\Psi(t) \tag{11}$$

$$\Psi(0) = 0, \quad \max|\Psi(t)| = 1 \tag{12}$$

2. The Permanent Shape Wave Type Exact Solutions

In case of the Korteweg-de Vries equation and other mathematical models for the description of nonlinear waves a considerable attention has been paid to the analysis of waves, which propagate without any change of their shape [3]. Let us consider such type of solutions of the equation (7) in case of the input (11). We search them in the form

$$U(X,t) = H(\zeta)u(\zeta) \qquad ,\zeta = t-X/c \tag{13}$$

under restrictions

$$u(0) = 0 \quad , \quad v(0) = 0, \qquad |v^{\bullet}(\zeta)| < \infty \tag{14}$$

where $v(\zeta)=u^{\bullet}(\zeta)$.
The expression (13) satisfies the conditions (2). By substituting it into (7) and taking into consideration that

$$\int_0^t R(\tau)U^{\bullet\bullet}(X,t-\tau)d\tau = \int_{X/c}^{\zeta} R(\zeta-z)H(z)u^{\bullet\bullet}(z)dz =$$

$$= H(\zeta)\int_0^{\zeta} R(\zeta-z)v^{\bullet}(z)dz \tag{15}$$

we obtain the equation

$$\{1 - q[-v(\zeta)/c]\}v^{\bullet}(\zeta) + \varepsilon_0\int_0^{\zeta} R(\zeta-z)v^{\bullet}(z)dz = 0 \tag{16}$$

where $\zeta > 0$. According to the second assumption (14) the solution of this equation is to be constructed by making use of the condition $v(0)=0$. From the physical point of view the equation (16) requires complete mutual compensation of the influence of the nonlinear and hereditary effects. It might be proved easily that at the front $\zeta=0$ the equation (16) is satisfied due to the restrictions laid by us on the functions $R(t)$, $q(U')$ and $v(\zeta)$. If the function $q(U')$ is given in the form

$$q(U') = 1 + k_1 U'(X,t) \tag{17}$$

then we obtain from (16) the equation

$$k_1 v(\zeta) v^{\bullet}(\zeta) + \varepsilon_0 c \int_0^{\zeta} R(\zeta - z) v^{\bullet}(z) dz = 0 \tag{18}$$

Example: Let us consider the special case of the standard visco-elastic body defined with the aid of the kernel function [2]

$$R(t) = \frac{1}{\tau_0} \exp(-t/\tau_0) \quad , \tau_0 > 0 \tag{19}$$

Then the equation (20) has the exact solution

$$v(\zeta) = 2\varepsilon_0 (c/k_1)[\exp(-\zeta/2\tau_0) - 1] \tag{20}$$

and the permanent shape wave

$$U^{\bullet}(X,t) = H(\zeta) v(\zeta) \tag{21}$$

is caused by the input (11) if

$$\varepsilon = -2\varepsilon_0/k_1 \quad , \quad \Psi(t) = \exp(-\zeta/2\tau_0) - 1 \tag{21}$$

This input might be interpreted as a special kind of loading which causes at X=0 a smooth growth of the velocity from zero to a constant value.

3. <u>Method of Successive Integration of the Linear Inhomogeneous Wave Equations</u>

Let us assume that ε_0 in equation (7) and ε in boundary conditions (8), (9), (11) are small in the sense

$$\varepsilon_0' << 1 \quad , \quad |\varepsilon| << 1 \tag{23}$$

According to the physical considerations [2] the application of the model of the nonlinear hereditary medium is just of the main interest in case (23). We shall consider the inputs (8), (9) under the restrictions (23). Then the distortion of the pulse shape due to the nonlinear and hereditary effects during its propagation develops slowly and there exists a region $0<X<L$ for which the asymptotic presentations of the solution might be obtained with the aid of the method of successive integration of the linear inhomogeneous wave equations. The basic idea of this

method has been pointed out already by Lamb [4]. Further, it has been used by several authors [5,6] in special cases of the governing quasi-linear differential equation. The author of this paper has suggested an elaborated version of this method [7] and has applied it in analyses of problems where the governing equation is a quasi-linear differential or integro-differential equation [8-10]. We shall briefly reproduce here the scheme of calculations.

The governing equation is presented in the form

$$\ddot{U}(X,t) - c^2U''(X,t) = G(X,t) \tag{24}$$

In case of the equation (7) we have

$$G(X,t) = c^2U''(X,t)[q(U') - 1]- \\ -\varepsilon_0 H(\zeta)\int_0^{\zeta} R(\zeta-y)\ddot{U}(X,y + c^{-1}X)dy \tag{25}$$

The zero-order asymptotic solution $U_o(X,t)$ is constructed by the integration of the elementary linear wave equation

$$\ddot{U}_0(X,t) - c^2U_0''(X,t) = 0 \tag{26}$$

All the next asymptotic solutions $U_j(X,t)$, $j \in 1,2,...,$are calculated by the successive integration of the linear inhomogeneous wave equations

$$\ddot{U}_j(X,t) - c^2U_j''(X,t) = G_j(X,t) \tag{27}$$

where $G_j(X,t)$ denotes the result of the calculation of the function $G(X,t)$ by making use of the approximation $U(X,t) \approx U_{j-1}(X,t)$. The functions $G_j(X,t)$ describe waves. By carrying out the calculation on the basis of the expression (25) in case of the input (8) or (9) it might be shown[9,10]that the functions $G_j(X,t)$ have the structure

$$G_j(X,t) = H(\zeta)g_{1j}(X,t) - H(\zeta - t_0)g_{2j}(X,t) \tag{28}$$

where $g_{1j}(X,t)$ and $g_{2j}(X,t)$ denote the continuous functions which are explicitly expressed through the kernel function $R(t)$ and through the input function $\psi(t)$. Since we have introduced $R(t)$ and $\psi(t)$ as arbitrary functions satisfying certain conditions, then we have to carry out the integration of the equations (27) with the right-hand-sides (28), where $g_{1j}(X,t)$ and $g_{2j}(X,t)$ are arbitrary functions having the property

$$g_{1j}(X,t) = 0 \quad \text{at } \zeta=0, \quad g_{2j}(X,t) = 0 \quad \text{at } \zeta=t_0 \tag{29}$$

which follows from the assumed properties of the functions $R(t)$ and $\psi(t)$. The exact solution obtained with the aid of the Laplace integral transform is the following [8] :

$$U_j'(X,t) = \varepsilon[H(\zeta) - H(\zeta-t_0)]\psi(\zeta) + \\ + \tfrac{1}{2}H(\zeta)[-TY_{1j}^{(1)} - Y_{1j}^{(2)} + Y_{1j}^{(3)}] - \tfrac{1}{2}H(\zeta-t_0)[-TY_{2j}^{(1)} - Y_{2j}^{(2)} + Y_{2j}^{(3)}] \tag{30}$$

$$\dot{U}_j(X,t) = -\varepsilon c[H(\zeta) - H(\zeta-t_0)]\psi(\zeta) + \\ + \tfrac{c}{2}H(\zeta)[TY_{1j}^{(1)} + Y_{1j}^{(2)} + Y_{1j}^{(3)}] - \tfrac{c}{2}H(\zeta-t_0)[TY_{2j}^{(1)} + Y_{2j}^{(2)} + Y_{2j}^{(3)}] \tag{31}$$

$$Y_{ij}^{(1)} = \int_0^{a_i} g_{ij}(y,\zeta-c^{-1}y)\,dy, \quad Y_{ij}^{(2)} = \int_0^{X} g_{ij}(y,\zeta+c^{-1}y)\,dy$$

$$Y_{ij}^{(3)} = \int_0^{b_i} g_{ij}(y,t + c^{-1}X - c^{-1}y)\,dy \quad , \; i = 1,2 \tag{32}$$

$$a_1 = \frac{1}{2}(ct - X) \quad , \; a_2 = \frac{1}{2}(ct - X - ct_0)$$

$$b_1 = \frac{1}{2}(ct + X) \quad , \; b_2 = \frac{1}{2}(ct + X - ct_0) \tag{33}$$

Here and further $T=1$ in case of the input (8) and $T=-1$ in case of the input (9).

The formulae (30) - (33) are an important tool of application of the method being considered in general case of the functions $R(t)$ and $\psi(t)$. As a matter of fact we mention that the exact solution of the wave equation with a right-hand-side which describes a wave moving with decreasing velocity, is available in [11] .

4. Asymptotic Solutions

By making use of the formulae of the previous section and the expansion (6) the first-order and the second-order asymptotic solutions of the problem, formulated in section 1, have been derived by the author in [2,9,10].

We reproduce some results here.

The first-order asymptotic solution might be presented in the form

$$U'_1(X,t) = \varepsilon V'_{00}(X,t)+\varepsilon^2 V'_{10}(X,t) +\varepsilon\varepsilon_0 V'_{01}(X,t)$$

$$\dot{U}_1(X,t) = \varepsilon \dot{V}_{00}(X,t)+\varepsilon^2 \dot{V}_{10}(X,t) +\varepsilon\varepsilon_0 \dot{V}_{01}(X,t) \tag{34}$$

The first term in the formulae (34) is the solution of the equation (26) defined by the expression

$$V'_{00}(X,t) = -c^{-1}\dot{V}_{00}(X,t) = [H(\zeta) - H(\zeta -t_0)]\psi(\zeta) \tag{35}$$

The second term in (34) determines the first-order asymptotic correction due to the nonlinear effects through the functions

$$V'_{10}(X,t) = [H(\zeta)-H(\zeta -t_0)][-c^{-1}XA_{10}(\zeta)+(1-T)B_{10}(\zeta)] \tag{36}$$

$$\dot{V}_{10}(X,t) =c[H(\zeta)-H(\zeta -t_0)][c^{-1}XA_{10}(\zeta)+(1+T)B_{10}(\zeta)] \tag{37}$$

$$A_{10}(\zeta) = - \frac{1}{2}k_1\psi(\zeta)\psi^{\cdot}(\zeta) \ , \quad B_{10}(\zeta) = - \frac{1}{8}k_1\psi^2(\zeta) \tag{38}$$

According to this correction term the nonlinear effects cause a distortion of the pulse shape, but they do not change the length of the pulse.

The third term in (34) defines the first-order correction due to the hereditary effects with the aid of the functions

$$V'_{01}(X,t) = H(\zeta)[-c^{-1}XA_{01}(\zeta) + (1-T)B_{01}(\zeta)]-$$

$$-H(\zeta -t_0) \ [-c^{-1}X\tilde{A}_{01}(\zeta) + (1-T)\tilde{B}_{01}(\zeta)] \tag{39}$$

$$\dot{V}_{01}(X,t) = cH(\zeta)[c^{-1}XA_{01}(\zeta) + (1+T)B_{01}(\zeta)] -$$

$$-H(\zeta -t_0)[c^{-1}X\tilde{A}_{01}(\zeta) \ +(1+T)\tilde{B}_{01}(\zeta)] \tag{40}$$

$$A_{01}(\zeta)= \frac{1}{2}\int_0^{\zeta}R(\zeta -y)\psi^{\cdot}(y)dy \ , \quad \tilde{A}_{01}(\zeta)= \frac{1}{2}\int_{t_0}^{\zeta}R(\zeta-y)\psi^{\cdot}(y)dy \tag{41}$$

$$B_{01}(\zeta)= \frac{1}{4}\int_0^{\zeta}R(\zeta -y)\psi(y)dy \ , \quad \tilde{B}_{01}(\zeta)= \frac{1}{4}\int_{t_0}^{\zeta}R(\zeta-y)\psi(y)dy \tag{42}$$

It follows from (39) - (42) that the distortion of the pulse shape due to the hereditary effects is connected with the growth of the length of the pulse during its propagation. Let us consider the shape of the pulse in a fixed point X=const > 0. Then in the interval $0 < \zeta < t_0$ the first term in the formulae (39), (40) describes the distortion of the shape of the pulse of initial length and the second term is zero, but at $\zeta > t_0$ the sum of the two terms is not zero and describes the formation of the tail of the pulse. Due to the fact that we have assumed fading memory, the amplitude of the tail quickly decreases with the growth of ζ . In case of the semi-infinite input (11) in (39) and (40) the terms containing $H(\zeta - t_0)$ are to be dropped, T=-1, and $\psi(\zeta)$ is to be replaced by $\Psi(\zeta)$. In this case the sum of the nonlinear and hereditary correction terms is zero, if to choose $-\varepsilon c\Psi(\zeta)$ equal to the function $v(\zeta)$, which satisfies the equation (16).

In the general case of the input (8) or (9) the first order nonlinear and hereditary correction terms in comparison with the zero-order term might be correspondingly estimated by quantities $\eta_1 L$ and $\theta_1 L$, if to define

$$\eta_1 = \frac{1}{2}\varepsilon k_1 \Omega_1 , \quad \theta_1 = \frac{1}{2}\varepsilon_0 R_0 \tag{43}$$

$$\Omega_1 = \max|\psi^{\bullet}(t)| , \quad R_0 = \max\left|\int_0^t R(t-\tau)\psi^{\bullet}(\tau)d\tau\right| \tag{44}$$

Therefore, the asymptotic solutions are valid in region

$$0 < X < L, \quad \eta_1 L < 1 , \quad \theta_1 L < 1 \tag{45}$$

The second-order asymptotic solution has the structure[2,9,10]

$$U_2'(X,t) = U_1'(X,t) + \varepsilon^3 V_{20}'(X,t) + \varepsilon^2 \varepsilon_0 V_{11}'(X,t) + \\ + \varepsilon\varepsilon_0^2 V_{02}'(X,t) \tag{46}$$

The functions through which the second-order correction terms are expressed have been derived in [2,9,10]. Here we draw attention to the fact that the leading terms of these functions are proportional to X^2.

5. Evolution Equations

In the theory of nonlinear waves considerable attention has

been paid to the derivation of the simplified evolution equations from the given governing equation (or the system of equations).

Let us define

$$\tilde{U}(X,\zeta) \equiv U(X,t) \;, W(X,\zeta) \equiv U_\zeta(X,\zeta) \;, \; \zeta = t - X/c \tag{47}$$

and denote derivatives by the indices. Then

$$U'(X,t) = \tilde{U}_x(X,\zeta) - c^{-1}W(X,\zeta), \quad \dot{U}(X,t) = W(X,\zeta) \tag{48}$$

By making use of these formulae and the expansion (6) we obtain from (7) the following equation [10]

$$K_1(X,\zeta) + K_2(X,\zeta) + \cdots = 0 \tag{49}$$

where

$$K_1(X,\zeta) = k_1 W(X,\zeta)W_\zeta(X,\zeta) + 2c^2 W_x(X,\zeta) + + \varepsilon_0 cH(\zeta)\int_0^\zeta R(\zeta-y)W_\zeta(X,y)\,dy \tag{50}$$

$$K_2(X,\zeta) = c^3\tilde{U}_{xx}(X,\zeta) + 2k_1 cW(X,\zeta)W_x(X,\zeta) + + k_1 c\tilde{U}_x(X,\zeta)W_\zeta(X,\zeta) + k_2 c^{-1}W^2(X,\zeta)W_\zeta(X,\zeta) \tag{51}$$

As shown in [10] $K_1(X,\zeta)$ is the sum of the first-order terms and $K_2(X,\zeta)$ is the sum of the second-order terms. The proof given in [10] is based on derivation of the asymptotic presentations of $\tilde{U}_x(X,\zeta)$, $W(X,\zeta)$, $\tilde{U}_{xx}(X,\zeta)$, $W_x(X,\zeta)$, and $W_\zeta(X,\zeta)$ by making use of the expressions (48) and the asymptotic solutions described in the previous section.

Therefore, the equations

$$K_1(X,\zeta) = 0 \tag{52}$$

$$K_1(X,\zeta) + K_2(X,\zeta) = 0 \tag{53}$$

might be interpreted as the first-order and the second-order evolution equations. We present here the following first-order asymptotic expressions valid in the region (45):

$$\tilde{U}_x(X,\zeta) = 2\varepsilon H(\zeta)[\varepsilon B_{10}(\zeta) + \varepsilon_0 B_{01}(\zeta)] - - 2\varepsilon H(\zeta-t_0)[\varepsilon B_{10}(\zeta) + \varepsilon_0 \tilde{B}_{01}(\zeta)] \tag{54}$$

$$W(X,\zeta) = \varepsilon cH(\zeta)[-\psi(\zeta) + \varepsilon c^{-1}XA_{10}(\zeta) + \varepsilon(1+T)B_{10}(\zeta) +$$
$$+ \varepsilon_0 c^{-1}XA_{01}(\zeta) + \varepsilon_0(1+T)B_{01}(\zeta)] -$$
$$- \varepsilon cH(\zeta - t_0)[-\psi(\zeta) + \varepsilon c^{-1}XA_{10}(\zeta) + \varepsilon(1+T)B_{10}(\zeta) +$$
$$+ \varepsilon_0 c^{-1}X\tilde{A}_{01}(\zeta) + \varepsilon_0(1+T)\tilde{B}_{01}(\zeta)] \qquad (55)$$

$$W_X(X,\zeta) = \varepsilon H(\zeta)[\varepsilon A_{10}(\zeta) + \varepsilon_0 A_{01}(\zeta)] -$$
$$- \varepsilon H(\zeta - t_0)[\varepsilon A_{10}(\zeta) + \varepsilon_0 \tilde{A}_{01}(\zeta)] \qquad (56)$$

The expression (55) is the asymptotic solution of the equation (52) in the region (45). For obtaining a better asymptotic solution of the equation (7) in this region one has to use the equation (53). Let ε_* denote the greatest of the constants ε_o and $|\varepsilon|$. Then on the basis of the expressions (54) - (56) and of the corresponding second-order expressions [10] one can prove the estimations

$$\tilde{U}_X(X,\zeta) \sim \varepsilon_* W(X,\zeta) \quad , \; \tilde{U}_{XX}(X,\zeta) \sim \varepsilon_*^2 W_\zeta(X,\zeta),$$
$$W_X(X,\zeta) \sim \varepsilon_* W_\zeta(X,\zeta) \qquad (57)$$

If to introduce these estimations as assumptions (as it is often done in the theory of nonlinear waves) then it is possible to derive the evolution equations (52) and (53) from(7) without making use of the asymptotic expressions of the previous section.

6. Evolution of the Sine Pulses

Let us consider the evolution of the pulse shape in case of the input

$$U(0,t) = - \varepsilon c\Omega^{-1}[H(t)P(t) - H(t-t_0)P(t-t_0)]\sin \Omega t$$
$$\Omega = 2n\pi/t_0 \qquad (58)$$

where n is an integer and P(t) denotes a function introduced for obtaining continuous $U^{\cdot}(0,t)$. We shall assume that P(t) is zero at t=0 and quickly (practically with an interval $0 < t < a$, $a << t_o$) grows to the constant value, equal to unity. For example $P(t)=1-\exp(-t/r)$, $r << t_o$ might be used. Then in $U^{\cdot}(0,t)$ dominates the cosine pulse of the duration t_o.

Let us define
$$f(t) = \int_0^t R(\tau)\cos[(t-\tau)\Omega]d\tau \qquad (59)$$

Then, dependently on the required accuracy of calculations and on the given kernel function, it is possible to establish such a constant b that

$$f(t) \approx f(\infty) \quad \text{if} \quad t > b \tag{60}$$

We shall consider the case $b < t_0$. Under the described restrictions the solution might be presented in the form

$$U(X,t) = H(\zeta)F_1(X,\zeta) + H(\zeta - t_0)F_2(X,\zeta) +$$
$$+ [H(\zeta) - H(\zeta - t_0)]F_0(X,t) \tag{61}$$

The functions in this presentation might be characterized by their following properties at X=const: $F_1(X,\zeta)$ describes the nonstationary part of the solution which decreases in the near--front region $0 < \zeta < b$ and is negligible at $\zeta > b$; $F_2(X,\zeta)$ describes the tail of the pulse at $\zeta > t_0$ and is negligibly small when $\zeta > t_0 + b$; $F_0(X,t)$ describes the stationary part of the solution, what is a periodic function of t in the interval $0 < t < t_0$ and might be separately calculated by replacing the equation (7) with the following equation

$$\ddot{F}_0(X,t) - c^2 q[F_0'(X,t)]F_0''(X,t) + \varepsilon_0 \int_0^\infty R(\tau)\ddot{F}_0(X,t-\tau)d\tau = 0 \tag{62}$$

We shall briefly outline a scheme of derivation of the asymptotic presentations of the functions $F_1(X,\zeta)$, $F_2(X,\zeta)$, $F_0(X,t)$. In this section we shall consider the solution of the corresponding linear problem $q(U')=1$ as the first-order asymptotic at $\varepsilon \to 0$ solution of the nonlinear problem. Let $F_{11}(X,\zeta)$, $F_{21}(X,\zeta)$, $F_{01}(X,t)$ denote correspondingly $F_1(X,\zeta)$, $F_2(X,\zeta)$ and $F_0(X,t)$ in the linear problem.

We shall point out two equivalent methods of constructing them.

According to the first method $F_{01}(X,t)$ is constructed as the solution of the linear equation obtained from (62). It belongs to a type of investigated equations[12] and has the exact solution

$$F_{01}(X,t) = -\varepsilon c\Omega^{-1}\sin\{\Omega[t-\beta(\Omega)X/c]\}\exp[-\alpha(\Omega)\Omega X/c] \tag{63}$$

$$\alpha(\Omega) = \{\tfrac{1}{2}[-M(\Omega)] + N(\Omega)]\}^{1/2}, \quad \beta(\Omega) = \{\tfrac{1}{2}[M(\Omega) + N(\Omega)]\}^{1/2} \tag{64}$$

$$M(\Omega) = 1 + \varepsilon_0 R_c(\Omega) \quad , \quad N(\Omega)=\{[M(\Omega)]^2+[\varepsilon_0 R_s(\Omega)]^2\}^{1/2} \tag{65}$$

$$R_c(\Omega) = \int_0^\infty R(t)\cos \Omega t dt \quad , \quad R_s(\Omega) = \int_0^\infty R(t)\sin \Omega t dt \tag{66}$$

This solution describes the propagation of the sine waves in the linear hereditary medium.

By making use of the expansion of this solution in the near--front region, where ζ is small, and by subtracting it from the asymptotic solution of section 4 for the linear case, the asymptotic at $\varepsilon_o \to 0$ presentation of $F_{11}(X,\zeta)$ is obtained. The function $F_{21}(X,\zeta)$ is constructed in the similar manner.

The other method of obtaining the equivalent results is based on making use of the Laplace integral transform $U(X,t) \div U^L(X,s)$. The solution might be constructed as the sum of two components: first, the contribution of the poles $s=i\Omega$, $s=-i\Omega$; second, the contribution of the branch points and of the function $P^L(s)$. It might be shown easily that the former is equal to $F_{01}(X,t)$ defined by (63). In general case of $R(t)$ the latter cannot be calculated straightforwardly with the aid of the method of deformation of the integration path in the s-plane, often used [2,13] in special cases of $R(t)$. However, the asymptotic at $s \to \infty$ inversion gives the formulae of the section 4 for the linear problem. Therefore, the calculation of $F_{11}(X,\zeta)$ and $F_{21}(X,\zeta)$ might be carried out in the same way as in case of the first method.

The second-order asymptotic solution, described by the functions $F_{o2}(X,t)$, $F_{12}(X,\zeta)$, $F_{22}(X,\zeta)$, might be constructed in the following way. By making use of the function of nonlinearity (17) it is possible to construct $F_{o2}(X,t)$ as the solution of the equation (62) in the form of a sum of the "first harmonic" (63) and of the "second harmonics". That has been done by M. Itskovitsch and by the author in a recent study, the results of which will be published elsewhere. Further, $F_{12}(X,\zeta)$ and $F_{22}(X,\zeta)$ might be constructed by making use of the asymptotic expressions described in the section 4.

In the nonlinear acoustics is of special interest the case when in the estimation (60) $b \ll t_o$ and the stationary part of the

solution $F_0(X,t)$ dominates in the large "central" part ($b < \zeta < t_0$) of the travelling pulse and is the subject of recording and analyses. According to (63) - (66) the "first harmonic" is dependent on $R(t)$ only through its Fourier sine and cosine transforms $R_s(\Omega)$, $R_c(\Omega)$ at the given value of Ω , and similarly the "second harmonics" are dependent on $R_s(2\Omega)$, $R_c(2\Omega)$. Consequently, the measurement of the stationary part of the solution at a single value of Ω gives rather poor information on $R(t)$. However, if $R(t)$ is known, then the recording of the "second" and "third harmonics" is correspondingly sufficient for calculation of the values of the coefficients of nonlinearity k_1 and k_2.
This conclusion is important from the point of view of the subject of the next section.

7. Inverse Problem of Acoustic Evaluation

Let us accept the following assumptions:

1) The shape of the pulse has been recorded at X=0 in the form (11) and at $X=L_0$ in the form

$$U^{\bullet}(L_0,t) = -\varepsilon c H(\zeta)\Phi(\zeta) \qquad (67)$$

where $\zeta=t-\zeta_0$, $\zeta_0=L_0/c$, and $\Phi(0)=0$

2) The value of L_0 is given and the value of ζ_0 has been extracted from the data (67), so that the values of c and ε might be calculated on the basis of the formula $c=L_0/\zeta_0$ and the condition (12).

3) The equation (7) together with the expansion (6) is a proper mathematical model of the medium.

Then the inverse problem of acoustic evaluation is in calculation of the function $K(t)\equiv\varepsilon_0 R(t)$ and of the coefficients k_1, k_2, ..., on the basis of the given data. This problem is a subject of the current study in nonlinear acoustics. Two basic types of pulse technique might be specified for the nonlinear hereditary media (we shall not concern here the stationary technique described, for example, in [12]).

Technique 1. Two experiments are carried out. In the first

experiment a weak broad-band input is used, in case of which the nonlinear effects are negligible but hereditary effects are well pronounced and the linear model of hereditary medium is valid for calculation of the kernel function $K(t)$.
In the second experiment a strong input is used, in case of which the nonlinear effects are well pronounced and the model of the nonlinear hereditary medium is used for calculation $k_1, k_2, \ldots$ by making use of the known $K(t)$. It is sufficient to use nearly sinusoidal pulses and record the higher order "harmonics" in the second experiment (see section 6).

Technique 2. One experiment is carried out by making use of a strong broad-band input, in case of which both the nonlinear and hereditary effects are well pronounced. The interpretation of the experimental data is carried out by making use of an approximation of the kernel function in a priori given form.

In connection with the Technique 1 we shall briefly concern the problem of calculation $K(t) \equiv \varepsilon_o R(t)$ in case of the linear medium.

By making use of the Laplace transform we obtain

$$\Phi^L(s)\exp(-\zeta_0 s) = \Psi^L(s)\exp\{-\zeta_0 s[1+K^L(s)]^{1/2}\} \tag{68}$$

It follows from (68) that

$$K^L(s) = \{[\ln A(s)]/\zeta_0 s\}^2 - 2[\ln A(s)]/\zeta_0 s \tag{69}$$

$$A(s) = \Phi^L(s)/\Psi^L(s) \tag{70}$$

Similar formulae might be derived with the aid of Fourier transform and applied for calculation $K(t)$ by making use of the fast numerical Fourier transform technique or other numerical methods [14].

If a moderate distortion of the pulse shape takes place during the propagation of the pulse from the point $X=O$ into the point $X=L_o$, then

$$\Phi(t) = \Psi(t) - y(t) \quad ,|y(t)| \ll |\Psi(t)| \leqslant 1 \tag{71}$$

$$A(s) = 1 - z(s) \quad , z(s) = y^L(s)/\Psi^L(s) \tag{72}$$

and by making use of the expansion

$$\ln(1-x) = -x - \frac{1}{2}x^2 - \frac{1}{3}x^3 - \cdots \tag{73}$$

we obtain the aaymptotic presentation

$$K^L(s) = 2(s\zeta_0)^{-1}[z(s) + \frac{1}{2}z^2(s) + \frac{1}{3}z^3(s) + \cdots] + \tag{74}$$
$$+ (s\zeta_0)^{-2}[z^2(s) + z^3(s) + \cdots]$$

If $\Psi^L(s)$ is a simple function then the expression (74) may happen to be useful for calculation $K(t)$.

Example. Let us consider the input $U(0,t) = -\varepsilon c\Omega^{-1}H(t)\sin\Omega t$. Then $\Psi(t)=\cos\Omega t$, $\Psi^L(s)=s/(s^2+\Omega^2)$ and on the basis of (74) we obtain the following expression:

$$K(t) = -2(\zeta_0)^{-1}\ln[1-y(0)] +$$
$$+ 2(\zeta_0)^{-1}[y(t) - y(0)][1 + \frac{1}{2}y(0) + \frac{1}{3}y^2(0) + \cdots] +$$
$$+ 2(\zeta_0)^{-1}[\Omega^2 y_2(t) + \frac{1}{2}y^{\cdot}(t)*y(t) + \Omega^2 y_1(t)*y(t) +$$
$$+ \frac{1}{2}\Omega^4 y_2(t)*y_1(t) + \frac{2}{3}y(0)y^{\cdot}(t)*y(t) + \frac{1}{3}y^{\cdot}(t)*y^{\cdot}(t)*y(t) +$$
$$+ \Omega^2 y(t)*y(t)*y(t) + \cdots] + (\zeta_0)^{-2}[y(t)*y(t) +$$
$$+ 2\Omega^2 y_1(t)*y_1(t) + \Omega^4 y_2(t)*y_2(t) + \cdots] \tag{75}$$

where

$$y_1(t) = \int_0^t y(\tau)d\tau \qquad , \quad y_2(t) = \int_0^t y_1(\tau)d\tau \tag{76}$$

and the symbol $*$ is defined by the formula

$$f(t)*g(t) = \int_0^t f(\tau)g(t-\tau)d\tau \tag{77}$$

At t=0 the formula (75) gives the exact result[2]

$$K(t) = -2(\zeta_0)^{-1}\ln[1-y(0)] \tag{78}$$

and at small values of t it is valid for calculation $K(t)$. Since we assume the fading memory, it is an important information on $K(t)$. On the basis of (75) we obtain also the rough estimation $K(t) \approx 2y(t)/\zeta_0$.

Other examples have also been analysed by the author. Here we draw attention to the fact that with the aid of the asymptotic presentation (75) one can establish how the kernel function $K(t)$ is dependent on the distortion function $y(t)$ and on its derivatives in case of the given input function $\Psi(t)$.

This information is important for making a proper choice of $\Psi(t)$ in experimental investigations.

Anyway, there are several methods for solving the inverse problem within the framework of the linear theory, i.e. for interpretation of the experimental data of the first experiment in case of the Technique 1. By carrying out the measurement of the higher order harmonics in the second experiment and by making use of the asymptotic formulae of the section 6, it seems to be possible to solve the inverse problem within the framework of the nonlinear theory.

As far as the Technique 2 is concerned, sẹveral algorithms for interpretation of the experimental data are being developed. The space available for publication of the report permits us only to mention that in certain cases it seems to be promising to apply the asymptotic presentations described in sections [4-6].

According to our opinion the further progress in the field of acoustic evaluation of the hereditary media will be dependent on the cooperation of efforts of the scientists carrying out the theoretical and experimental investigations.

References

1. Rabotnov, Yu.N.: Elements of the hereditary mechanics of solid bodies. Moscow: Nauka 1977 (in Russian).

2. Nigul, U.: Nonlinear acoustic diagnostics. Leningrad: Sudostroyenie 1981 (in Russian).

3. Eckhaus, W.; Van Harten, A.: The inverse scattering transformation and the theory of solitons. Amsterdam, New-York, Oxford: North-Holland Mathematics Studies 50 1981.

4. Lamb, H.: The dynamic theory of sound. London: Edward Arnold 1931.

5. Blackstock, T.D.: Propagation of plane sound waves of finite amplitude in nondissipative fluids. J. Acoust. Soc. Amer. 34 (1962) 9-30.

6. Zarembo, L.K.; Krasilnikov, V.A.: Introduction to nonlinear acoustics. Moscow: Nauka 1966 (in Russian).

7. Nigul, U.: Deviation of the solution of the quasi-linear wave equation from the solution of the linear wave equation in the region, where the first derivatives are continuous. Prikl. Mat. Mekh. 37 (1973) 434-447 (in Russian).

8. Nigul, U.: Echo-signals from elastic objects I. Tallinn: Valgus 1976 (in Russian with a summary in English).

9. Nigul, U.: Asymptotic description of a process of weak distortion of the shape of a longitudinal wave in nonlinear hereditary elastic media. Dokl. Akad. Nauk SSSR 247 (1979) 60-63 (in Russian).

10. Nigul, U.: The method of successive integration of the linear inhomogeneous wave equations in the theory of transient wave propagation in non-linear hereditary elastic media. Int. J. Non-Linear Mechanics 15 (1980) 469-483.

11. Nigul, U.: Plane stress waves caused in membranes by an arbitrary convex pressure wave. Archives of Mechanics (Poland) 23 (1971) 393-403.

12. Nunziato, J.W.; Walsh E.K.; Schuler, K.W.; Barker, L.M.: Wave propagation in nonlinear viscoelastic solids.

Handbuch der Physic. Band VI/4. Berlin, Heidelberg, New--York: Springer-Verlag (1974) 1-108.

13. Nigul, U.: The correct application of the method of deformation of the integration path for inversion of the Laplace integral transform in problems of the propagation of visco-elastic waves. Dokl. Akad. Nauk SSSR 248 (1979) 56-59 (in Russian).

14. Sachse, W.; Pao, Y.S.: On the determination of phase and group velocities of dispersive waves in solids. J. Appl. Phys. 48 (1978) 4320-4327.

Parameter Identification of Mechanical-Mathematical Models Using the Data Obtained by Following an Experimental Wave Method

A.Baltov, I.Nicolova

Institute of Mechanics and Biomechanics, Bulg.Acad.of Sciences, Sofia, Bulgaria

Summary

The following research is a continuation of the work[1] where some ideas connected with parameter identification of mechanical-mathematical models based on data of an experimental wave method are given. We consider experiments made by means of different experimental technique, which creates stress pulses on one side of a given plane specimen. A stress wave propagates through the specimen and reaches its other side, where it is recorded.

The methods are based on registered data concerning the wave propagation and the changes in the pulse shape after it has propagated through the material. Our aim is to determine the most appropriate model among the typical representatives of a given class of models, which gives the best description (in some sense) of the mechanical behaviour of the material. Using the methods mentioned one can obtain quantitative data on the parameters of the model chosen.

1. Proposed methods

We propose methods with the following stages:

Stage I. Determination of some mechanical properties typical to the models of the class given before. Thus we define several vast classes of mechanical-mathematical models K_J,J=1,2,...Q. For each class we choose a number of typical representatives K_J -I, I =1,2,...N_j.

Stage II. At this stage the experiment is modelled. The wave problem for the propagation of a plane stress wave in the direction of Ox(x/O,L/, where L is the thickness of the specimen in this direction), is formulated. The type of the pattern problem depends on the shape of the pulse p(t), produced on the x=O side (or group of pulses characterised by their envelope), and on the model adopted. As an example we consider the case of pulses which satisfy the condition p(t) = O at the initial

Nonlinear Deformation Waves
IUTAM Symposium Tallinn 1982
Editors U Nigul, J Engelbrecht

moment. In this case we describe the process of wave propagatior in terms of the following system of equations.

$$\partial\sigma/\partial x = \rho\,\partial v/\partial t \qquad \text{(equation of motion)}$$

$$\partial\varepsilon/\partial t = \partial v/\partial x \qquad \text{(geometry relation)}$$

$$\Phi(\sigma,\varepsilon,\dot{\sigma},\dot{\varepsilon},\ldots) = 0 \qquad \text{(constitutive equation)}$$

where $\sigma(x,t)$ is the normal stress on the planes with normals parallel to Ox; $\varepsilon(x,t)$ - the linear strain in the same direction; $v(x,t)$ - the particle velocity in the direction of Ox; ρ = const - the density of the material.The specimen is initially unstressed and unstrained. Then the initial conditions are:

$$t = t_0, \quad \sigma(x,t_0) = 0, \quad \varepsilon(x,t_0) = 0, \quad u(x,t) = 0.$$

The pulse p(t) acts on the x=0 plane, the x=L plane is fixed. The corresponding boundary conditions are:

$$x = 0, \qquad \sigma(0,t) = -p(t)$$

$$x = L, \qquad v(L,t) = 0$$

We have to solve the problem for all typical representatives of the class K_J-I, I = 1,2...N_j, and for each model we have to choose a number of characteristic values of its material constants. This implies a repeated solving of the same initial boundary problem for various constitutive laws. Hence the finite system method, proposed by one of the authors [2] is appropriate.

After solving these problems we obtain the pulse $\overline{p}(t) = \sigma(L,t)$, which has passed through the specimen.
Stage III. For the functions $\overline{p}(t)$ we fix M indications,clearly differing for the models K_J-I. The choice of the indications is connected with the choice of an appropriate pulse. Then those indications define discernable areas of belonging of models in the M-dimensional space.

Stage IV. A statistically necessary number of experiments with specimens of a fixed material is carried out. We determine the function p(t) and its characteristic indications. For each area of models there exist R basic points in the M-dimensional space, obtained by solving the pattern problem for a typical number of values of the model materials constants.

we determine the first S minimal distances to the basic points of the model areas. On this basis we determine the most appropriate model among the typical representatives of the considered class.

Stage V. Using data for the indications of the chosen model we may determine the values of the model parameters.

It is advisable to use other types of experiments (for instance, vibrations of cantilever beams, etc.) to justify the choice of the model.

If the examined material shows properties different from those of the representatives of the class, then we must look for a more precise model of the same class. It must satisfy well enough the condition for closeness in the M-dimensional space of indications and predict data of the control experiments well.

2. Discussion on the methods

We will discuss some questions concerning the stages of the methods.

For instance, we could admit the following characteristic properties distinguishing some classes of some mechanical-mathematical models:

A) For the class of nonlinear elastic and elasto-plastic bodies - the nonlinear velocity of propagation of the plane longitudinal wave. That can be easily checked by examining specimens with different thickness.

B) For the class of inelastic bodies of Maxwellian type- the linear velocity of propagation of the longitudinal plane wave, which is the same as the one in the elastic case ($c^2 = \frac{E}{\rho}$, E = const), and the dissipation of the energy. It is easily checked by means of the data obtained by following the non-destructive method.

We shall consider the class of models of Maxwellian type. The corresponding constitutive equations are:

$$\sigma = E(\varepsilon - \varepsilon^a), \quad \dot{\varepsilon}^a = F(\varepsilon, \varepsilon^a)$$

where ε^a is the inelastic part of the linear strain. The function on f, giving the change of ε^a, determines the models of that class. We will consider the following representatives:
1) $f \equiv 0$ (linear elastic body); 2) $f = \frac{E}{\eta}(\varepsilon - \varepsilon^a)$, E=const, η_a= const (Maxwellian linear visco-elastic body);
3) $f = E\frac{(\varepsilon - \varepsilon^a)}{\eta_0} \cdot \exp E\frac{(\varepsilon - \varepsilon^a)}{m_0}$, η_0= const, m_0 = const (Maxwell-Gurevitch-Rabinovitch nonlinear visco-elastic body);
4) $f = \gamma_p\{\frac{E(\varepsilon - \varepsilon^a)}{\sigma^0} - 1\}^q$, γ_p= const, E = const, σ^0 = const, q = const (visco-plastic body of Malvern-Perzyna type);
5) $f = A \exp\{-D_0/z\}$, $A = A(\sigma, \varepsilon^a)$, $D_0 = D_0(\sigma, \varepsilon^a)$,
$\Delta z = \frac{1}{E}(\sigma^2 - \sigma_s^2) + \sigma\varepsilon^a - \sigma_s\varepsilon_s^a + (\sigma - \sigma_s)\alpha_T\theta_0$ where σ_s is the normal stress corresponding to the static yield condition, and ε_s^a is the corresponding inelastic strain, E = const, α_T=const, θ_0 =const (visco-plastic model with hardening of Baltov[4]).

The repeated solution of the pattern problem for the models mentioned with various characteristic values of their material constants is carried out by means of the finite system method, with the algorithm of the wave problem, described in [1]. The method combines a finite difference discretization with an iteration procedure, representing body's work as a cybernetic system. Using that method one can easily change the part, containing the constitutive equations, i.e. in our case, the function f. When we choose the steps Δx and Δt we must satisfy the convergence condition of the iteration procedure, which depends on the differential scheme adopted for the differential operators for Δx and for Δt.

The solution is performed in a nondimensional form. Nondimensional variables are introduced; $X = x/L$ and $\tau = t/t_a$ where t_a is the duration of the initial pulse (or group of pulses); nondimensional functions: $S = \frac{\sigma}{\sigma_p^0}$, $U = u/L$, and parameters:

$$\mu = \frac{\rho \dot{L}^2}{t_a^2 \sigma_p^0} \qquad \text{and } E1 = E/\sigma_p^0.$$

An investigation of the elastic case, f = 0, has been done to work out an appropriate scheme of discretization in view of minimum time expenses.

We adopted a scheme of central differences for x, and for t: I) a scheme forward; II) a predictor-corrector scheme with a scheme forward; III) a Cranck-Nicholson scheme; IV) a predictor-corrector scheme with a Cranck-Nicholson scheme. The schemes were compared to the exact solution for the following data: $E = 2\cdot 10^{7} N/cm^{2}$; $\sigma_p^0 = 2\cdot 10^4 N/cm^2$, $L=5cm$, $t_a = 10^{-5}sec$; $\rho = 7.85\cdot 10^{-5} Nsec^{2}/cm^{4}$ and for an input pulse $p(\tau) = \frac{p}{\sigma_0} = 10^3(\tau-1), \tau/0,1/$. On reaching a maximum error about 1.5% by scheme (III) we spent 354 MU, by scheme (IV) - 407 MU, by scheme (I) - much more. Scheme (III) was adopted as a basic one for the computations.

The methods were tested for models[1],[2],[3],[4] for various data on their material constants.

In Fig.2 the changed pulse for models [3],[4] compared to [1] is given. It is seen that the visco-elastic model does not lead to a change of the pulse shape, but the maximum stress max S decreases. The visco-plastic model leads to a change of the pulse shape and decreasing of max S.

We adopt the following indications in the methods for pattern recognition: max S, the norm in Sobolev space of the first rang $||\overline{P}||_W = \int_0^{\tau_1} \overline{P}^2 d\tau + \int_0^{\tau_1} \dot{\overline{P}}^2 d\tau$ where τ_1 is the time of action of the passing pulse; the characteristics of $\overline{p}(\tau)$ - the points and the values of the local maximums and the points of zeros with respect to the frequency ω[5]. After determining the most appropriate model in the sense of Sect.1, we can determine its material constants using the values of the indications. On Fig.3 the dependences of max S on various parameters for the models considered from the class of inelastic Maxwellian bodies are shown.

References

1. Baltov,A.I.; Nicolova,I.N.: On the question of model identification using data obtained by following a nondestructive wave method, Theor. and Appl. Mech.l 12(1981) 69-75.

2. Baltov,A.: Application of mathematical system theory to dynamical thermoplasticity, ZAMM, 128(1978) 197-199.

3. Tou,J.T.; Gonzales,R.C.: Pattern Recognition Principles, Addison-Wesley Publ. Company, Mass. 1974.

4. Baltov,A.I.: Application of the finite system method in structural dynamics, Theor. and Appl. Mech., 12(1981)51-63.

5. Otnes,R.K.; Enochson,L.: Applied Time Series Analysis, vol.1, Wiley Intersciences Publ., New York 1978.

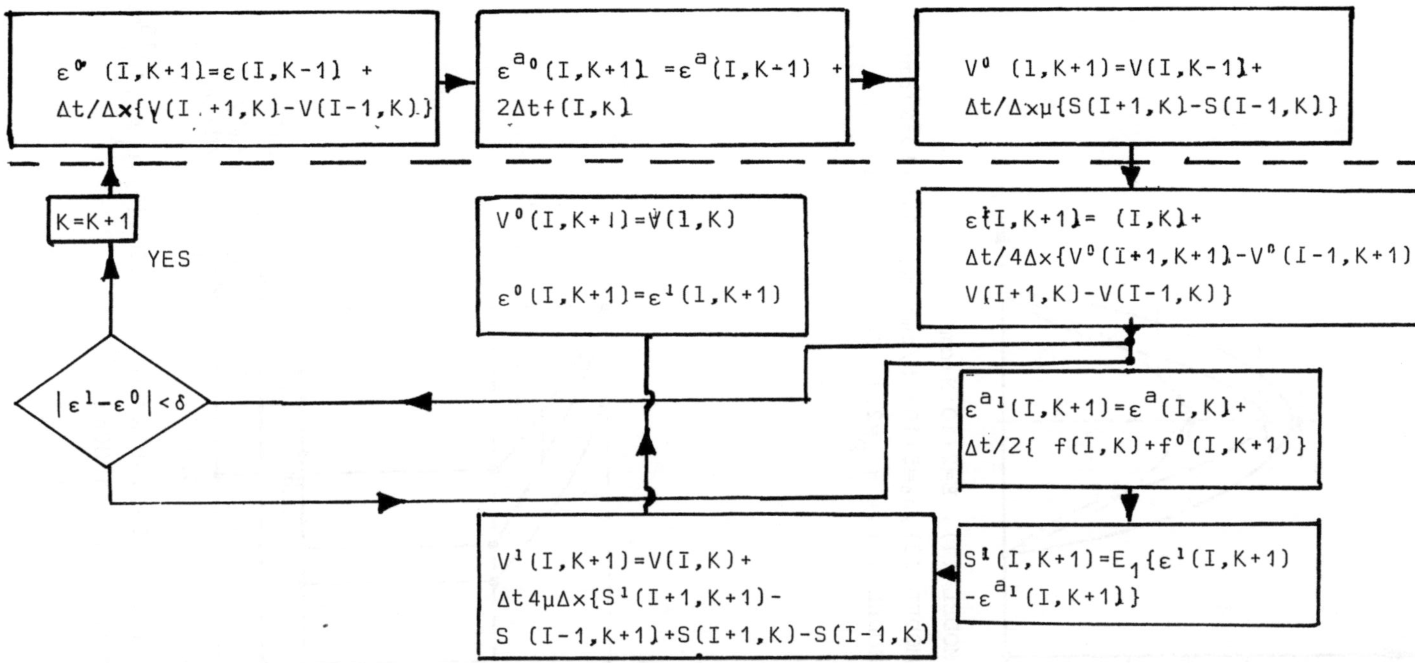
ε⁰ (I,K+1)=ε(I,K-1) +
Δt/Δx{V(I+1,K)-V(I-1,K)}
ε^a0(I,K+1) =ε^a(I,K-1) +
2Δtf(I,K)
V⁰ (I,K+1)=V(I,K-1)+
Δt/Δxμ{S(I+1,K)-S(I-1,K)}
K=K+1
YES
V⁰(I,K+1)=V(I,K)
ε⁰(I,K+1)=ε¹(I,K+1)
ε¹(I,K+1)= (I,K)+
Δt/4Δx{V⁰(I+1,K+1)-V⁰(I-1,K+1)+
V(I+1,K)-V(I-1,K)}
|ε¹-ε⁰|<δ
ε^a1(I,K+1)=ε^a(I,K)+
Δt/2{ f(I,K)+f⁰(I,K+1)}
V¹(I,K+1)=V(I,K)+
Δt4μΔx{S¹(I+1,K+1)-
S (I-1,K+1)+S(I+1,K)-S(I-1,K)
S¹(I,K+1)=E1{ε¹(I,K+1)
-ε^a1(I,K+1)}

Fig.1

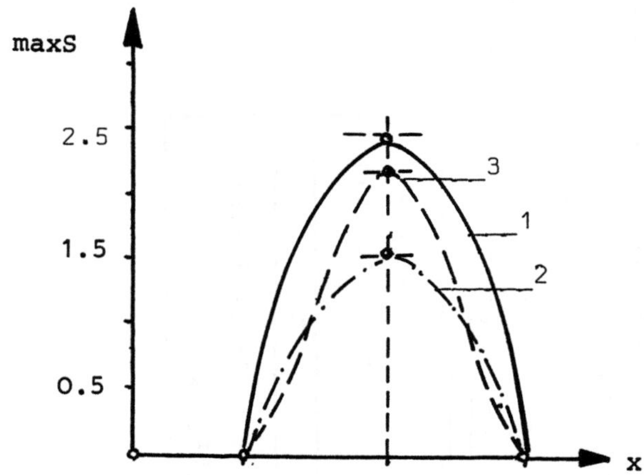

Fig.2. 1- MODEL (1), $E=2\cdot 10^{7}N/cm^{2}$;
2- MODEL (3), $\eta_0=5\cdot 10^{11}N/cm$; $m_{02}=1.5\cdot 10^{3}N/cm^{2}$;
3- MODEL (4), $\sigma_p=2.1\cdot 10$ N/cm, $\gamma_p=45$.

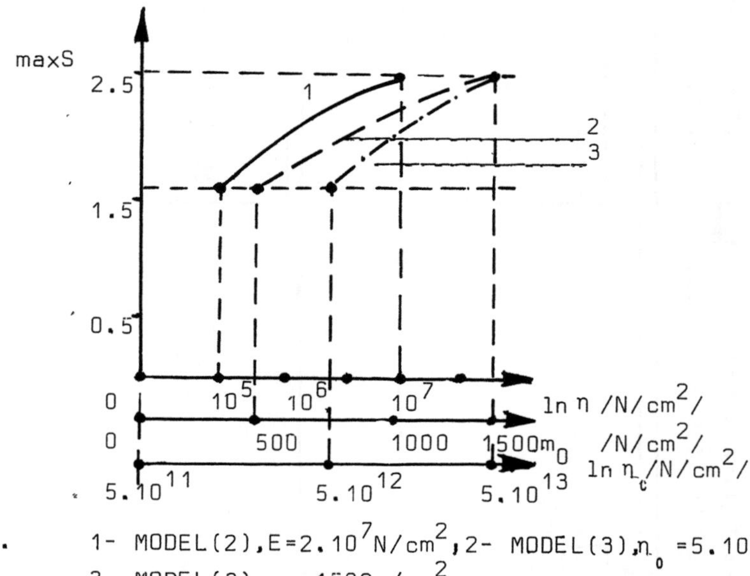

Fig.3. 1- MODEL(2), $E=2.10^{7}N/cm^{2}$, 2- MODEL(3), $\eta_0=5.10^{13}N/cm$
3- MODEL(3), $m_0=1500n/cm^{2}$

On the Use of Ultrasonic Velocity Measurements for the Study of Stress Waves in Solids

N G OHLSON

Royal Institute of Technology
Materials Research Center
S-100 44 STOCKHOLM
Sweden

Summary.
Experiments for determining the effect of superimposed stresses on sound velocity are described. Material characteristics may be obtained from velocity measurements. The stress state may also be analysed by studying the damping of the transversal wave.

Introduction.

Ultrasonic waves have become widely used for investigating materials with respect to possible interior defects in engineering components. To a first approximation, this may be considered as a linear problem in wave mechanics, the material being assumed as linearly elastic and the ultrasonic pulses propagating either as transversal or as longitudinal waves with a single, well-defined velocity.

However, a close examination reveals certain deviations from linearity. Mechanical properties of the material may differ from the Hooke'ian model of material. Inhomogeneities as well as non-isotropic elastic properties also affect the propagation. The boundary conditions governing the problem may differ from one part of the body to another, thereby affecting the degree of triaxiality of the stress state and, hence, the velocity of propagation. Closely related to this second reason for deviation from linearity is the occurrence of superimposed states of stress in the body.

The present paper is aimed at illustrating the possibility of tracing the state of stress in a solid by studying velocity of propagation and damping of ultrasonic pulses.

Nonlinear Deformation Waves
IUTAM Symposium Tallinn 1982
Editors U Nigul, J Engelbrecht

Velocity of sound in linearly elastic medium.

Elementary calculations show that the sound velocity for longitudinal waves in a one-dimensional medium may be written as

$$C_{11} = \sqrt{E/\rho}, \tag{1}$$

and in a three-dimensional body

$$C_{31} = \sqrt{(E/\rho) \cdot (1-\nu)/(1+\nu)(1-2\nu)} \tag{2}$$

Propagation according to eq. (1) takes place when the extension of the body in directions perpendicular to the direction of the wave are small in comparison with the wave-length λ.

For commercial use in ultrasonic flaw detection a typical frequency is 4 MHz which corresponds to a wave-length of the order of 0.001 m in aluminium and steel. According to [1], the one-dimensional velocity of sound dominates when the cross-section d of the body fulfills the inequality

$$d \lesssim 5\lambda \tag{3}$$

i.e. in thin rods.

In technical applications of ultrasonics, however, several types of waves are present simultaneously. A surface wave may also appear. This leads to certain distorsion of the original ultrasonic pulse emitted.

The transversal sound wave velocity is

$$C_t = \sqrt{G/\rho} \tag{4}$$

Transversal waves inevitably appear in connection with the reflection of longitudinal waves at interfaces unless the incident wave follows the normal of the interface.

Evaluation of elastic constants from ultrasonic measurements.

It is realized from (1), (2) and (4) that the elastic constants of the solid may be evaluated from velocity measurements. A very simple set-up may be used (fig. 1), the same crystal acting as transmitter and receiver. To the right in fig. 1 is shown a device for obtaining shear (transversal) waves: the effective distance of propagation of the shear wave is l_t, irrespective of the precise position of the crystal.

Complete material characteristics in the form of a stress-strain curve may be obtained from wave propagation measurements [2].

Normally, velocities are evaluated from the horisontal position of the reflections ("echoes") observed from the CRT (cathode ray tube) screen of the ultrasonic device used, fig. 2. The horisontal axis of the screen is proportional with time elapsed from the emission of each ultrasonic pulse.

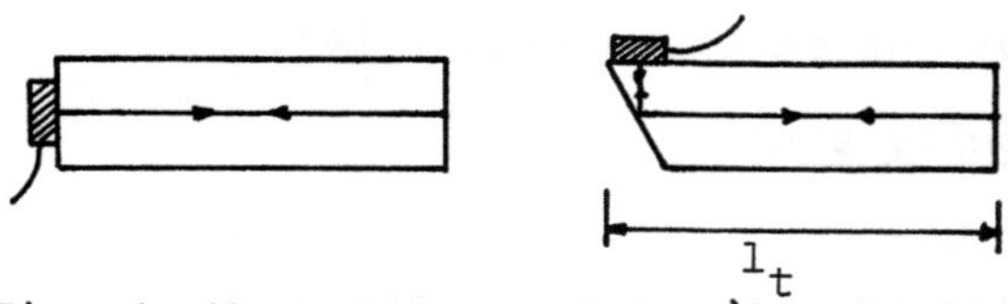

Fig. 1. Measurement of longitudinal (left) and transversal (right) wave speed.

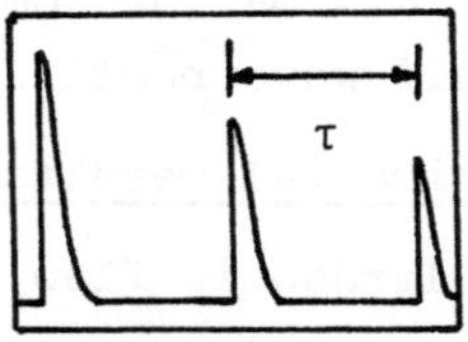

Fig. 2. CRT display of echoes.

Improvement of resolution.

For the purpose of studying the influence of superimposed stresses in the material, exceedingly small velocity changes should be resolved. Recent progress in the development of accurate electronic equipment for detecting ultrasonic waves [3] has led to a measuring technique for wave speeds which can be used for determining stationary states of stress.

In order to be able to study transient stresses, two experimental conditions should be fulfilled, namely,

1° the device for recording the time τ between consecutive echoes should be fast enough and

2° the repetition rate of the emitted pulses should be high enough.

The recording device may be designed according to fig. 3. It consists of a clock-pulse generator, an electronic gate which opens and shuts by the echoes received by the ultrasonic unit, and a counter which registers the number of pulses which have passed through the gate during time τ.

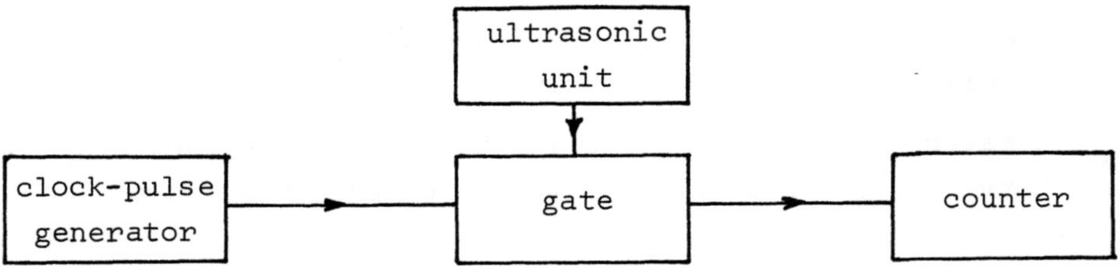

Fig. 3. Principal diagram for accurate recording of τ.

The resolution is enhanced if the echo can be given a very steep flank, i.e. a short rise-time. In a commercial ultrasonic unit this may be sometimes achieved by switching from full wave to half wave rectification of the received signal [4].

Experimental results of velocity studies.

The ultrasonic flaw detector used was a Krautkrämer USM 2, with crystals having resonance frequencies of 2 and 4 MHz. The material tested was an aluminium alloy (Swedish standards no. 4102) available in rods and bars. If a dual-crystal set-up is used (fig. 4) across a hexagonal rod, loaded in compression in its axial direction, one observes a velocity change according to fig. 5. An average was taken over 10 echoes.

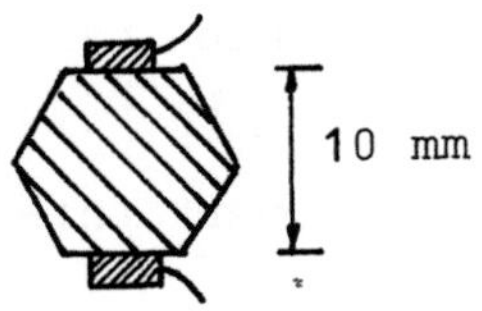

Fig. 4. Velocity measurement across a rod.

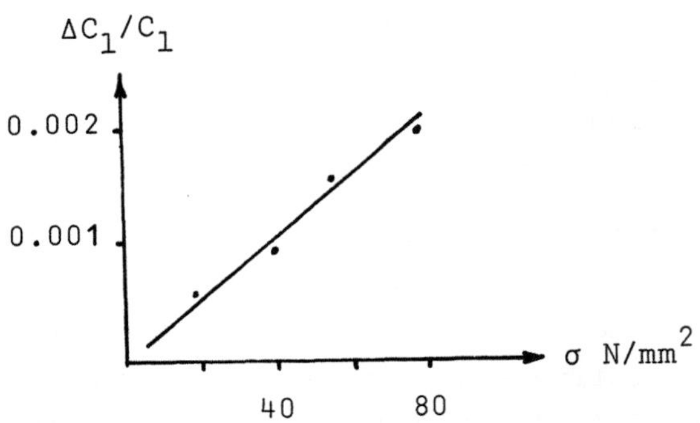

Fig. 5. Relative change in velocity due to uniaxial stress state.

Damping investigations.

A typical picture of the CRT screen is shown in fig. 6. The rugged "hills" appearing between the distinct echoes obtained from the other side of the rod are caused by scattered transversal waves. Experiments show that the amplitude of these intermediate echoes also varies with superimposed stresses. An experimental set-up for tracing bending waves as well as longitudinal waves is seen in fig.7.

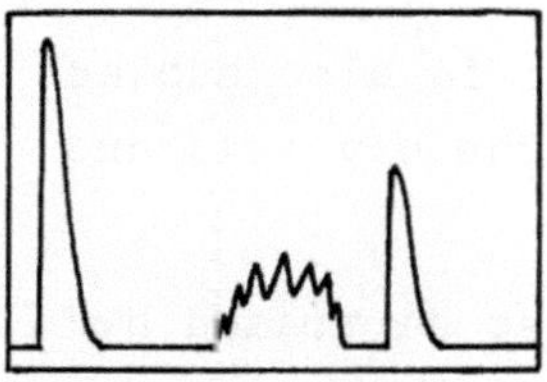

Fig. 6. CRT display showing transversal wave echoes.

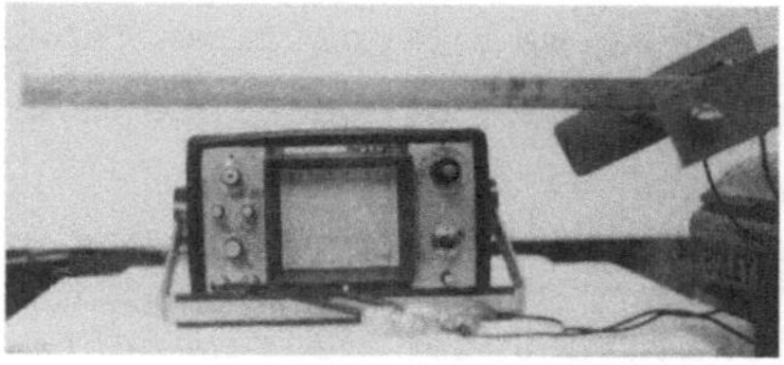

Fig. 7. Experimental set-up for the study of vibrating beam.

The intensity - which is the square of the amplitude registered on the screen - may be written as a function of the distance x which the wave has travelled:

$$I = I_o \, e^{-\alpha x}, \qquad (5)$$

α being the coefficient of absorption and is a measure of the damping. For stress-free aluminium, α is of the order of 0.01 m^{-1} but increases to about 0.7 in uniaxial compression near the yield stress.

The above results apply in the far-field of the transmitter crystal,

$$x > r^2/\lambda \qquad (6)$$

where r denotes the radius of the crystal. The intensity function in the far-field depends also on the direction from the crystal. Measurements were taken in the central lobe, limited to a cone of top angle

$$\theta = 2 \arcsin (0.61\frac{\lambda}{r}) \quad (7)$$

Experimental results of damping studies.

Impact tests were carried out on a cantilever beam of aluminium according to fig. 7. The magnitude of the stress amplitude was 100 N/mm^2. This causes an amplitude drop of the intermediate echoes of the order of 30 per cent. The drop seems to be rather linear with the average stress in the cross-section where the ultrasonic measurements were taken.

This method can be made sensitive. However, it is also subject to disturbances caused by possible motion of the crystals and should be applied with caution.

Other changes in mechanical properties may also be traced by means of ultrasonic speed measurements, e.g. creep damage [5]. Therefore, only comparative studies should be made, since the absolute value of sound speed as well as damping is affected by many factors.

References.

[1] Filipczynski, L., Pawlowski, Z. & Wehr, I., "Ultrasonic methods of testing materials". Butterworths, London, 1966.

[2] Birch, F., "The Effect of Pressure Upon the Elastic Parameters of Isotropic Solids, According to Murnaghan's Theory of Finite Strain". J. Appl. Phys., 9, 279 -288, 1938.

[3] Hsu, N. N., "Acoustical Birefringence and the Use of Ultrasonic Waves for Experimental Stress Analysis". Exp. Mech., 31, 169 - 176, 1974.

[4] Krautkrämer GMBH, "Ultrasonic flaw detector USM 2, Manual 23906 - 2.930", Krautkrämer, Köln 1972.

[5] Bråthe, L., "Macroscopic Measurements of Creep Damage in Metals". Scand. J. Metallurgy, 7, 199 - 203, 1978.

Numerical Analysis of Hopkinson Split Pressure Bar Test

J. BUCHAR, Z. BÍLEK and Z. KNÉSL

Institut of Physical Metallurgy, Czechoslovak Academy of Science, Brno, Czechoslovakia

Introduction

The split Hopkinson pressure bar (SHPB) is widely used to measure the mechanical properties of structural materials at high strain rates $\dot{\varepsilon} \in (10^2, 10^4)s^{-1}$ [1]. To verify the validity of current procedures for SHPB experiments evaluation the extensive numerical calculations based on finite difference method (fdm) [2] and on finite element method [3] were performed. In this paper fem [4] is applied to simulate elastic and elastic--plastic SHPB specimen behaviour and to investigate the influence of loading stress pulse duration and specimen length. Our basic problems are shown in Fig.1. In case (a) we look for two dimensional elastic response of semi-infinite bar loaded uniformly thorough the cross section by the stress pulse $\sigma_{zz}(0,t)$ and in case (b) we solve rather simplified SHPB set up for loading stress pulses $\sigma_I^1(t)$, $\sigma_I^2(t)$ (Fig.1), chosen according Jahsman [5]. The fem results are compared with one dimensional solution [5,6] and relevant implications for more precise evalution of dynamic mechanical properties are worked out.

Semi-infinite elastic circular bar

The numerical computations were carried out for, Young modulus $E = 2.1$ Mba, density $\rho = 7.75 \times 10^3$ Kgm^{-3} and Poisson ratio $\gamma = 0.299$. The bar was loaded as shown in Fig.1 (a) with the stress pulse amplitude $\sigma_{Im} = 10^3$ MPa and duration $t_I = 10$, 5 and 2 μs. Time dependence of the axial stress σ_{zz} is plotted in Fig.2 together with the asymptotic solu-

Nonlinear Deformation Waves
IUTAM Symposium Tallinn 1982
Editors U Nigul, J Engelbrecht

tion [6] and the values of σ_{zz} determined from the radial displacement u_z following the relation

$$\sigma_{zz} = E \, \nu \, \frac{u_z}{a} \tag{1}$$

used to evaluate the records obtained by radial capacity gauges [1]. From Fig.2 and additional results summarized in [7] one can conclude that: (1) Two dimensional (2D) solution performed by dynamic fem code BKDYN has identical features with (2D) solution recently obtained by fdm [2]. (2) The stress state described by the ratio max σ_{ij}/σ_{zz} ($\sigma_{ij} = \sigma_{rr}, \sigma_{\varphi\varphi}, \sigma_{r\varphi}$) differs from (1D) solution only at the region of stress wave arrival as it is also obvious from Fig.2. Therefore Eq. (1) is quite applicable for our bars having 20 mm in diameter and can corrected for (2D) effects as suggested in Fig.2 considering the ratio $\sigma_{zz}/ou_z/\partial t$).

The split Hopkinson pressure bar

Elastic-plastic calculations are made within the framework of rate independent incremental plasticity theory for Mises yield criterion. Fig.3 shows that the boundaries of elastic and plastic unloading and loading regions are dependent on radial coordinate and somewhat larger extend of plastic flow is observed for 2D calculations. Large difference between 1D and 2D stress data is apparent from Fig.4. Knowledge of σ_I, σ_R, σ_T makes possible to find out the strain rate $\dot{\varepsilon}$

$$\dot{\varepsilon}(t) = \left(\frac{\rho^2}{E}\right)^{1/2} \left\{\sigma_I(t) - \sigma_R(t) - \sigma_T(t)\right\} \tag{2}$$

and the stress-strain curves $\sigma - \varepsilon$ [1] plotted in Figs.5, 6. It immediately follows that the validity of usual assumptions made to analyze SHPB tests increases with stress pulse duration and specimen length. The increase of elastic stress lavel for 2D is related to the first stress tensor invariant. As the results of this effect the simple SHPB experiment interpretation may lead to the apparent occurrence of strain rate influence on mechanical properties, namely yield stress σ_y. The complete understanding of SHPB technique requires further investigation e.g. to include polynomial hardening law, strain rate dependence of σ_y , heat generation, friction and etc. Some calculations concerning the above mentioned problems are in progress.

References

[1] Kolsky, H.: An Investigation of the Mechanical Properties of Materials at Very High Rates of Loading. Proceedings of the Physical Society, 62 (1949) 676-700.

[2] Bertholf, L.D.: Feasibility of Two-Dimensional Analysis of the Split-Hopkinson Pressure Bar System. Journal of Applied Mechanics, 41 (1974) 137-144.

[3] Kawata, K.: Hashimoto, S., Kurokawa, K.; Kanayama, N.: Mechanical Properties at High Rates of Strain. Institute of Physics Conference Series Number 47. The Institute of Physics,Bristol and London 1979, pp. 71-80.

[4] Bílek, Z., Knésl, Z., Buchar, J., Polcar, P.: Dynamic finite element code BKDYN. Research Report Institute of Metallurgy C.A.S., Brno, 1982 (In Czech.).

[5] Jahsman,W.E.: Reexamination of the Kolsky Technique for Measuring Dynamic Material Behaviour. Journal of Applied Mechanics, 38 (1971) 75-82.

[6] Kennedy, L.W., Jones, O.E.: Longitudinal Wave Propagation in a Circular Bar Loaded Suddenly by a Radially Distributed Stress. Journal of Applied Mechanics, 36 (1969) 470--478.

[7] Buchar, J., Bílek, Z., Knésl, Z., Polcar, P.: Numerical experiments on elastic-plastic wave propagation in bars. Research Report Institute of Metallurgy C.A.S. Brno 1982 (In Czech.).

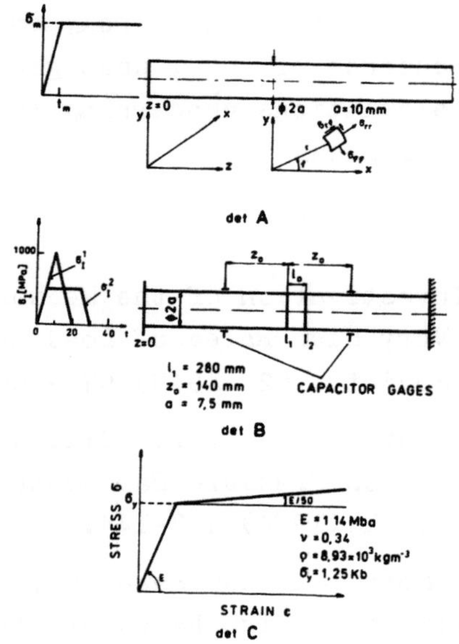

Fig.1. Diagram of computational research program. Detail C shows the stress-strain curve used for the computer simulations

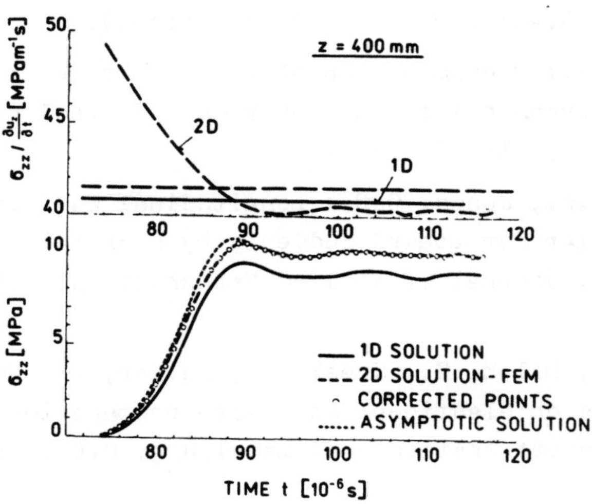

Fig.2. Depedence of axial stress σ_{zz} and the ratio $\sigma_{zz}/(\partial u_z/\partial t)$ on time at the distance of 40 bar radii from the loaded end

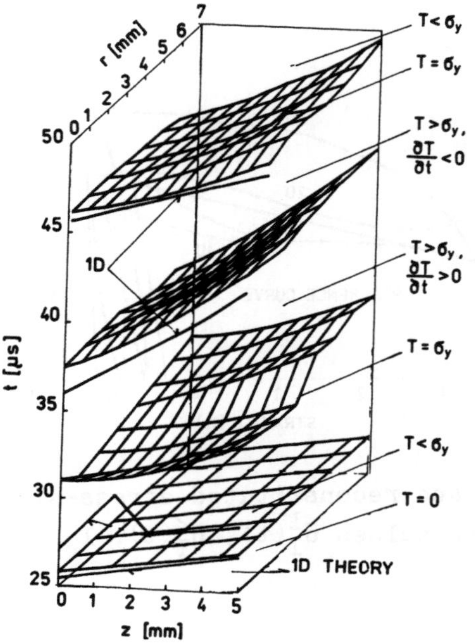

Fig.3. Distribution of deviatoric stress T and its comparison with yield stress σ_y in the specimen with length l_o = 5mm

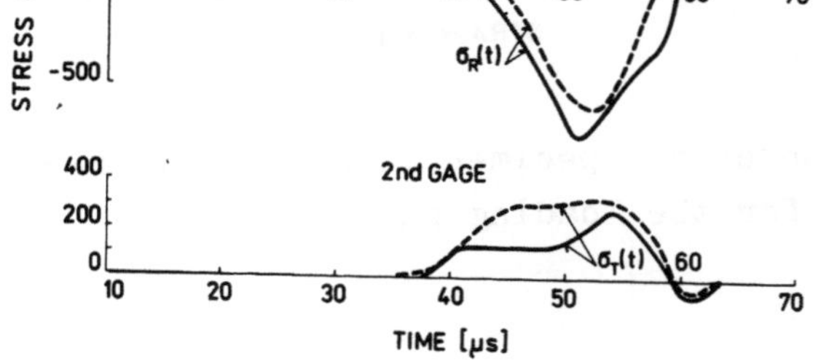

Fig.4. Incident (σ_I), transmitted (σ_T) and reflected (σ_R) stress pulses at the capacity gauges sites, l_o = 5mm

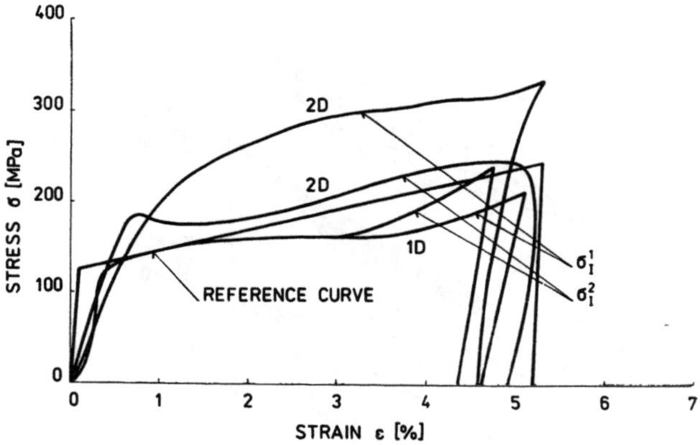

Fig.5. Reference and reconstituted stress-strain curves for two loading stress pulses $\sigma_I^1(t)$, $\sigma_I^2(t)$, l_o = 5mm

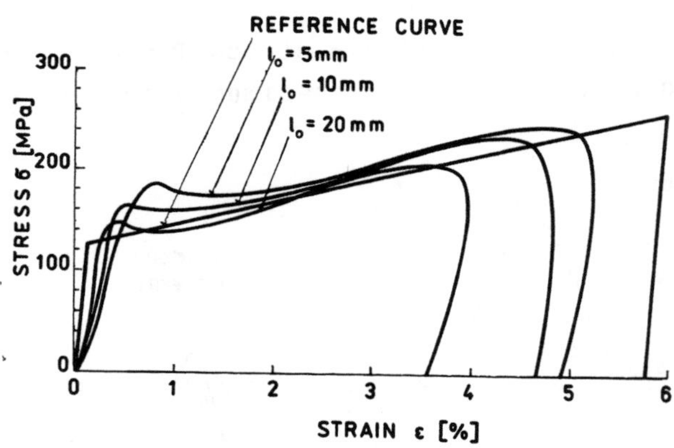

Fig.6. The influence of specimen length on stress-strain curve behaviour for the loading pulse $\sigma_I^2(t)$

Nonlinear Deformation Waves in the Previously Stressed Elastic Media

A.G.Kulikovskiy[1], E.I.Sweshnikova[2]

1) Steklov Mathematical Institute, Acad.of Sciences USSR, Moscow
2) Moscow State University,USSR

Isotropic elastic media is specified by elastic potential $\Phi = \rho_0 U(\varepsilon_{ij}, S)$ where ρ_0 - density in nondeformed state, U and S - internal energy and entropy of mass unity, $\varepsilon_{ij} = \frac{1}{2}\left(\frac{\partial w_i}{\partial \xi_j} + \frac{\partial w_j}{\partial \xi_i} + \frac{\partial w_k}{\partial \xi_i}\frac{\partial w_k}{\partial \xi_j}\right)$ - components of deformation tensor, w_i - components of displacement vector, ξ_i - material coordinates. One dimensional simple waves and shock waves of low intensity in weakly deformed media are studied. (Such waves in nondeformed media were investigated previously [1]). All the waves are divided into quasi-longitudinal and quasi-transversal waves.

Quasi-longitudinal simple and shock waves demonstrate the behaviour usual for many problems of continuous media which is result of second order nonlinearity.

Here only the quasi-transversal waves are investigated as less studied. So far as the motion with small deformations is considered the elastic potential is expanded into a series with respect to deformation powers. For quasi-transversal waves it is enough to consider the terms up to fourth powers

$$\Phi = \frac{1}{2}\lambda I_1 + \mu I_2 + \beta I_1 I_2 + \gamma I_3 + \delta I_1^3 + \xi I_2^2 + \rho_0 T_0 (S - S_0) + const$$

$$I_1 = \varepsilon_{ii}, \quad I_2 = \varepsilon_{ik}\varepsilon_{ik}, \quad I_3 = \varepsilon_{ik}\varepsilon_{kj}\varepsilon_{ji}$$

Here are written only the terms which are important for investigation.

The coordinate system is choosen in such a way that the plane $\xi_3 = const$ represents the wave front. The only quantities which variate in simple and shock waves are the following $\partial w_i / \partial \xi_3 = u_i(\xi_3, t)$. For quasi-transversal waves it is pos-

Nonlinear Deformation Waves
IUTAM Symposium Tallinn 1982
Editors U. Nigul, J. Engelbrecht

sible [2] to eliminate u_3 and express u_3 as function of $u = u_1(\xi_3, t)$ and $v = u_2(\xi_3, t)$ which describe the shear along the directions tangent to wave front. Further we investigate the wave behaviour in uv plane. The initial deformation is given by initial values of u and v and by the components of deformation tensor which are constant in the waves (ε_{11} and ε_{22}).

Previous deformations lead to anisotropy of initial state and to splitting of quasi-transversal waves both simple and shock into "fast" and "slow" waves. The characteristic speeds are

$$\rho_0 C_{1,2}^2 = \frac{1}{2}(\alpha_1+\alpha_2) - \varkappa\left[u^2+v^2 \pm \frac{1}{2}\sqrt{(u^2-v^2+G)^2+4u^2v^2}\right]$$

$$\alpha_i = \mu + (\lambda+\mu+\beta+\frac{3}{4}\gamma)I_1^0 - (\mu+\frac{3}{4}\gamma)\varepsilon_{ii}, \qquad i=1,2$$

$$G = \frac{1}{\varkappa}\,(2\mu+\frac{3}{2}\gamma)(\varepsilon_{22}-\varepsilon_{11})$$

$$\varkappa = \mu + \frac{1}{\lambda+\mu}\,(\mu+\beta+\frac{3}{2}\gamma)^2 - 2\xi$$

For fast waves the sign "-" corresponds to media with $\varkappa>0$ and sign "+" - to media with $\varkappa<0$. The dependence of characteristic speeds on current values u and v makes possible the simple wave overturning. The connection between u and v in simple waves is represented at the uv plane by two mutually ortogonal families of curves which correspond to fast and slow waves. Their equations are

$$\frac{dv}{du} = \frac{v^2-u^2 \pm\sqrt{(v^2-u^2+G)^2+4u^2v^2}}{2uv}$$

By the choice of axis numeration it is possible to make $G \geq 0$.

The integral curves of above written equations are represented at fig.1. They are symmetric with respect to u and v axis and have singular points $u=0$, $v=\sqrt{G}$ and $u=0$, $v=-\sqrt{G}$. For $u^2+v^2 >> G$ the integral curves are close to circles and to rays. For $\varkappa>0$ fast waves correspond to closed curves. Characteristic speeds along integral curves variate in such a way that for $\varkappa>0$ fast waves in which $|u|$ increases and slow waves in which $|u|$ decreases have the tendency to overturn. At the fig.1 the arrows

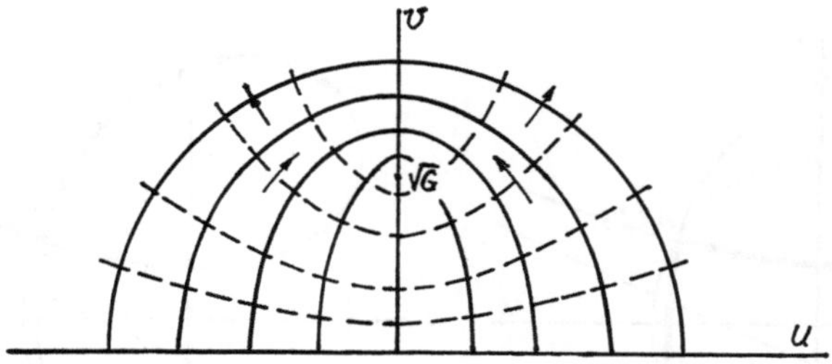

fig.1

characterise the directions of magnitude change in nonoverturning waves for media with $\varkappa > 0$. For media with $\varkappa < 0$ the directions change to the opposite ones and closed curves correspond to the slow waves.
For $G \to 0$ (in the limit $\mathcal{E}_{22} - \mathcal{E}_{11} = 0$) all integral curves turn into circles and rays. The wave which corresponds to the circle has constant characteristic speed and moves without change of its form [1]. The wave may also have the form of the step-function and represent the discontinuous solution (shock) with continuous entropy.
For quasi-transversal shocks it is possible to obtain the relation between state u , v behind shock and state U, V in front of shock from shock conditions.
This relation (shock adiabate) in uv plane is represented by the third order curve

$$(u^2+v^2-R^2)(Uv-Vu) + 2G(u-U)(v-V) = 0, \quad R^2 = U^2+V^2 \quad (1)$$

This curve is drawn at fig.2.
The condition of entropy nondecreasing must be satisfied for real shocks

$$8\rho_0 T_0 (S^+ - S^-) = -\varkappa[(u-U)^2+(v-V)^2](u^2+v^2-R^2) \geq 0 \quad (2)$$

For $\varkappa > 0$ this inequality takes place for that parts of shock adiabate which lie within the circle with centre at coordinate origin and passing through initial point $A(U, V)$. For $\varkappa < 0$ the condition of entropy nondecreasing is satisfied out of the circle.

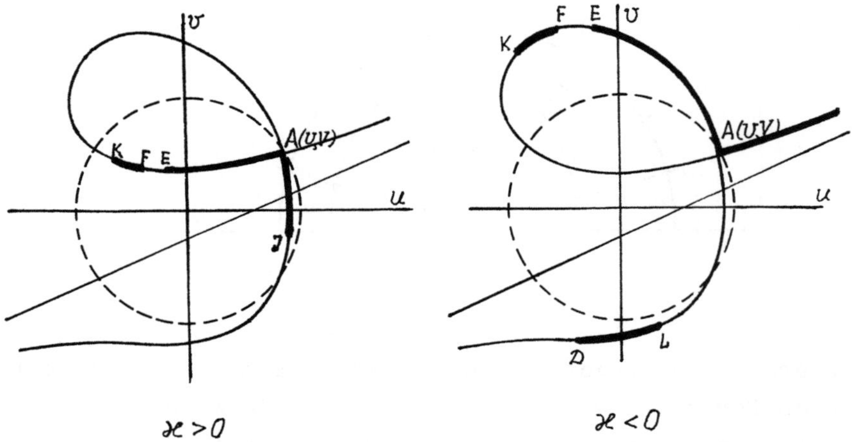

fig.2

Besides the evolution conditions [3] (necessary conditions for correctness of boundary relation at the shock) must be satisfied. These conditions are represented by the following inequality systems

$$\begin{cases} C_2^- \leq W \leq C_2^+ \\ C_1^+ \leq W \leq C_3^- \end{cases} \qquad \begin{cases} C_1^- \leq W \leq C_1^+ \\ 0 \leq W \leq C_2^- \end{cases} \tag{3}$$

where $W = d\xi_3/dt$ - shock speed, C_i^-, C_i^+, $i = 1,2,3$ - characteristic speeds of two quasi-transversal ($i = 1,2$) and one quasi-longitudinal ($i = 3$) waves. Upper subcripts + and - correspond to the states behind and in front of the shock. The numeration of C_i is so chosen that $C_1 \leq C_2 \leq C_3$. If the first inequality system is satisfied shock is called a fast one, in the second case shock is called a slow one.

The shock conditions permit to represent the shock speed W as a function of some parameter x which monotonous changes along adiabate (fig.3). The "real" parts of adiabate with the conditions (2) and (3) satisfied are shown at fig.2 and 3 (solid curves). For different values of coefficients in elastic potential and initial deformations U, V, G two, three or four such parts of shock adiabate are possible. Two of such parts go out of initial point U, V and two other

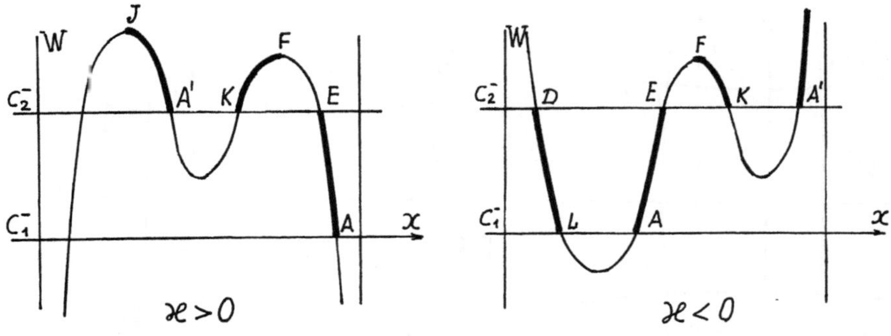

fig.3

are separated.
The real parts of adiabate are terminated by the points in which the shock speed coincides with one of characteristic speeds in front of or behind shock (Jouget points) or go to the infinity. The above results are applied to "piston problem". Let the medium occupying the half space $\xi_3 \geqslant 0$ is at a rest for $t < 0$ and has homogeneous deformations which are described by the values U_0, V_0, G. Let the constant tangential stress corresponding to the values u_*, v_* be applied to the plane $\xi_3 = 0$ at $t \geqslant 0$. For simplicity the case $U_0^2 + V_0^2 \ll G$ and $u_*^2 + v_*^2 \ll G$ is considered. The above problem is a self-similar one and its solution consists of time-independent shock and simple waves. Their relative location is determined by their speeds.

The variation of medium parameters is represented at uv plane by segments of adiabate and integral curves of simple waves. The sequence of these segments must connect the point U_0, V_0 with the point u_*, v_*. The wave sequence representing the solution for different points u_*, v_* is written at fig.4. The letters R_1, R_2 denote slow and fast simple waves and the letters S_1, S_2 - slow and fast shock waves. Subscript J by S means that the shock obeys the Jouget condition. At the fig.4 the initial point U_0, V_0 is fixed and point u_*, v_* can be arbitrary located. The sequence of waves representing solution is shown according to the location of this point. The bo-

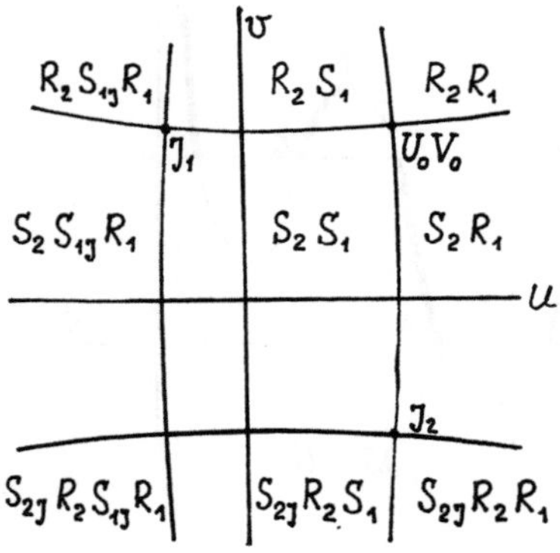

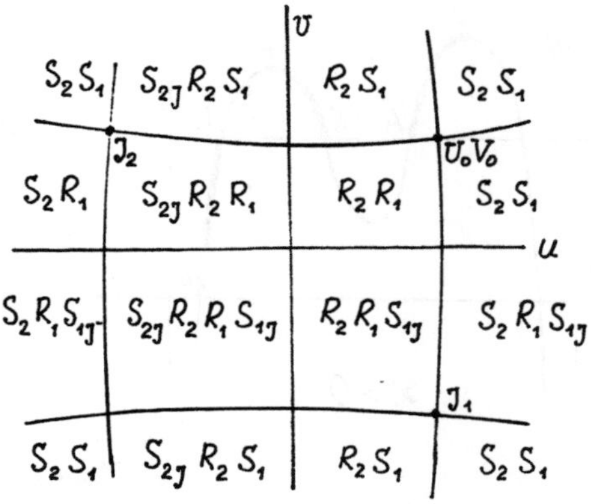

fig.4

undaries of the regions with the same sequence of waves consist of shock adiabates and segments of simple waves integral curves, drawn through the initial point and Jouget points J_1 J_2. The weak shocks in deformed elastic media were considered also in [4]. In the case under consideration the weakness condition of [4] requires

$$|u-U| \ll U, \quad |v-V| \ll V$$

and is fulfilled only in the small vicinity of initial point.

References

1. Bland, D.R.; Nonlinear dynamic elasticity. Waltham, Massachusetts Toronto. London, 1969.

2. Kulikovskiy, A.G.; Sweshnikova, E.I.: On the shock propagating in strained isotropic elastic media. PMM, 1980, v.44, No.3.

3. Lax, P.D.: Hyperbolic Systems of Conservation Low II. Communs Pure and Appl. Math., 1957, vol.10, No.4.

4. Wesolowski, Z.: Strong discontinuity wave in initially strained elastic medium. Arch. Mech. Stos., 3, 30, 1978.

PART 4
NONLINEAR ACOUSTICS

Spectral Interactions in Nonlinear Acoustics – A Review

DAVID T. BLACKSTOCK

Applied Research Laboratories
The University of Texas at Austin
Austin, Texas 78712, USA

Summary

The nonlinear interaction of two collinear sound waves gives rise to a variety of interesting phenomena. The discussion here is limited to two-tone interactions, that is, the interaction of two progressive primary waves, angular frequencies ω and Ω. First, basic solutions are briefly reviewed. A classification based on the value of the frequency ratio $r = \omega/\Omega$ is given. The applications for closely spaced primary frequencies ($r \doteq 1$) are the parametric transmitting array and self-demodulation. Widely separated primary frequencies are used for suppression of sound by sound ($r \gg 1$) and the parametric receiving array ($r \ll 1$). Finally when r is a simple rational number, such as 1/2, degenerate parametric amplification is feasible.

INTRODUCTION

This paper is a review of two-tone spectral interactions in nonlinear acoustics. Only collinear waves are considered. The review is descriptive, not exhaustive, and although not intended to be limited to, is certainly colored by, the author's own experience in the field.

Propagation distortion of sound from a two-frequency source (angular frequencies ω and Ω) is qualitatively different from that from a single-frequency source. A single-frequency tone suffers only harmonic distortion, i.e., only up conversion of energy takes place. A two-frequency tone is subject not only to harmonic distortion but also to intermodulation distortion, that is, the production of components at frequencies $m\omega \pm n\Omega$, where m and n are integers. Thus down conversion as well as up conversion takes place. Several interesting applications, of which the parametric array [1] is the most famous, are possible. Before describing the applications, however, we first review some basic theory of two-tone interaction.

A. PLANE WAVE THEORY

A generalized Burgers equation [2] may be used to describe the propagation of plane waves of finite amplitude in a fluid,

Nonlinear Deformation Waves
IUTAM Symposium Tallinn 1982
Editors U Nigul, J Engelbrecht

$$u_x - (\beta/c_0^2) u u_{t'} = L(u) \quad , \tag{1}$$

where u is the particle velocity, x is distance, t is time, $t' = t - x/c_0$ is retarded time, c_0 is small-signal sound speed, β is the parameter of nonlinearity (for gases $\beta = \frac{\gamma+1}{2}$, where γ is the ratio of specific heats; for liquids $\beta = 1 + \frac{B}{2A}$ where B/A is a coefficient of the quadratic term in an isentropic pressure-density relation), and the operator L(u) describes the attenuation and dispersion properties of the fluid. For a two-frequency source the boundary condition is

$$u(0,t) = U_0 \sin(\Omega t + \Phi) + u_0 \sin(\omega t + \phi) \quad , \tag{2}$$

where U_0, u_0 and Φ, ϕ are the amplitudes and phase angles, respectively of the two tones. The two signals are usually called primaries. When one primary is much stronger than the other, our practice is to represent the strong one, which is usually called the pump, by the first term (amplitude U_0) in Eq. 2. In the case of the parametric transmitting array, u_0 and U_0 are normally of the same order of magnitude.

Fenlon [3] gave an exact solution of the two-tone problem for the case of a lossless fluid [$L(u) = 0$]. Very useful for some of the applications we shall discuss, this solution is as follows:

$$u = \sum_{m=0}^{\infty} \sum_{n=-\infty}^{\infty} u_{m,n} \sin[m(\omega t' + \phi) + n(\Omega t' + \Phi)] \quad , \tag{3}$$

where the amplitude of the component at frequency $m\omega + n\Omega$ is

$$u_{m,n} = \frac{\varepsilon_m u_0}{(mr+n)\sigma/r} J_m[(mr+n)\sigma/r] \, J_n[(mr+n)\Sigma] \quad . \tag{4}$$

Here J_m and J_n are Bessel functions of the first kind, $r = \omega/\Omega$ is the ratio of the two frequencies, $\sigma = \beta u_0 \omega x/c_0^2$ and $\Sigma = \beta U_0 \Omega x/c_0^2$ are the distortion range variables appropriate for the two tones, and ε_m is the Neumann factor ($\varepsilon_0 = 1$; $\varepsilon_m = 2$, $m \neq 0$). The shock formation distance $\bar{x}$ depends on the phase angles ϕ and Φ. When both are zero, $\bar{x}$ has its minimum value, which may be found from the relation $\sigma + \Sigma = 1$, i. e.,

$$\bar{x} = \frac{c_0^2}{\beta(u_0\omega + U_0\Omega)} \tag{5}$$

In the limit as one of the tones disappears, say $u_0 = 0$, Eq. 3 reduces to a single series (only the $m = 0$ term survives), and the familiar Fubini solution [4], in which the harmonic amplitudes are $2u_{0,n} = \frac{2}{n\Sigma} J_n(n\Sigma)$ is recovered.

The problem is more difficult when losses are included. For a thermoviscous fluid the loss term in Eq. 1 is $L(u) = Au_{t't'}$, where the constant A includes the coefficients of viscosity and heat conduction. For this case the equation has an exact solution. Unfortunately, this solution for a two-tone source is not easy to interpret. A second-order solution was obtained by Naugol'nykh, Soluyan, and Khokhlov [5] (1963) for both the two-frequency wave and an amplitude modulated wave. Using weak shock theory, they also investigated the problem of the amplitude modulated wave in the postshock region and discovered the phenomenon of self-demodulation, about which more later. Tjøtta and Tjøtta [6] (see also the references in their paper) analyzed the solution of Burgers' equation for a two-frequency source (with $\phi = \Phi = \frac{\pi}{2}$) and obtained a Fay-like solution for the difference frequency wave and its harmonics,

$$u = \frac{2\alpha_- c_0^2}{\beta\omega_-} \sum_{m=1}^{\infty} \frac{(-1)^m \sin m\omega_- t'}{\sinh m(\alpha_- x + \Gamma_\omega^{-1} + \Gamma_\Omega^{-1})} , \qquad (6)$$

where $\omega_- = \omega - \Omega$ is the difference frequency; $\alpha_- = A\omega_-^2$, $\alpha_\omega = A\omega^2$, and $\alpha_\Omega = A\Omega^2$ are the small-signal absorption coefficients at frequencies ω_-, ω, and Ω, respectively; and $\Gamma_\omega = \beta u_0\omega/c_0^2\alpha_\omega$ and $\Gamma_\Omega = \beta U_0\Omega/c_0^2\alpha_\Omega$ are the Gol'dberg numbers for the two primary waves. Equation 6 is an asymptotic solution, valid for strong primary waves whose frequencies are close together, at distances such that $\alpha_\omega x >> 1$ and $\alpha_\Omega x >> 1$. It is interesting to note the weak dependence of the difference frequency harmonic amplitudes on the primary amplitudes. When the distance is such that $\alpha_- x >> \Gamma_\omega^{-1} + \Gamma_\Omega^{-1}$, the dependence disappears altogether, and we observe a saturation phenomenon similar to that for single-tone sources [7].

The applications that are now discussed may be classified according to the proximity of the two primary frequencies, that is, the size of the frequency ratio r, and the relative amplitudes U_0 and u_0. When $r \doteq 1$, the difference frequency wave has a much lower frequency than either of the two primaries. Such an arrangement is generally sought for the parametric transmitting array and self-demodulation. For suppression of sound by sound, however, one usually chooses $r >> 1$. Finally, degenerate parametric amplification of sound requires that r be a simple ratio, for example, $r = \frac{1}{2}$.

If r is not a rational number, the phases ϕ and Φ are unimportant. Accordingly, ϕ and Φ are assumed to be zero for Sections B and C.

B. CLOSELY SPACED PRIMARY FREQUENCIES ($r \doteq 1$)

1. Parametric transmitting array. The main application when the primary frequencies are close together is the parametric transmitting array. The signal of interest is the difference frequency component $u_{1,-1}$, which is low in frequency because $\omega \doteq \Omega$. The two primary waves are narrow collinear beams, not infinite plane waves, and an expression for the difference frequency sound is desired for the entire half space in front of the source. This means that the plane wave solutions discussed in the previous section are not directly applicable. The usual analytical approach is to solve the (three-dimensional) inhomogeneous wave equation [1]. It suffices here to give a simple physical explanation of the process. The interaction of the two primary beams produces virtual sources of difference frequency sound along the path of the two primary beams. The phasing of the virtual sources is such that they constitute an endfire line array. The array is tapered in strength because of absorption of the primary beams by the medium. Westervelt [1] (1960) derived an expression for the difference freauency sound in the farfield of the array for the case in which the primary beams are perfectly collimated plane waves. The difference frequency amplitude $u_{1,-}$, may be expressed as follows:

$$u_{1,-} = \frac{\beta u_0 U_0 \omega_-^2 S_0 e^{-\alpha_- R}}{4\pi c_0^3 R \alpha_T} D(\theta) \quad , \tag{7}$$

where $D(\theta)$ is the amplitude directivity function (θ is the angle relative to the axis of the primary beams),

$$D(\theta) = \frac{\alpha_{T_0}}{\alpha_T^2 + 4k_-^2 \sin^4\theta/2} \tag{8}$$

In these formulas S_0 is the cross sectional area of the primary beams, R is the radial distance from the source of the primary waves, $\alpha_T = \alpha_\omega + \alpha_\Omega - \alpha_- \cos\theta$ and $\alpha_{T_0} = \alpha_\omega + \alpha_\Omega - \alpha_-$ are combination absorption factors, and $k_- = \omega_-/c_0$ is the difference frequency wave number. The other main components of the wave -- the primaries and the sum and harmonic components -- are rapidly absorbed by the medium because of their hiah frequency. After a certain distance, therefore, the difference freauency wave is the surviving signal.

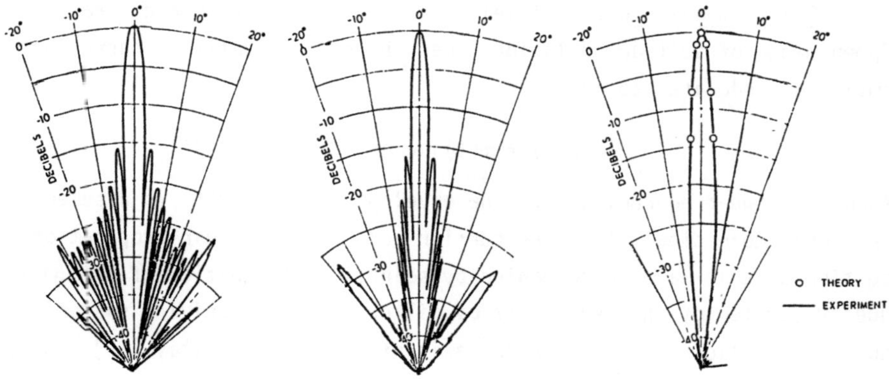

Fig. 1 Parametric transmitting array. Experiment of Muir [8] showing beam patterns of the two primary (carrier) waves and the difference frequency wave.

Two of the remarkable features of the difference frequency wave may be seen in Fig. 1, which shows the results of an early parametric array experiment done in fresh water by Muir [8]. The primary beams, 418 kHz and 482 kHz, were produced by a 7.6 cm diameter piston and are narrow but have many sidelobes. The 64 kHz difference frequency beam is also very narrow but has negligible sidelobes. The narrowness of the difference freauency beam is due to the length of the array of virtual sources. Had the beam been produced directly by the 7.6 cm diameter piston, it would have been about an order of magnitude broader. The lack of minor lobes is due to the exponential taper of the array. Another useful feature of the parametric array is its broad bandwidth. A small percentage change in one of the primary frequencies results in a large percentage change in the difference frequency. Thus the relative bandwidth at the difference frequency may be larae even though the transducer used to produce the primary beams is a narrowband device.

The parametric receiving array has also been the subject of considerable theoretical and experimental study. a pioneering investigation was done by Berktay and Al-Temimi [9]. We shall not attempt to analyze the parametric receiver here. It suffices to say that the pump frequency Ω is generally chosen much higher than the signal frequency ω, i.e., $r \ll 1$. The distortion components of interest are the sum and difference frequency signals, which are thus sidebands about the pump.

2. Self-demodulation. Closely related to the parametric array is the phenomenon of self-demodulation. Let the source radiate a narrow beam of amplitude modulated sound

$$u(0,t) = f(t) \cos \Omega t \qquad (9)$$

instead of a two-frequency wave. The envelope function f(t) is assumed to vary much more slowly than the carrier $\cos \Omega t$. Nonlinear propagation distortion of this wave, accompanied by selective attenuation of the high frequency carrier by the medium (or by filters in the receiving system), produces the following signal on the axis of the beam (Berktay [10], 1965):

$$u = \frac{\beta S_0}{16\pi c_0^3 R \alpha_\Omega} \frac{\partial^2}{\partial t^2} [f^2(t - R/c_0)] \quad . \qquad (10)$$

Because this signal is associated with the modulation envelope, not the carrier, self-demodulation is said to have occurred. If the primary radiation (Eq. 9) is an infinite plane wave, or uniform spherical wave, rather than a beam, the demodulation signal is proportional to the first derivative, rather than the second derivative, of f^2, a result consistent with the findings of Naugol'nykh, Soluyan, and Khokhlov [5], although the domains of validity of the two analyses are quite different.

Figure 2 shows an experimental verification by Moffett [11] of the demodulation process for a 10 MHz pulse in carbon tetrachloride. Because of the high absorption of the medium at 10 MHz, the carrier wave is virtually extinct by a distance of 19 cm. All that remains are the transients associated with the rise and fall of the envelope. The final waveform agrees very well with the theoretical prediction, Eq. 10.

Because the second derivative of f^2, not f itself, is produced by the self-demodulation process, it has been suggested that for actual applications, such as speech transmission, the envelope function should be predistorted prior to radiation so that the received signal is the desired signal f.

C. WIDELY SPACED PRIMARY FREQUENCIES ($r \gg 1$ AND $r \ll 1$)

1. Suppression of sound by sound. Returning now to Fenlon's solution, we consider the circumstance in which $\Omega \ll \omega$ and $U_0 \gg u_0$. The first term in

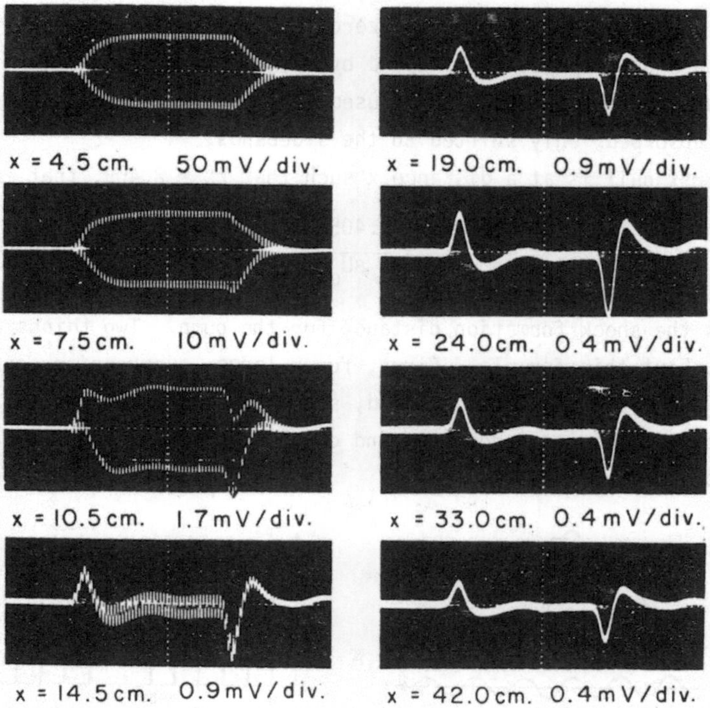

Fig. 2. Experiment of Moffett [11] showing self-demodulation of a pulsed carrier.

Eq. 2 thus represents a powerful low frequency pump, the second term a weak high frequency signal. Because of the smallness of u_0 (more accurately, the smallness of σ), only the $m = 0$ and $m = 1$ series in Eq. 3 are significant. The components of the $m = 0$ series constitute a harmonic spectrum of the Fubini type representing the pump and its harmonics. The components of the $m = 1$ series form a high frequency spectrum composed of the weak signal and its pairs of sidebands at frequencies $\omega \pm n\Omega$. The high frequency spectrum is the more interesting. Its components have amplitudes (note that for small arguments, $J_1(x) = x/2$)

$$u_{1,n} = u_0 J_n[(r+n)\Sigma] \qquad (11)$$

In particular, the expression for the amplitude of the weak primary,

$$u_{1,0} = u_0 J_0(r\Sigma) \quad , \qquad (12)$$

shows that the weak wave vanishes at zeros of the Bessel function. This is the process of suppression of sound by sound. Although the term "absorption of sound by sound" is sometimes used, the energy in the weak primary is not really absorbed, only shifted to the sidebands.

The first null is at a distance x such that $r\Sigma = 2.405$, that is,

$$x_{null} = \frac{2.405\bar{x}_\Omega}{r} = \frac{2.405c_0^2}{\beta U_0\omega} , \qquad (13)$$

where $\bar{x}_\Omega$ is the shock formation distance for the pump. Two things are noteworthy about this formula. First, for r large, suppression occurs long before the pump forms a shock. Second, suppression depends on the amplitude of the pump, but not its frequency, and on the frequency of the weak wave, but not its amplitude.

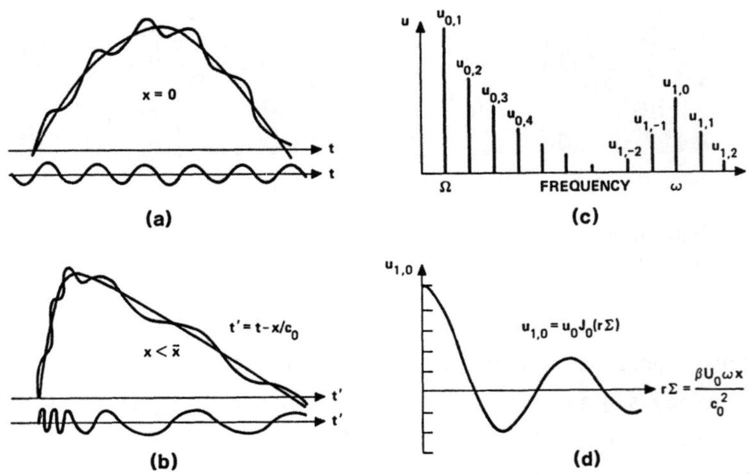

Fig. 3. Suppression of sound by sound, a frequency modulation process.

2. Interpretation: frequency modulation. The physical explanation for suppression is quite simple and is illustrated by Fig. 3. Part (a) of the figure shows the source signal ($x = 0$): several cycles of the weak signal superposed as a ripple on a half cycle of the pump signal. At the bottom of part (a) is the signal obtained by subtracting the pump. After the wave has propagated a distance x, the waveform is given by part (b). The compression section of the pump half cycle tends to form a shock and squeezes

the high frequency ripples together. Conversely, the ripples in the expansion section are stretched out. If the distorted pump half cycle is mentally filtered out of the waveform, the result is the weak signal shown at the bottom of part (b). This is a frequency modulated signal. The combined spectrum of both pump and frequency modulated weak signal is shown in part (c). The amplitude of the weak signal component at frequency ω is plotted as a function of the distance variable $r\Sigma$ in part (d).

3. Historical survey. Some comments about the history of this problem are in order. Mrass and Brinkman [12] (1964) analyzed the effect of a strong low frequency pump on a weak high frequency wave by an approximate treatment. They discovered the frequency, or phase, modulation nature of the interaction and carried out a confirming experiment. They proposed that the phase change provides a way to obtain an absolute measurement of sound pressure. Although their results included Eq. 12 and an approximate form of Eq. 11, they did not mention suppression as an application. Early proponents of the use of the interaction for suppression were Westervelt [13] and Rudenko and Chirkin [14], but in both cases the main interest was the suppression of a low frequency wave by isotropic high frequency noise. To the author s best knowledge the first observation and correct interpretation of suppression in a liquid occurred in 1972; see Moffett, Konrad, and Carlton [15]. Soon afterward Schaffer [16] carried out experiments on suppression in air in a plane wave tube. Independently, Rouvaen, et al [17] analyzed and measured the same phenomenon in a solid.

4. Two experiments. Figure 4 shows a spectrogram from one of Schaffer's experiments. The distance was $x = 7.6$ m and the pump and weak wave frequencies were 0.775 kHz and 3.5 kHz (the point marked by the arrow on the abscissa), respectively. Figure 4 is really the superposition of two spectrograms, one taken with the pump turned off, the other with it on. With the pump off, the spectrum consisted of the single line for the weak wave, namely the upper peak at 3.5 kHz. When the pump was turned on (SPL $\doteq$ 147 dB at $x = 0$), all the other spectral lines appeared: the harmonics at 0.775n kHz and the sidebands at $3.5 \pm 0.775n$ kHz. Simultaneously the weak wave component at 3.5 kHz dropped 25 dB to the lower peak shown.

An experiment very similar to Schaffer's has been reported recently by Gong, Zhu, and Du [18]. Their plane wave tube arrangement and suppression curve are shown in Fig. 5. A maximum suppression of 47 dB was achieved. It will be noticed that the null occurs at a value of $r\Sigma = 2.83$

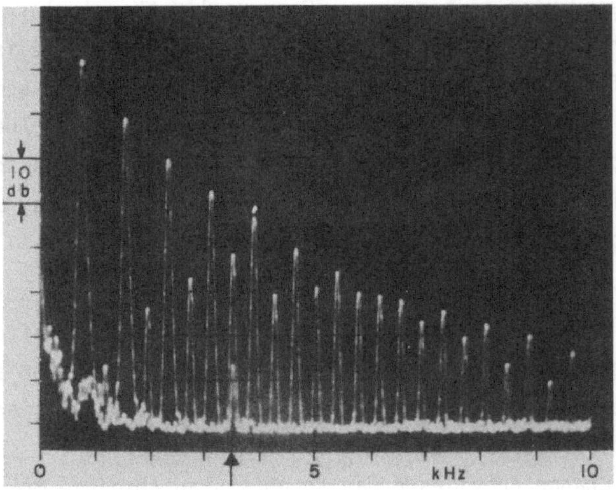

Fig. 4. Spectrogram showing the pump (0.775 kHz) and its harmonics and the weak wave (3.5 kHz) and its multiple sidebands [16].

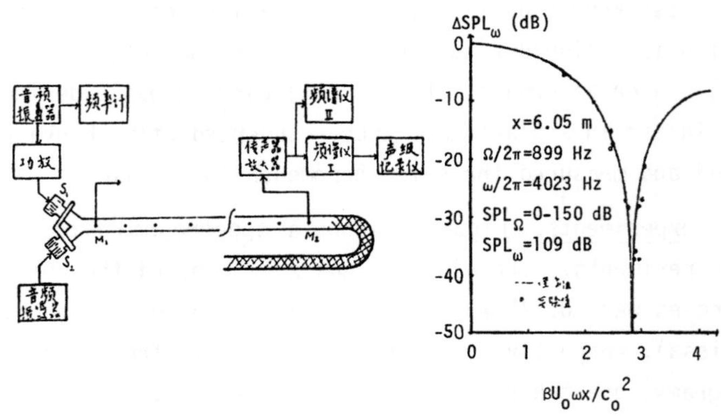

Fig. 5. Experiment of Gong, Zhu, and Du [18] (1979), showing 47 dB suppression.

rather than the value 2.405 given in Eq. 13. The reason is that ordinary tube wall absorption of the pump lowers its amplitude as it travels through the tube and thus slows down the modulation of the weak wave. The average amplitude $\overline{U}_o$ of the pump over the distance x is [16]

$$\overline{U}_o = \frac{1}{x} \int_0^x U_o e^{-\alpha_\Omega x} dx = U_o \frac{1 - e^{-\alpha_\Omega x}}{\alpha_\Omega x} = U_o F. \qquad (14)$$

The numeric 2.405 in Eq. 13 is thus to be divided by F to account approximately for the effect of the absorption. In this particular experiment 2.405/F = 2.405/0.85 = 2.83.

5. Practical aspects. Thus far, despite the apparent attractiveness of suppression for specialized uses in noise control, no practical application has been found. Although considerable suppression may be achieved with modest pump levels and at reasonable distances, the spectrum in the neighborhood of the weak wave frequency is heavily polluted by the sidebands and, in some cases, by the pump harmonics. See, for example, Fig. 4. The reappearance at nearby frequencies of the energy lost by the weak wave is a distinct disadvantage.

An obvious way to avoid the pollution is to reverse the frequencies of the weak wave and pump, i.e., use a high frequency pump ($r \ll 1$). The sidebands then appear near the pump and its harmonics, which are far removed in frequency from the weak wave. Unfortunately very little suppression is achieved with this arrangement [19]. As Eq. 12 shows, when $r \ll 1$, the zero of the Bessel function cannot be reached because the argument is small even when $\Sigma \to 1$. The physical explanation is that the low frequency wave always modulates the high frequency wave, regardless of their respective amplitudes, and so it is the high frequency wave, in this case the pump, that furnishes most of the energy to produce the sidebands. Using weak shock-theory, Willshire [19] investigated suppression (for $r \ll 1$) in the postshock region ($\Sigma > 1$), where Fenlon's solution is not applicable. No significant increase in suppression was found.

Practical application of the arrangement $r \ll 1$ thus appears to be limited to the parametric receiving array.

D. "HARMONICALLY" RELATED PRIMARY FREQUENCIES

In this section rational values of the frequency ratio r are considered, in particular, simple fractions such as r = 1/2, 1/3, etc. Because the two primary waves are locked together in this case, their relative phase $\phi-\Phi$ is important. Without restriction we set $\Phi = 0$.

1. Degenerate parametric amplification: $r = 1/2$. The possiblity of amplifying a weak signal by mixing it with a pump of twice the frequency, i.e., $\Omega = 2\omega$, was described to the author by Naugol'nykh in 1964 [20]. It soon became clear, however, that the phase ϕ must be carefully chosen, or the interaction produces loss instead of gain. Rudenko and his coworkers [21,22,23] have thoroughly analyzed the problem, and measurements have recently been reported by Sakagami, Aoki, Kamakura, and Ikegawa [24].

The gain at the low frequency may easily be found by referring once more to Fenlon's solution. When $\Omega = 2\omega$ (and $\Phi = 0$), Eq. 3 becomes

$$u = \sum_{m=0}^{\infty} \sum_{n=-\infty}^{\infty} u_{m,n} \sin[(m+2n)\omega t' + m\phi] \quad . \tag{15}$$

The components u_ω of frequency ω are found by setting $m + 2n = \pm 1$. The pair values that satisfy this relation are m,n = 1,0; 3,-1; 5,-2;... and 1,-1; 3,-2; 5,-3; .... . If $\sigma << 1$ only the two $m = 1$ terms are important, and we obtain

$$u_\omega = u_{1,0} \sin(\omega t' + \phi) - u_{1,-1} \sin(\omega t' - \phi)$$

$$= u_0[J_o(\Sigma/2)\sin(\omega t' + \phi) - J_1(\Sigma/2)\sin(\omega t' - \phi)] \quad . \tag{16}$$

The signal gain G at frequency ω is thus

$$G(\phi) = \frac{|u_\omega|}{u_o} = \sqrt{J_o^{\,2} - 2J_oJ_1\cos 2\phi + J_1^{\,2}} \quad , \tag{17}$$

where the argument of the Bessel function is $\Sigma/2$. This result was derived by Novikov and Rudenko [23] by another method. The maximum gain is realized when $\phi = \pi/2$; at the shock formation distance $G(\pi/2) = 1.18$. The least gain occurs when $\phi = 0$ and has the value $G(0) = 0.70$ at $\Sigma = 1$. The difference in the two gains is 4.6 dB.

The measurements of Sakagami, et al were made with an air-filled plane wave tube. Amplitude modulated waves as well as simple two-frequency waves were studied. The results are shown in Figs. 6 and 7, which are taken from

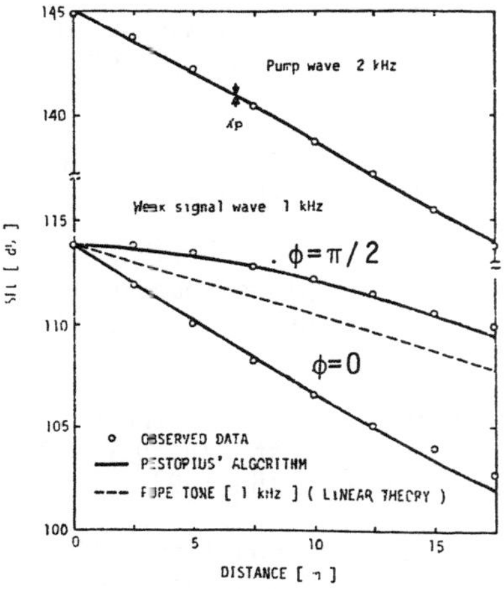

Fig. 6 Propagation curves for the pump and weak signal for phases $\pi/2$ and 0 [24].

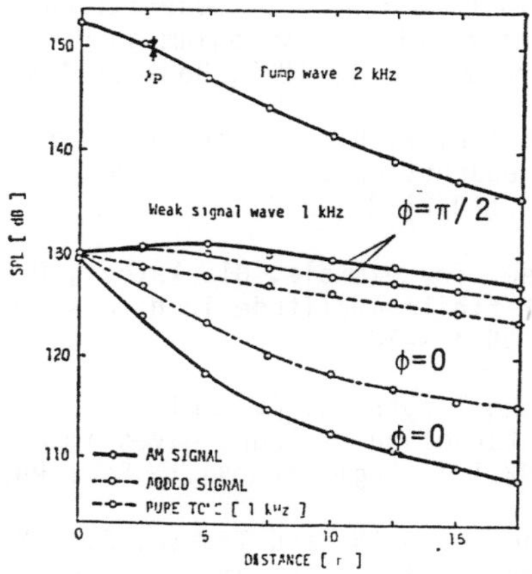

Fig. 7 Comparison of propagation curves for two-frequency signal (denoted "added signal") and AM signal [24]

[24]. Figure 6 (two-frequency case) shows that when ϕ changes from 0 to $\pi/2$, the weak signal increases by about 4 dB at the shock formation point (X_p in the figure). The authors' theoretical curves were obtained by using Pestorius' computer algorithm [25], which includes the effect of tube wall losses. Figure 7 shows a comparison of the gains using the AM signal and the two-frequency signal. The AM signal may be obtained by adding an upper sideband (frequency 3ω) to the pump.

References

1. Westervelt, P. J.: Parametric end-fire array. J. Acoust. Soc. Am. 32 (1960) 934-935 (A). Also: Parametric acoustic array. J. Acoust. Soc. Am. 35 (1963) 535-537.

2. Blackstock D. T.: Generalized Burgers equation for plane waves. Ninth International Symposium on Nonlinear Acoustics, University of Leeds, England, July 1981, Book of Abstracts, p. 7.

3. Fenlon, F. H.: An extension of the Bessel-Fubini series for a multiple-frequency cw acoustic source of finite amplitude. J. Acoust. Soc. Am. 51 (1972) 284-289.

4. See, for example, Blackstock, D. T.: Propagation of plane sound waves of finite amplitude in nondissipative fluids. J. Acoust. Soc. Am. 34 (1962) 9-30.

5. Naugol'nykh, K. A.; Soluyan, S. I.; and Khokhlov, R. V.: Nonlinear interaction of sound waves in an absorbing medium. Akust. Zh. 9 (1963) 192-197 [Engl. transl.: Sov. Phys. - Acoust. 9 (1963) 155-159].

6. Tjøtta, J. N. and Tjøtta, S.: Nonlinear equations of acoustics with application to parametric acoustic arrays. J. Acoust. Soc. Am. 69 (1981) 1644-1652.

7. See, for example, Blackstock, D. T.: Thermoviscous attenuation of plane, periodic, finite-amplitude sound waves. J. Acoust. Soc. Am. 36 (1964) 534-542.

8. Muir, T. G.: An analysis of the parametric acoustic array for spherical wave fields. Tech. Rep. ARL-TR-71-1, Applied Research Laboratories, The University of Texas at Austin (1971) (AD 723 241). See, also Muir, T. G. And Willette, J. G.: Parametric acoustic transmitting arrays. J. Acoust. Soc. Am. 52 (1972) 1481-1486.

9. Berktay, H. O.; and Al-Temimi, C. A.: Virtual arrays for underwater reception. J. Sound Vib. 9 (1969) 295-307.

10. Berktay, H. O.: Possible exploitation of non-linear acoustics in underwater transmitting applications. J. Sound Vib. 2 (1965) 435-461.

11. Moffett, M. B.: Large-amplitude pulse propagation, a transient effect. T. G. Muir (ed.) Nonlinear Acoustics, Proceedings of the Second International Symposium on Nonlinear Acoustics, November 1969, Applied Research Laboratories, The University of Texas at Austin (1970) 143-163. See also, Moffett, M. B.; Westervelt, P. J.; and Beyer, R. T.: Large-amplitude pulse propagation - a transient effect. J. Acoust. Soc. Am. 47 (1970) 1473-1474 (L).

12. Mrass, H. and Brinkmann, K.: Phasenmodulation einer Ultraschallwelle in Luft, verursacht durch die Einwirkung einer zweiten Schallwelle. Acustica 14 (1964) 205-211.

13. Westervelt, P. J.: Absorption of sound by sound. J. Acoust. Soc. Am. 53 (1973) 385(A). Also: J. Acoust. Soc. Am. 59 (1976) 760-764.

14. Rudenko, O. V. and Chirkin, A. S.: The theory of nonlinear interaction of monochromatic and noise waves in weakly dispersive media. JETP, 67 (1974) 1903-1912.

15. Moffet, M. B.; Konrad, W. L.; and Carlton, L. F.: Experimental demonstration of the absorption of sound by sound in water. J. Acoust. Soc. Amer., 63 (1978) 1048-1051.

16. Schaffer, M. E.: The suppression of sound with sound. Tech. Rep. ARL-TR-75-64, Applied Research Laboratories, The University of Texas at Austin (1975) (ADA 023 128). See also Schaffer, M. E. and Blackstock, D. T.: Modulation of a small-signal wave by a finite-amplitude wave of lower frequency. J. Acoust. Soc. Amer. 57 (1975) S73(A).

17. Rouvaen, J. M.; Bridoux, E.; Moriamez, M.; and Torguet, R.: Modulation of hypersound by ultrasound via strong collinear interaction. Appl. Phys. Lett. 23 (1973) 499-501.

18. Gong Xuifen, Zhu Zhemin, and Du Gonghuan: Nonlinear interaction of a finite amplitude wave with a small-signal wave in air. J. Nanjing University (1979) 19-28, 76 (in Chinese).

19. Willshire, W. L.: The suppression of sound by sound of higher frequency. Tech. Rep. ARL-TR-77-22, Applied Research Laboratories, The University of Texas at Austin (1977) (ADA 040 008).

20. Naugol'nykh, K. A.: personal communication (1964).

21. Rudenko, O. V.: Parametric interaction of traveling sound waves. Akust. Zh. 20 (1974) 108-111 [Engl. transl.: Sov. Phys.-Acoust. 20 (1974) 63-64].

22. Lyakhov, G. A. and Rudenko, O. V.: Parametric amplification of weak signals in nonlinear acoustics. Akust. Zh. 20 (1974) 738-744 [Engl. transl.: Sov. Phys.-Acoust. 20 (1975) 447-450].

23. Novikov, B. K. and Rudenko, O. V.: Degenerate parametric amplification of sound. Akust. Zh. 22 (1976) 461-462 [Engl. transl.: Sov. Phys.-Acoust. 22 (1976) 258-259].

24. Sakagami, K.; Aoki, S.; Kamakura, T.; and Ikegaya, K.: Experimental investigations on the degenerate parametric amplification. J. Acoust. Soc. Japan 37 (1981) 504-509 (in Japanese).

25. Pestorius, F. M.: Propagation of plane acoustic noise of finite amplitude. Tech. Rep. ARL-TR-73-23, Applied Research Laboratories, The University of Texas at Austin (1973) (AD 778 868).

Asymptotic Description of Nonlinear Interaction of Surface and Bulk Waves

L.M.Liamshev, K.A.Naugolnykh, S.A.Rybak

Acoustical Institute of the Academy of Sciences
Moscow, USSR

1. The interaction of waves propagating through a condensed medium with surface waves forms a basis for a variety of physical phenomena. Wave phenomena in the ocean can serve as a good example. The free oceanic boundary happens to be greatly excited because of environmental effects. As a result of this interaction, surface wave energy can be transmitted to the bulk waves forcing the internal modes of the system to be excited. This mechanism seems to be a cornerstone of low-frequency oceanic noise generation [1].

Nonlinear surface wave interaction contributes significantly to the generation of internal waves in the stratified ocean.

Energy exchange as a result of bulk and surface wave interaction in solids, as well as waves in liquids with that in piezosemiconductors interfacing with liquids, makes possible the application of wave generation, gain and transformation in different acousto-electronic devices [2-5]. The analysis of bulk and surface wave interaction requires the nonlinear solutions, which are usually difficult to obtain.

Approximate asymptotic approaches and ideas developed in nonlinear wave physics seem therefore effective as applied to the study of weakly nonlinear interactions. The paper by Benjamin and Feir [6] can serve as a good example.

Modern methods have stimulated good understanding of such classical phenomena as water waves and the detection of their in-

Nonlinear Deformation Waves
IUTAM Symposium Tallinn 1982
Editors U. Nigul, J. Engelbrecht

stability.

Although the nonlinear interactions are weak, they nevertheless appreciably affect the process. Two problems arise: 1) the condition of the origin of resonance interactions; 2) the efficiency of these interactions, determined by potential values of the interaction and dispersion. It is precisely resonance interactions that lead to the cumulative effects that we single out.

This paper deals with various cases of interaction of bulk waves in condensed media with surface waves on a free boundary.

By no means do we intend to comprehensively cover the problem; we try rather to present the results of our investigation systematically.

The interaction of sound in a liquid with surface gravity-capillary waves is the object of the 1st part of the paper. A low frequency acoustic wave can parametrically generate two surface waves at a frequency equal to half the sound frequency. On the other hand, two acoustic waves of similar frequency can resonate with capillary waves. This leads to induced combination sound scattering of surface waves. Gain increments of interacting waves and stabilized amplitude values have been achieved.

If the wavelength of the bulk wave is long enough, when we analyze its interaction with a surface wave, the bulk wave can be considered as an inhomogeneous flow. This is dealt with in the 2nd part of the paper. Also discussed are such questions as dispersion and instability of surface waves interacting with horizontally or vertically inhomogeneous flows. Conditions for the generation of negative energy waves are also analyzed.

2. Parametric generation of waves

The interaction of sound and surface waves results in the formation of a resonant wave system from a pair (incident and reflected components) of surface and sound waves.

A system of waves in the horizontal plane of the free liquid surface corresponds to a resonance triplet, which is described by the following equations:

$$\vec{q}_1 + \vec{q}_2 = \vec{k}_\rho , \qquad (1)$$

$$\omega_1 + \omega_2 = \omega ,$$

where $\vec{q}_i$, ω_i - wave vectors and surface wave frequencies; $\vec{k}_\rho$ - wave vector projection of an acoustic wave on the horizontal plane; ω - sound frequency.

As can be seen from this relation, the sound frequency is of the order of the surface wave frequency; at the same time the wave vectors of the sound and surface waves differ greatly.

In fact, because of the considerable difference in the velocities of the sound and surface waves, the sound wave vector $|\vec{k}|$ is small compared to the surface wave vector $|\vec{q}|$ at the same frequency.

Fig.1 shows the location of wave vectors for the simple case of normal incidence of sound on a free liquid surface (here $\vec{k}_1$ - a wave vector of the reflected sound wave, $\vec{k}_\rho$=0).

The problem of parametric generation of surface waves by sound, considered as the instability of free liquid surface, has been investigated by Rayleigh [7] and other authors [8-11]. This effect has been analyzed in works [12-13] from a "three-wave" interaction point of view.

It is necessary to retain nonlinear terms to describe this process in boundary conditions for the free liquid surface. Let the axis Z be vertically up and the plane Z=0 coincide with the free surface of an unexcited liquid. Then the boundary conditions for the free surface can be expressed this way:

$$\frac{\partial \zeta}{\partial t} + \frac{\partial \zeta}{\partial x_i}\frac{\partial \varphi}{\partial x_i} = \frac{\partial \varphi}{\partial z} , \qquad i = 1,2 , \qquad (2)$$

$$\frac{\partial \varphi}{\partial t} + \frac{1}{2}(\nabla \varphi)^2 + g\zeta - \alpha \Delta_\rho \zeta = 0,$$

where

φ - velocity potential,
ζ - surface liquid displacement,
g - gravity acceleration,
σ - surface tension factor,
ρ - liquid density.

Let us consider the simplest case, normal sound incidence, when a resonance triplet is formed by the wave system, as shown in the Fig.1 . The sound field in this case can be described by

$$\varphi_a = e^{-i\omega t}\left[\varphi_0 e^{ikz} + (\varphi_0 + \varphi_3) e^{-ikz}\right], \qquad (3)$$

which satisfies the linear wave equation always at $z < 0$, where amplitude φ_3 is independent of the coordinates. Here φ_0 - is a time constant amplitude of an incident wave, φ_0' - is a reflected wave amplitude, derived in the linear approximation, also independent of time, and φ_3 - an addition to the reflected wave, born in the process of nonlinear interaction with capillary waves near the liquid surface. The potentials of the capillary waves can be expressed by

$$\varphi_{iq} = \varphi_i(t)\, e^{i\vec{q}\vec{\rho} - q_i z - i\omega_i t}.$$

From expression (1) and the Fig.1 it follows that the vectors $\vec{q}_1$ and $\vec{q}_2$ must be antiparallel and equal $|\vec{q}_1| = |\vec{q}_2| = |\vec{q}|$, and the interacting wave frequencies should satisfy the relation

$$\omega_1 = \omega_2 = \frac{\omega}{2}.$$

From equation (2) we shall have:

$$\left(\frac{\partial^2}{\partial t^2} + \alpha \frac{\partial^3}{\partial z \partial x^2}\right)\varphi = -\frac{1}{2}\frac{\partial}{\partial t}(\nabla\varphi)^2 + \frac{\partial}{\partial t}\left(\frac{\partial \varphi}{\partial z}\right)^2 + \qquad (4)$$

$$+ \frac{\alpha\partial^2}{\partial x^2}\left(\frac{\partial\varphi}{\partial x}\right)\left(\frac{\partial\varphi}{\partial x_i}\right) - \left(\frac{\partial^2}{\partial z^2} + \alpha\frac{\partial^3}{\partial z\partial x^2}\right)\varphi\frac{\partial\varphi}{\partial z} = 0 .$$

Grouping and equating to zero the terms in equation (3) whose exponents are the same, we shall get time-dependent equations of the 1st order in relation to the slowly varying triplet amplitudes:

$$\dot{\varphi}_1 = -kq\,(2\varphi_0 - \varphi_3)\,\varphi_2^+ ,$$

$$\dot{\varphi}_2 = -kq\,(2\varphi_0 - \varphi_3)\,\varphi_1^+ , \qquad (5)$$

$$\dot{\varphi}_3 = -i\,\frac{4q^2}{\omega}\,\varphi_1\varphi_2 .$$

In coefficients of the right sides of equations (5), that is, in the interaction potentials, only the principal terms in the small parameter $\varepsilon = k/q$ are preserved. First ignoring sound wave variations, we omit the 2nd member in the right sides of the 1st two equations. Then solving the equation in a form $\dot{\varphi}_1 = -i\varphi_2^+$, we shall get the expression

$$\varphi_i = \varphi_{i0}\, e^{\delta t} ,$$

indicating the infinite growth of capillary waves under the influence of sound. Here $\delta = 2qk\varphi_0$. Along with the dissipative losses, the damping of the excited capillary waves occurs as a result of random interaction with the capillary, wave background, which is dependent on time [14],

$$\tau^{-1} = V^2\,\frac{C_q^2}{\omega_q} \sim \frac{q^4}{\omega_q}\,\varphi_q^2 , \qquad (6)$$

where V - capillary wave interaction potential,
C_q - their amplitude.

The steady shift in amplitude of the capillary wave can be calculated at $\tau = \delta^{-1}$; we obtain

$$q\zeta = (q\zeta_s)^{\frac{1}{2}} , \qquad (7)$$

where ζ_s - displacement amplitude in sound wave. We shall give a numerical estimation for the case in which a sound wave of $f = 10^2$ Hz generates capillary waves on a water surface

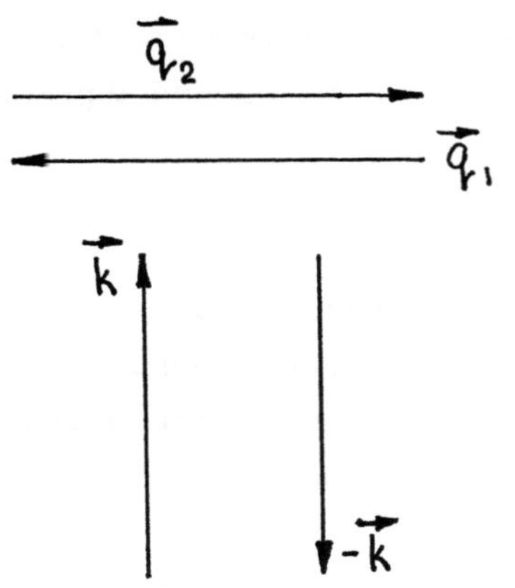

Fig.1

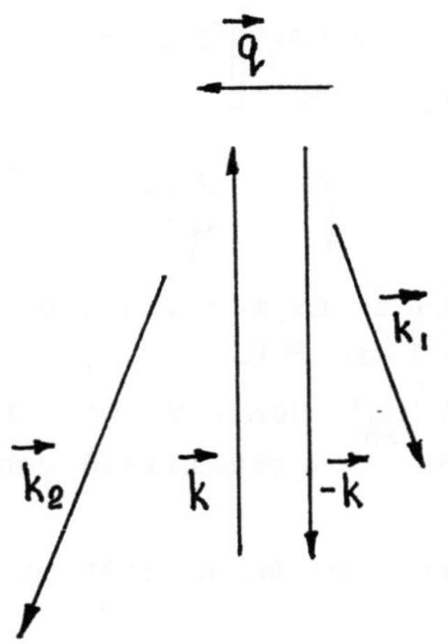

Fig.2

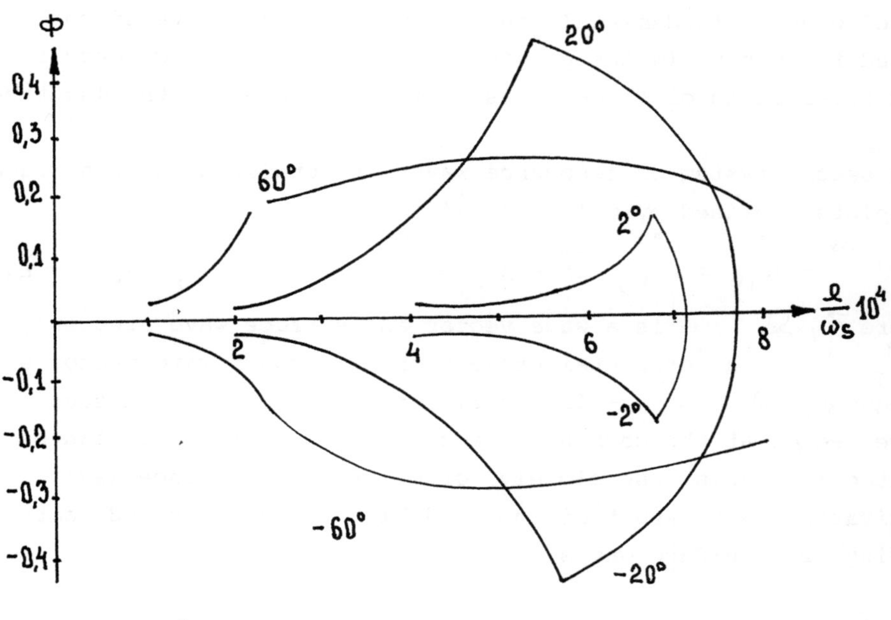

Fig.3

($\alpha = 74$ dyne $cm^2 g^{-1}$; $c = 1.5 \cdot 10^5 \frac{cm}{s}$) The ratio of gain time for a capillary waves τ to the period T of the sound equals to

$$\frac{\tau}{T} = \frac{k}{qM} \approx M^{-4}\ 4\ 10^{-4},$$

according to expression (16) the threshold value is $M_K = \nu q/\alpha c = 10^{-6}$, corresponding to the pressure amplitude $10^4 \frac{dyne}{cm^2}$ Hence we obtain $\tau/T = 10^2$ and a displacement amplitude in a stabilized capillary wave equals to $\zeta = 10^{-2}$ cm.

3. Combination sound scattering

When sound interacts with surface waves, two types of resonance wave systems can be produced: the 1st is of a pair of surface waves and a sound wave (consisting of an incident and reflected components) the 2nd is composed of a single surface wave, a sound wave of fundamental tone, and a scattered wave of combined frequency. In the previous part of the paper we described the interaction of three waves forming a system of the 1st type.

The second system of resonance waves is consistent with surface triplets, defined by relations [14]:

$$\vec{q} + \vec{k}_{1\rho} = \vec{k}_\rho,\quad \vec{q} + \vec{k}_{2\rho} = \vec{k}_\rho,\quad \omega_1 = \omega - \Omega,\quad \omega_2 = \omega + \Omega,$$

where $\vec{q}$, Ω - is a wave vector and surface wave frequency; $\vec{k}_\rho$ - is a horizontal projection of a sound wave vector at frequency ω ; $\vec{k}_{i\rho}$ - is a horizontal projection of a sound wave vector at the combined frequency ω_i (Fig.2). It can be stated that this case closely correlates with surface wave activation as a result of combined high frequency sound scattering from surface waves.

Indeed even for the shortest surface waves the frequency of the corresponding sound waves occurs rather high. For example, for the surface waves with Λ = 1.73 cm (corresponding to the transfer of gravity waves to capillary waves and to the minimum velocity of surface wave propagation) we shall get:

$$q = 3{,}69 \text{ cm}^{-1}, \quad \omega \sim c_s q = 5{,}53\ 10^6 \text{ s}^{-1}.$$

The differing mechanism of interaction inside the resonance triplet should be marked. When the initial sound wave is at the highest in triplet frequency, it then, in the process of disintegrating, results in the growth of surface waves. In the 2nd triplet, where the dispersed wave of a combined tone has the maximum frequency, the interaction between the surface and initial sound waves causes the damping of the surface one. Thus, the growth and the damping of the surface wave depend on the relative contribution made by the 1st and the 2nd triplets. As I will show below this relative contribution is dependent on the angle of incidence of the initial sound wave.

Let us consider a process of induced sound wave scattering on the surface wave.

The process of scattering in linear approximation is characterized by the appearance of difractional spectra - that is plane waves, propagating in the direction defined from the condition:

$$k \sin\theta - k_n \sin(\theta - \theta_n) = nq, \quad (8)$$

where k_n - a wave number of the n difractional spectrum; θ_n - the angle of the relative specular reflection, at which the n difractional spectrum is propagating; θ - the angle of sound wave incidence. It implies that the angle θ is positive, if the horizontal project of the sound wave vector and the surface wave vector are directed likewise. The angle is negative in case these vectors are directed oppositely.

Let us proceed to the case of acoustic plane wave scattering with velocity potential

$$\varphi_n = \varphi_0 \exp(ik \sin\theta + ikz \cos\theta - i\omega t) \quad (9)$$

on the surface wave

$$\zeta = a \sin(qx - \Omega t), \quad (10)$$

which velocity potential has a form:

$$\varphi_q = a_q \exp\left(iqx + qz - i\Omega t\right),\ a_q = -a c_q. \qquad (11)$$

For a further solution in approximation of slowly varying amplitudes

$$\dot{a}_q \ll \Omega a_q$$

we make use of the joint boundary condition (4).

Grouping and equating to zero the members in equation (4) at the same exponents we get the 1st order differential equation of a relatively slow varying amplitude of the surface wave:

$$\dot{a} = a\,\delta(M, q, k, \omega) \qquad (12)$$

It is obvious that the solution of the equation has a form

$$a = a_0 \exp(\delta t),$$

where

$$\delta = M^2 \frac{c_s}{c_q}\frac{\omega}{2}\cos\theta\left\{\left[1-\left(\frac{\sin\theta - \frac{q}{k}}{1-\frac{\Omega}{\omega}}\right)^2\right]^{\frac{1}{2}} - \left[1-\left(\frac{\sin\theta + \frac{q}{k}}{1+\frac{\Omega}{\omega}}\right)^2\right]^{\frac{1}{2}}\right\}. \qquad (13)$$

Here M - Mach number for a sound that satisfies the relation

$$k\varphi_0 = M c_s.$$

Depending on the sign of the expression in the parentheses, that is on the ratio of sound and surface wave frequencies and on the angle of sound incidence, the process of nonlinear interaction of a sound with a wave on liquid surface may be accompanied by either the time dependent growth or the damping of surface waves with increment (or decrement) δ.

Fig.3 shows a family of frequency relations of

$$\Phi = \frac{\cos\theta}{2}\left\{\left[1-\left(\frac{\sin\theta - \frac{q}{k}}{1-\frac{\Omega}{\omega}}\right)^2\right]^{\frac{1}{2}} - \left[1-\left(\frac{\sin\theta + \frac{q}{k}}{1+\frac{\Omega}{\omega}}\right)^2\right]^{\frac{1}{2}}\right\}, \qquad (14)$$

derived for frequency $f = 2,3\ kHz$ and various values of the angle ϑ.
It is obvious, when the angle of the sound incidence increases (by its absolute value) the curve peak $\Phi(\Omega/\omega)$ widens, provided the peak amplitude grows initially and then goes down, reducing to zero at sliding incidence $(\theta \approx \pi/2)$. The value Φ proved positive at positive angles of incidence θ (excluding the narrowest region close to zero), that is induced combined sound scattering on the surface waves is accompanied by the wave growth on liquid surface; the negative angles of sound incidence make the Φ value also negative and the analysed process is characterized by the damping of surface waves.

4. Low frequency wave and high frequency wave train interaction

Let us consider the interaction of waves having different spacial and time scales. It may be gravity waves on internal ones, capillary waves, superimposed on gravity waves and so on.

Assign a slow wave in a form of a harmonic one:

$$\varphi_i = \varphi_{io}(z)\, e^{-i\Omega t + iqx}, \quad U_x = iq\varphi_{io}, \quad U_z = \frac{\partial \varphi_{io}}{\partial z}\Big|_{z=0}, \tag{15}$$

and a surface high frequency wave in a form of narrow spectrum wave train

$$\varphi_g = \varphi_{go}(x,t)\, e^{-i\omega_o t + ik_o x + k_o z}, \tag{16}$$

$\varphi_{go}(x,t)$ - is a slow varying amplitude. The below conditions must be followed

$$\omega_o \gg \Omega, \quad k_o \gg q. \tag{17}$$

The boundary is excited by the slow wave, thus changing the high frequency wave parameters

$$\frac{\partial \varphi_g}{\partial t} + u_x \frac{\partial \varphi_g}{\partial x} - g\zeta + \alpha \zeta_{xx} = 0, \tag{18}$$

$$\frac{\partial \zeta}{\partial t} + U_x \zeta_x = \frac{\partial \varphi}{\partial z} .$$

If solving the system (18) linearly at φ_{cg} = const, in the form (16) we obtain a dispersion equation, having Doppler frequency shift, induced by the slow U_x :

$$(\omega - U_x k)^2 = gk + \alpha k^3 \tag{19}$$

(Fig.4). The dispersion equation is greatly complicated by considering inhomogeneous flows with depth. Supposing that the function $U_x(z)$ has a stepped form we shall have instead of (19) [15,16] : $U(z) = U_0$ at $0 > z > -h$, $U(z) = 0$, $z < -h$

$$[(\omega - U_0 k)^2 - gk - \alpha k^3][(\omega - U_0 k)^2 + \omega^2] - \tag{20}$$
$$- [(\omega - U_0 k)^2 - gk + \alpha k^3][(\omega - U_0 k)^2 - \omega^2] e^{-2kh} = 0$$

(Fig.5, the calculation has been performed by S.Yu.Tretiakova).

As it is shown on Fig.5 group wave velocity turns to zero at point 1 and 3, and its phase velocity zeroing at points 2 and 4. Zeroing of the group wave velocity means that the flow stops the wave train. In inhomogeneous flow along the x-axis (for example on internal wave) blocking takes place at these points [17,18] . High frequency wave train is reflected noticeably growing in amplitude. On Fig.5 group wave velocity at points 1 and 2 turns to infinity. After passing the point $C_r = \infty$ the wave properties qualitatively change: in the region from zero to one (0-1) and from two to three (2-3) on Fig.5 wave energy becomes negative and such a wave is characterized by unusual properties: the presence of dissipation may lead it to its swinging.

Variations of parameters $\omega(x,t)$, $k(x,t)$ on inhomogeneous flow is determined from the following equations: the dispersion equation $Z(\omega, k) = 0$, and

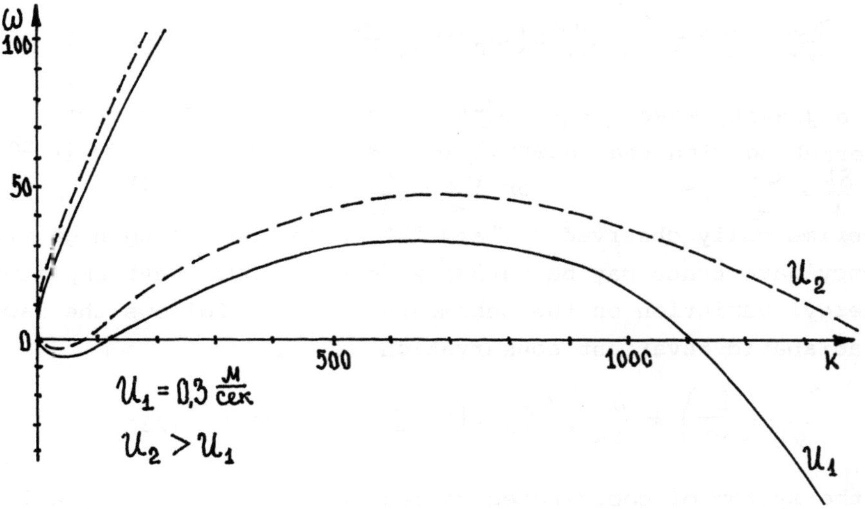
ω
100
50
0
500
1000
K
U_1 = 0,3 м/сек
U_2 > U_1
U_2
U_1

Fig.4

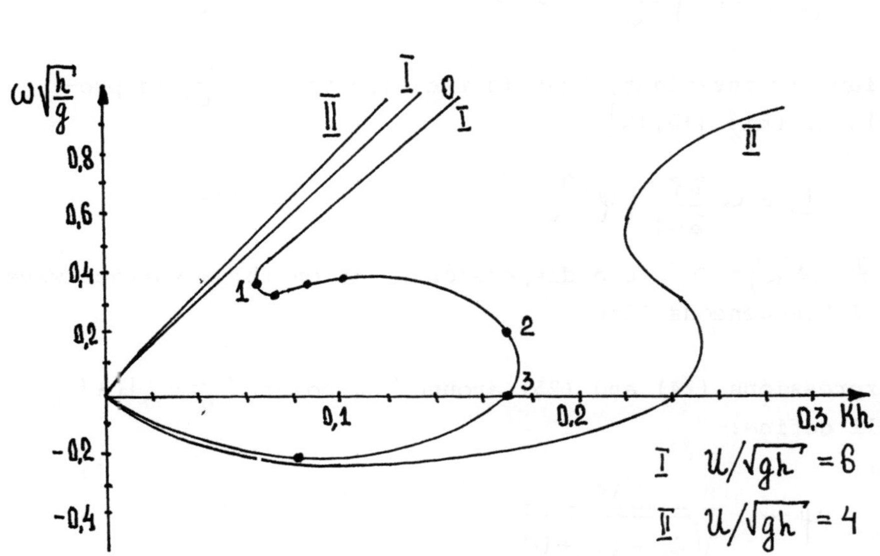
ω√(h/g)
0,8
0,6
0,4
0,2
-0,2
-0,4
0,1
0,2
0,3 Kh
II
I
0
I
II
1
2
3
I U/√gh = 6
II U/√gh = 4

Fig.5

$$\frac{\partial \omega}{\partial x} + \frac{\partial k}{\partial t} = 0, \quad \frac{\partial \omega}{\partial x} = (C_r + U_o)\frac{\partial k}{\partial x} + \frac{\partial u}{\partial x} k. \quad (21)$$

For a gravity wave $\varphi = \varphi_c \exp(-i\omega t + ikx)$, $\frac{\partial \omega}{\partial k} = C_r$, interacting with the internal wave we get [19]: if $C_r = C_i$, then

$$\frac{\delta k}{k} = \frac{q^2}{k} U_o t \quad \text{or} \quad \frac{\delta k}{k} = \frac{q^2}{k} \frac{U_o x}{C}. \quad (22)$$

Experimentally observed "ω" and "k" variations along high frequency wave trace may be rather wide up to 50%. Wave amplitude (energy) variation on the inhomogeneous flow follows the law of adiabatic invariant conservation [20].

$$\frac{\partial}{\partial t}\left(\frac{E}{\omega}\right) + \frac{\partial}{\partial x}\left[\left(C_r + U + C_i\right)\frac{E}{\omega}\right] = 0 \quad (23)$$

In the system of coordinates having the speed C_i, all the factors depend only on $\xi = x - C_i t$
Then

$$\left(C_r + U\right)\frac{E}{\omega} = \text{const} \quad (24)$$

(an adiabatic invariant, that is the flux vector $\frac{E}{\omega}$ is preserved). At this [15,18]

$$E = \omega \frac{\partial Z}{\partial \omega} |\varphi|^2, \quad (25)$$

where $Z(\omega, k) = 0$ is a dispersion equation for a surface wave on the inhomogeneous flow.

From expressions (24) and (25) around the point $C_r = -(U + C_i)$ we shall define:

$$\varphi \sim \frac{\varphi_o}{\sqrt{c_r - (c_i + U)}}. \quad (26)$$

In the regions from 2-3 to 4 on Fig.4 and from 1 to 0 (zero) and from 2 to 3 on Fig.5 $\omega > 0$, $\frac{\partial z}{\partial \omega} < 0$, it follows that according to (25) $E < 0$. Let us consider how the dissipation affects the negative wave energy [21]. Dissipation causes the imaginary adding to the frequency $\omega = \omega_o + i\omega_1$ of a harmonic wave, corresponding to the effective wave number k . The dispersion equation $Z(\omega, k, \alpha) = 0$ transforms into:

$$Z(\omega_0 + i\omega_1, k, \alpha_0 - i\alpha_0) = 0 .$$

Disintegrating it in small values α_0, ω_1 we obtain

$$\omega_1 = \alpha_1 \frac{\partial Z}{\partial \alpha} \Big/ \frac{\partial Z}{\partial \omega} \qquad (27)$$

For a wave of positive energy $\frac{\partial Z}{\partial \omega} > 0$ and $\omega_1 > 0$ that is the wave decays $e^{-\omega_1 t}$. For a wave of negative energy $\frac{\partial Z}{\partial \omega} < 0$ and at $\alpha_1 > 0$, a wave due to dissipation swings exponentially. It should be noted, that the ratio $\partial Z / \partial \omega$ is invariant under the Galilean transformation of the coordinates [22] and negative energy wave growth due to dissipation has obviously no connection with the choice of a coordinate system. A surface wave of negative energy is swung as a result of its bulk wave radiation as well as dissipation. In [22] an example of transverse wave swing in the membrane due to its radiation of sound waves into a liquid half-space, and relative to which the membrane is moving with speed $U > C_\varphi$, where C_φ is wave velocity in a free membrane, is given. Another case of radiation instabilities, that is a shift flow, may be instable in relation to the radiation of internal bulk waves.

High frequency wave can inversely affect low frequency wave. A high frequency wave may disintegrate, that is, for example, a pair of high frequency waves composing a high frequency wave train as a result of resonance triplet interaction can give birth to a low frequency wave with phase velocity close to sound velocity of a high frequency wave train. Further self-trapping is possible resulting in moving of the above waves with equal velocity, and their envelopes form coupled solitons [24,25] .

References

1. Brekhovskikh L.M. On sound wave generation in liquid by surface waves. - Acoust. Zh. 1966, v.12, N 3, p.376-378.

2. Liamshev L.M. On ultrasonic wave gain in non-conducted liquid at the interaction with fine elastic piezosemiconductor - Acoust. Zh. 1969, 15, 3, p.460-462.

3. Liamshev L.M. On the non-specularly reflected wave gain. - Acoust. Zh. 1969, 15, 2, p.303-304.

4. Liamshev L.M. On acoust. wave gain and absorption when reflected by a crystal CdSe in water. - DAN SSSR, 1970, 194, 5, p.1049-1051.

5. Liamshev L.M., Kurilkin Ju.V. On sound reflection by crystals in liquid. - Acoust. Zh. 1970, 16, 4.

6. Benjamin T.B. , Feir J.E. The disintegration of wave trains on deep water. Part 1. Theory. J.Fluid Mech. 1967, 27, 417.

7. Rayleigh G. Sound theory. M., GITTL., 1955.

8. Brand R.P., Nyberg W.L. Parametrically exited surface waves. - J.Acoust. Soc. Am. 1965, 37, 3, 509-515.

9. Eisenmenger W. Dynamic properties of the surface tension of water waves and aqulous solution of surface active agents with standing capillary waves. Acustica, 1959, 9, 327-334.

10. Gavrikov V.K., Lukashova N.V. Induced sound scattering on liquid surface. - Acoust. Zh. 1973, v.19, N 4, p.512-518.

11. Bozhkov A.I. Instability of the two media interface in a sound field. - Acoust. Zh., 1974, v.20, N 5, p.690-699.

12. Huges B. Some theoretical aspects of the generation of surface ripples by parametric subharmonics resonance with sound waves. - J. Acoust. Soc. Am., 1977, 61, 2, 407-412.

13. Goncharov V.V., Naugolnykh K.A., Rybak S.A. On surface wave excitation by sound. - AN SSSR FAO, 1977, v.4, p.431-434.

14. Naugolnykh K.A., Rybak S.A. Low linear wave attenuation. - Acoust. Zh., 1976, v.22, N 5, p.735-741.

15. Voronovich A.G., Rybak S.A. Implosive instability of stratified flows. - DAN SSSR, v.239, N 6, 1978.

16. Voronovich A.G., Lobanov E.V., Rybak S.A. On gravity-capillary wave stability in the presence of inhomogeneous flow. Izv. AN SSSR FAO, 1980, N 3, p.329-331.

17. Basovich A.Ja., Talanov V.I. Adiabatic wave interaction. - In "Nonlinear waves", ed. by Gaponov-Grekov A.V., M., 1981,

p. 147-165.

18. Fillips O.N. Dynamics of oceanic upper layer. - L., Gidrometeoizdat, 1980.

19. Lewis J.E., Lake B.M., Ko D.R.S. On the interaction of internal waves and surface gravity waves. - J. Fluid Mech., 1974, 63, 773-800.

20. Whitham G.B.Waves with dispersion and variational principles. - In "Nonlinear waves" edited by Leibovich S.and Seebass A., M., Mir, 1977, chap.5.

21. Lighthill J. Waves in liquids. - M., Mir, 1981.

22. Rybak S.A. Sound waves with negative energy. - Acoust. Zh., 1980, v.26, N 2, p. 248-256.

23. Ostrovsky L.A., Tsimring L.Sh. Radiative instability of shear flows in stratified liquid. - AN SSSR FAO, 1981, v.17, N 7, p. 766-768.

24. Karamzin Ju.N., Sukhorukov A.P., Filipchuk T.S. On a new class of connected solitons in a dispersive medium with square nonlinearity. - Vest. Mosk. Univ. Fiz., astr., v.19, N 4, p.91-98, 1978.

25. Petrov V.V. On non-linear surface wave dynamics in stratified ocean. - AN SSSR FAO, v.15, N 7, p. 740-749, 1979.

Experimental Determination of the Nonlinear Properties of Elastic Solids

MACK A. BREAZEALE

Department of Physics
The University of Tennessee
Knoxville, TN 37996-1200 USA

Summary

Since an ultrasonic wave propagating through a solid is phase matched with its harmonics, harmonic growth becomes a means of evaluating the nonlinear properties of solids. By generating a finite amplitude ultrasonic wave in crystalline solids at 30 MHz and measuring absolute amplitudes as small as 10^{-4} Angstroms at 30 MHz and 60 MHz, one can determine the nonlinear properties of solids and characterize them with third-order elastic (TOE) constants.

Most TOE constants are negative; however, a few are positive. This means that in most cases the (displacement) second harmonic is generated in phase with the fundamental, but in a few it is generated out of phase. Since the phase of the second harmonic is not readily determined in routine measurements, we have used a phase sensitive detector at 60 MHz to determine relative phase of fundamental and second harmonic and hence unambiguously to determine TOE constants.

The experimental technique is described and a review is given of data on fused silica, copper single crystals, and other crystals of cubic symmetry.

I. Introduction

An adequate description of the propagation of an ultrasonic wave through a solid can only be made in terms of a nonlinear differential equation, because in general the response of a solid to a deformation wave is nonlinear. This is obvious, and has been for some time. What is less obvious is where we go after admitting that linearity is a figment of our physical imagination.

Whether viewed from a macroscopic perspective or a microscopic perspective, many properties of the solid can only be described in terms of a nonlinear (anharmonic) theory. Such properties as thermal expansion, wave interaction, pressure and temperature

Nonlinear Deformation Waves
IUTAM Symposium Tallinn 1982
Editors U Nigul, J Engelbrecht

dependence of the elastic constants, coupling among the lattice phonon modes, and the interaction between sound waves and thermal vibrations require nonlinear (anharmonic) theories. Thus, the results of experiments which measure the nonlinear properties of solids are of fundamental interest to solid state physics.

The most convenient form for data on the nonlinear properties of solids is in terms of the coefficients of a Taylor expansion of the strain energy in terms of the strains. This leads to the definition of the second-order (ordinary) elastic constants and the third-order elastic (TOE) constants. The second-order elastic constants come directly from the square of the sound velocity and enter into virtually any theory which describes the linear mechanical properties of solids. The TOE constants which result from measurement of the nonlinearity parameter enter into virtually any theory which describes their nonlinear (or anharmonic) mechanical properties. Thus, accurate values of TOE constants are very important.

I should like to review some recently published results on TOE constants of solids taken by the harmonic generation technique. Before I do, however, Section II is devoted to a short analysis of the theory, Section III is a description of the experimental arrangement, and Section IV describes two experimental observations which appear to be unique to our experiment. Finally, Section V is a review of some recent results.

II. Theory

A formal description of the propagation of a deformation wave can be made in terms of Lagrangian variables a_i (i = 1,2,3) and the time t. The a_i are Cartesian coordinates of a point in the solid at time t = 0. At a later time t, the Cartesian coordinates of the point in the strained solid are x_i (i = 1, 2,3). Then, the a_i and t are the independent (Lagrangian) variables upon which the variables x_i depend. These variables enter into the Jacobian matrix as matrix elements in the form

$$J_{k\ell} = \frac{\partial x_k}{\partial a_\ell} \quad . \tag{1}$$

In turn, the Jacobian matrix is used to define the strain matrix [1]

$$\eta = \frac{1}{2} (J^*J - I) \tag{2}$$

where I is the unit matrix and J* is the transpose of J. The elements of the strain matrix then enter into the definition of the elastic potential energy per unit unstrained volume. The elastic potential energy is the function which contains all the material properties of the medium, and the definition of the elastic potential energy is the starting point for the derivation of the nonlinear wave equation. Once we define the elastic potential energy, the only task remaining is to substitute it into Lagrange's equations, then out will come the nonlinear wave equation [2]. The elastic potential energy can be written in the form

$$\phi(\eta) = \sum_{ijk\ell} C_{ijk\ell}\eta_{ij}\eta_{k\ell} + \sum_{ijk\ell mn} C_{ijk\ell mn}\eta_{ij}\eta_{k\ell}\eta_{mn} + \ldots \tag{3}$$

where the $C_{ijk\ell}$ are the ordinary (second-order) elastic constants and $C_{ijk\ell mn}$ are the third-order elastic (TOE) constants. It enters directly into the Lagrangian function $\mathcal{L}$:

$$\mathcal{L} = \frac{1}{2} \sum_{i=1}^{3} \rho \left(\frac{\partial x_i}{\partial t}\right)^2 - \phi(\eta) \tag{4}$$

where ρ is the unstrained mass density. In turn, the Lagrangian function enters into Lagrange's equation

$$\frac{d}{dt}\left(\frac{\partial \mathcal{L}}{\partial\left(\frac{\partial x_i}{\partial t}\right)}\right) + \sum_{k=1}^{3} \frac{d}{da_k}\left(\frac{\partial \mathcal{L}}{\partial\left(\frac{\partial x_i}{\partial a_k}\right)}\right) = 0 \tag{5}$$

with the result that if we choose the a_1 axis along the direction of propagation Lagrange's equation becomes

$$\rho \frac{\partial^2 x_i}{\partial t^2} = \frac{d}{da_1} \sum_{k=1}^{3} J_{ik} \frac{\partial \phi(\eta)}{\partial \eta_{1k}} \tag{6}$$

where i = 1,3,3.

Equation 6 points out clearly the central role played by the elastic potential energy expansion $\phi(\eta)$. For purposes of reducing the equations to a form which can be applied to the propagation of deformation waves in cubic crystals, we will use the expansion given in Eq. 3; however, one is not limited to this expansion. In situations in which there are other processes than mechanical nonlinearity, the effect of that new process would change the form of the potential energy expansion, and hence the form of the final differential equation.

For the present purposes, however, we restrict our attention to mechanical nonlinearities in cubic crystals. If one chooses the x_k axes parallel to the a_k axes, one can write the particle displacement

$$\xi_k = x_k - a_k \, . \tag{7}$$

One may then perform the operations appropriate to the crystal symmetry and obtain

$$\rho \frac{\partial^2 \xi_i}{\partial t^2} = \sum_{j=1}^{3} A_{ij} \frac{\partial^2 \xi_i}{\partial a_1{}^2} + \sum_{j,\ell=1}^{3} B_{ij\ell} \frac{\partial \xi_j}{\partial a_1} \frac{\partial^2 \xi_\ell}{\partial a_1{}^2} \tag{8}$$

where the A_{ij} are known linear combinations of second-order elastic constants and the $B_{ij\ell}$ are known linear combinations of second-order and TOE constants. Equation 8 is the general form taken by the nonlinear wave equation in a crystal; however, it can be further simplified by specializing to the principal directions in cubic crystals.

For comparison with experiment in which waves are excited along [100], [110], and [111] directions, it is convenient to separate out those terms in $B_{ij\ell}$ (in Eq. 8) which contain second-order elastic constants. If we let K_2 represent a linear combination of the appropriate second-order elastic constants and K_3 represent a linear combination of TOE constants, then the nonlinear equation for plane longitudinal waves propagating in either the [100], the [110], or the [111] direction in a cubic crystal takes the form [3]

$$\rho \frac{\partial^2 \xi}{\partial t^2} = K_2 \frac{\partial^2 \xi}{\partial a^2} + (3K_2 + K_3) \frac{\partial \xi}{\partial a} \frac{\partial^2 \xi}{\partial a^2} + \cdots \tag{9}$$

For the three principal directions in a cubic crystal the linear combinations of second-order elastic constants K_2 and the linear combinations of TOE constants are given in Table I.

TABLE I

K_2 AND K_3 PARAMETERS FOR THE THREE PRINCIPAL DIRECTIONS

Direction of Wave Propagation	K_2	K_3
[100]	C_{11}	C_{111}
[110]	$\frac{1}{2}(C_{11} + C_{12} + 2C_{44})$	$\frac{1}{4}(C_{111} + 3C_{112} + 12C_{166})$
[111]	$\frac{1}{3}(C_{11} + C_{12} + 4C_{44})$	$\frac{1}{9}(C_{111} + 6C_{112} + 12C_{144} + 24C_{166} + 2C_{123} + 16C_{456})$

For example, for the [100] direction Eq. 9 becomes

$$\rho \frac{\partial^2 \xi}{\partial t^2} = C_{11} \frac{\partial^2 \xi}{\partial a^2} + [3C_{11} + C_{111}] \frac{\partial \xi}{\partial a} \frac{\partial^2 \xi}{\partial a^2} + \ldots \tag{10}$$

If one produces a sinusoidal oscillation at the end of the solid, defined by a = 0, one finds that a perturbation solution of Eq. 15 takes the form

$$\xi = A \sin(ka - \omega t) - (3 + \frac{C_{111}}{C_{11}}) \frac{A^2k^2a}{8} \cos 2(ka - \omega t) + \ldots \tag{11}$$

with a similar solution for the other directions, as can be obtained by substitution of the appropriate expressions for K_2 and K_3 from Table I. These solutions form the basis of the measurement technique described in Section III.

III. Experimental Arrangement

As can be seen from Eq. 11, one can obtain a value of the TOE constant C_{111} by generating a sinusoidal plane wave and allowing the wave to propagate along the [100] direction in the sample of interest. The measurement consists of determining the fundamental amplitude A, the second harmonic amplitude A_2, the sample length a, the propagation constant $k = \omega/c$. These quantities are related by

$$C_{111} = -(3 + \frac{8A_2}{A^2k^2a}) C_{11} , \tag{12}$$

with similar expressions for the [110] and the [111] directions. For a fundamental of 30 MHz, typical values of A_2 and A are, respectively, 10^{-9} cm and 10^{-7} cm, which means that a very sensitive technique is required for accurate measurements.

To make measurements we use the experimental arrangement shown in Fig. 1. A pulsed oscillator drives a quartz crystal at 30 MHz. The sample has optically flat and parallel ends which are coated with a conductive layer if necessary. The detector, to be described shortly, produces an electrical signal which is proportional to the amplitude of either the 30 MHz fundamental component or the 60 MHz second harmonic. The two signals can be measured by use of the oscilloscope, or other more complicated devices.

The detector is shown in Fig. 2. The quartz transducer at the top of the sample generates a 30 MHz ultrasonic pulse which propagates through the sample and impinges on the bottom which is spaced approximately 10 μ from the detector button. A dc bias of ∿ 150 volts causes the space between the sample and the

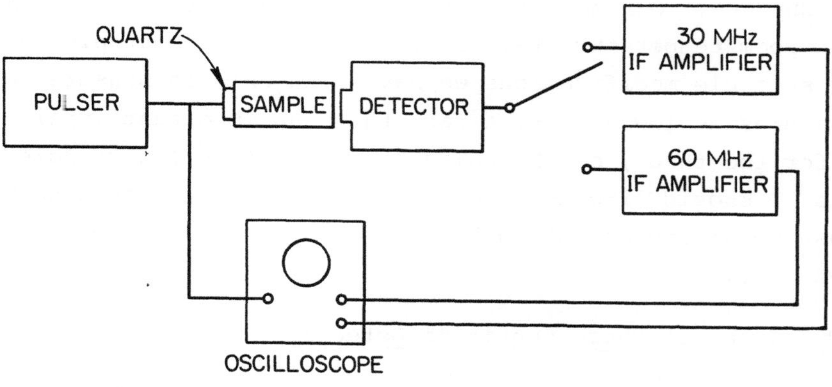

Fig. 1. Experimental arrangement.

detector button to function as a capacitive microphone which generates an electrical signal proportional to the amplitude of the ultrasonic wave. Since the proportionality constant is a known function of the physical parameters of the detector, measurement of the voltage generated gives a direct measure of the amplitude of the ultrasonic wave. We have been able to make

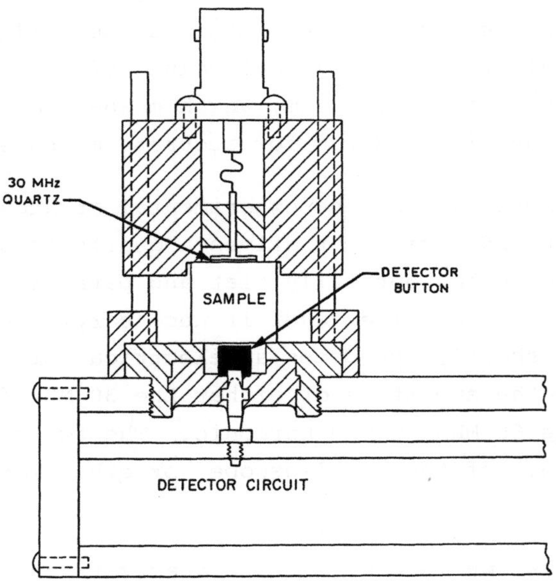

Fig. 2. Capacitive detector for measuring ultrasonic wave amplitudes.

measurements of waves with amplitudes as small as 10^{-12} cm with this technique. Our measurements have produced information not only on room temperature values of TOE constants, but also by making suitable modifications we have been able to measure to temperatures as low as 3°K, a fact that has important implications for the theory of the solid state. Before I give some results, I should like to describe points at which the concept of phase sensitive detection have entered into our experimental observations.

IV. Two Types of Phase Sensitive Detection

a. Pulse superposition of second harmonic. During the course of investigating possibilities to improve signal-to-noise ratio the pulse overlap technique of McSkimin [4] was applied to the harmonic generation technique with rather intriguing results. With the overlap technique a pulsed ultrasonic signal (1) is generated at the end of the sample as indicated in Fig. 3.

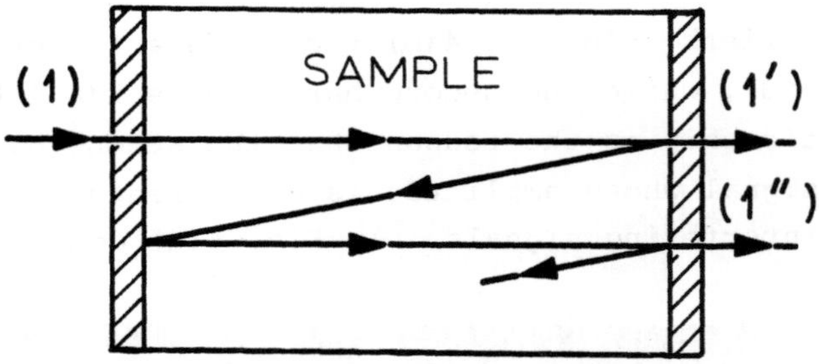

Fig. 3. Pulse overlap in sample.

If the pulse duration is less than the time required for the deformation wave to travel twice the length of the sample, a pattern of distinct pulses is sensed by the detector as (1'), (1''), etc. Each successive pulse echo is smaller than its predecessor, and an exponential decay pattern is observed. If the pulses are lengthened, a part of the signal (1') will still be at the detector when the reflection (1'') arrives. The frequency of the signal can be adjusted to give either constructive or destructive interference. The constructive interference provides enhancement of the signal. Thus, one is able to measure at one time the sum of several pulses.

If the interference of distorted (nonsinusoidal) waves is observed [5], the pulse decay patterns of the fundamental and the second harmonic are different, since the amplitude of the second harmonic depends on the square of the fundamental amplitude (cf. Eq. 16). Since the distorted wave passes through amplifiers tuned to the fundamental or the second harmonic, the two components can be displayed separately. Figure 4 gives the detected pulse-echo patterns obtained by overlapping distorted wave pulses. In each photograph the upper trace shows the fundamental and the lower trace the second harmonic. The distinct pulses are shown in Fig. 4(a), then single overlap in phase (Fig. 4(b) and out of phase (Fig. 4(c)). Double overlap in phase is shown in Fig. 4(d). Let us label the amplitudes of the distinct fundamental pulses A, B, and C. The amplitudes of the second harmonic pulses then are kA^2, kB^2, and kC^2, where the constant k depends on the relative amplification factors of the two amplifiers. In Fig. 4(b) the fundamental on overlap becomes (A + B), while the second harmonic is $k(A + B)^2$. This square relationship is the reason that the second harmonic produces a signal whose amplitude is much greater than the sum of the two interfering signals. In Fig. 4(d) double overlap

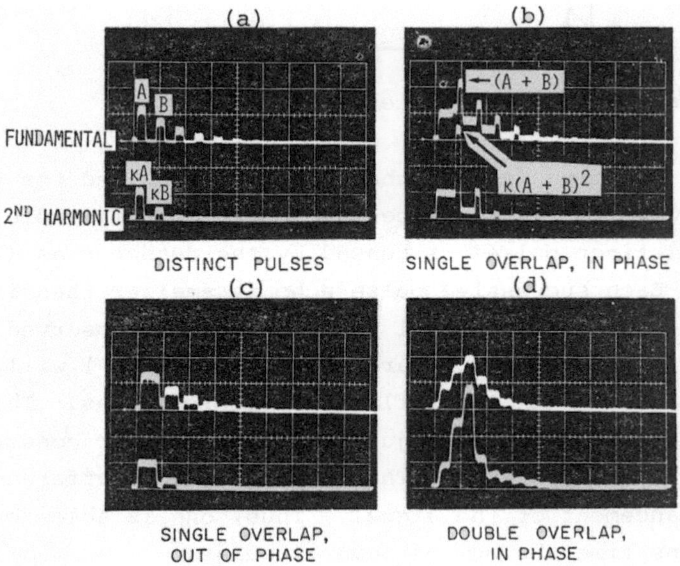

Fig. 4. Overlap of distorted deformation waves. Upper traces: fundamental; lower traces: second harmonic.

shows the same effect. The ledges for the fundamental are A, (A + B), and (A + B + C); those for the second harmonic are kA^2, $k(A + B)^2$, and $k(A + B + C)^2$. This interference phenomenon is the nonlinear deformation wave analogue of phase sensitive detection. It can be quite useful in situations in which the signals are weak because the square relationship on overlap most favorably enhances the second harmonic, the weaker signal.

b. Use of Phase Sensitive Detector to Exhibit Backward Distortion of Nonlinear Deformation Waves. In fluids and in solids as a sinusoidal nonlinear deformation wave propagates, the waveform becomes distorted. This distortion, characterized by the growth of harmonics of the initial disturbance, has been described for fluids as resulting from the fact that points of maximum condensation travel faster than points of minimum condensation. This results in a "forward-sloping wave" with the velocity profile shown in Fig. 5(a). Such a velocity profile is found in most materials, and it is tempting to assume that it exists in all materials, for such an assumption is valid for all fluids. For solids, however, the situation can be different. We have found that at least one exceptional material, fused silica, exists in which the waveform distortion results in a "backward-sloping wave" as indicated in Fig. 5(b). Such a wave profile presents an interesting physical situation in addition to presenting a potential problem in data interpretation. The fundamental and second harmonic Fourier components drawn in Fig. 5 clearly show that the difference between the two wave profiles is only that the second harmonic in Fig. 5(b) has a phase difference of 180° from that in Fig. 5(a). In ordinary measurements phase information is lost; however, a phase difference of 180° is readily demonstrated by an electrical phase sensitive detector. Such devices can be purchased for operation at low frequencies; however, the fact that we wanted to operate at 60 MHz made it necessary to build our own.

A quartz crystal is used to generate pulsed 30 MHz ultrasonic waves, which are detected by a capacitive detector as they impinge on the optically flat sample end. The output of the

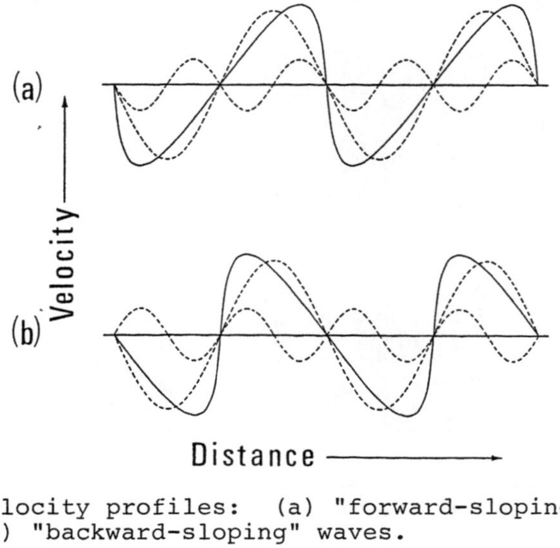

Fig. 5. Velocity profiles: (a) "forward-sloping" waves; (b) "backward-sloping" waves.

capacitive detector is fed to two tuned RF amplifiers—one for the fundamental and one for the second harmonic, as shown in Fig. 6. The fundamental frequency echo train is then frequency doubled and this signal and the second harmonic are fed into the phase-sensitive detector whose output is proportional to the cosine of the phase angle between the two 60 MHz signals.

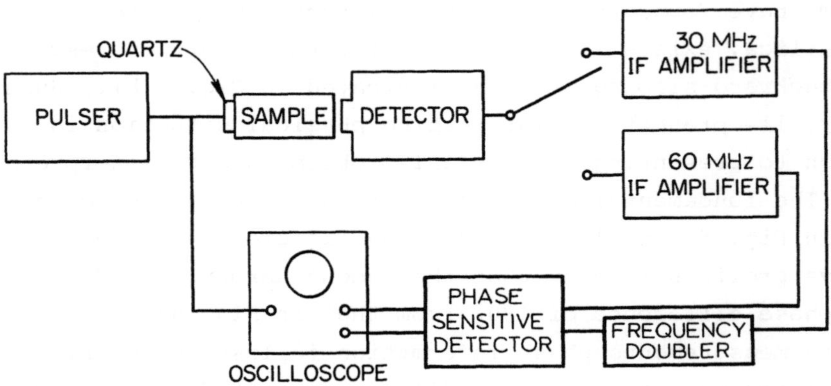

Fig. 6. Experimental arrangement for phase-sensitive detection.

With a sample of copper single crystal oriented in such a manner that the ultrasonic waves travelled along the [100] direction, the phase between the harmonic and the (frequency-doubled) fundamental signals in the phase-sensitive detector is adjusted to zero [6]. This produced a maximum positive-going echo train shown in the trace in Fig. 7(a).

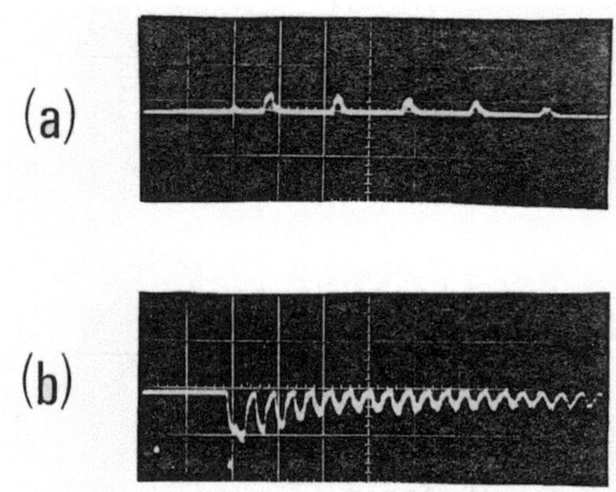

Fig. 7. Oscilloscope traces showing relative phases of second harmonic in (a) copper and (b) fused silica.

Without making any adjustment in the electronics, the copper sample was replaced by the fused silica sample. The signal was observed to be negative, as shown in the oscilloscope trace, Fig. 7(b). Further adjustment of the phase did not produce a more negative signal. Thus, within the accuracy of the apparatus there is exactly a 180° difference between the phase of the second harmonic with respect to the (frequency-doubled) fundamental in fused silica and that in copper.

The interpretation of this phase difference can readily be made by referring to the second harmonic amplitude which comes from Eq. 11:

$$A_2 = -(3 + \frac{C_{111}}{C_{11}}) \frac{A^2k^2a}{8} . \qquad (13)$$

All quantities in this expression are inherently positive except C_{111}. For copper it is inherently negative, with a magnitude greater than $3C_{11}$. Thus, A_2 is positive. For A_2 to be negative, one must have $C_{111} > -3C_{11}$. In fact, for fused silica C_{111} is positive, a most unusual situation. The positive C_{111} in fused silica means that A_2 is negative, which corresponds to distortion in the "backward" sense shown in Fig. 7(b). Fused silica becomes less stiff on compression, which causes the points of maximum condensation to travel more s l o w l y than the points of minimum condensation. Such a condition might not be observed for arbitrarily large amplitudes; however, it is clear from the results of our phase sensitive detector measurements shown in Fig. 7(b) that for the amplitudes used in our experiment this is a valid description.

V. Experimental Results

Measurement of the fundamental and the second harmonic amplitude of nonlinear deformation waves propagating through a sample allows one to calculate the combinations of TOE constants K_3 given in Table I: C_{111}, $\frac{1}{4}(C_{111} + 3C_{112} + 12C_{166})$ and $\frac{1}{9}(C_{111} + 6C_{112} + 12C_{144} + 24C_{166} + 2C_{123} + 16C_{456})$. One can evaluate simpler combinations, for example, simply by subtracting C_{111} from the K_3 in the [111] direction. This allows one to evaluate: C_{111}, $(C_{112} + 4C_{166})$, and $(C_{123} + 6C_{144} + 8C_{456})$. These are the combinations to be discussed.

<u>a. Fused Silica</u>. Fused silica, in which a "backward-sloping" deformation wave propagates, is the single substance known to date which has a positive TOE constant C_{111}. Since fused silica is an isotropic material, one can measure only one TOE constant C_{111} by the harmonic generation technique. Such measurements [7] prove interesting because they are very sensitive to structural characteristics of the samples. Measurements of four different types of fused silica samples have given the results shown in Fig. 8. The symbol W indicates that the sample

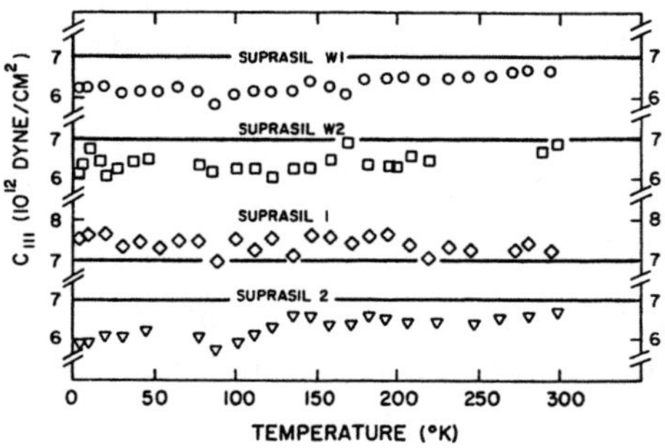

Fig. 8. Temperature dependence of C_{111} for the fused-silica samples.

has only 5 ppm of OH ions, while the other samples had 1200 ppm. The designation 1 indicates strong homogeneity in all directions, whereas 2 implies a guaranteed strong homogeneity only in the direction parallel to the direction of propagation of the ultrasonic waves.

The magnitudes of C_{111} for three of the four samples are nominally the same, as can be seen by noting that the horizontal line is drawn at 7×10^{12} dynes/cm^2 for each sample. The fourth sample, Suprasil 1 has a magnitude of C_{111} that is approximately 20% higher than the others. We have not been able to associate this difference with obvious differences in physical properties; however, the scatter of the data as the temperature changed seems to indicate structural changes of the samples as a function of temperature, for such scatter of the data does not show up with other materials. In large part, this structural change may be due to the asymmetrical nature of the void surrounding the oxygen atom. As the temperature is lowered there is evidence that the Si-O-Si bond angle changes because of the asymmetrical vibrations of the oxygen atoms. Thus, the tetrahedral network of fused silica below the quenching temperature is in a state of stress. The resulting states of strain are more accentuated for the weak impurity bonds and this makes them more efficient electron traps than the strong Si-O bonds. This means that the effect of the temperature change is to cause variations, especially in the anharmonicity-dependent effects since it is these effects that are most sensitive to variation in structural symmetry and strain perturbations introduced by impurity sites.

b. Strontium Titanate. The sensitivity of C_{111} to structural changes is especially noticeable in materials which undergo phase transitions. For example, strontium titanate, which undergoes a phase transition from cubic to tetragonal phase as the temperature is lowered to 102.5°K, exhibits the behavior in Fig. 9. These data, taken by Mackey and Arnold [8], show that C_{111} approaches very large negative values as the transition temperature is approached. The other two combinations of TOE constants also show anomalous behavior as the transition temperature is approached. They appear to be going to zero. The exact meaning of this is difficult to interpret in terms of a physical model, however, for the transition from

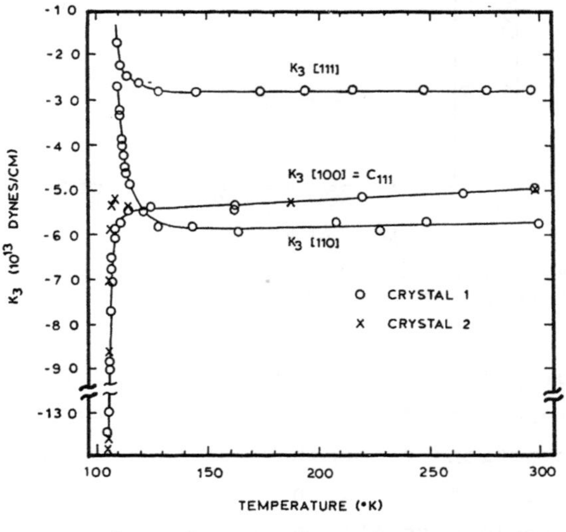

Fig. 9. Temperature dependence of K_3 in strontium titanate.

cubic to tetragonal phase involves a change in the total number of TOE constants. Hence, as the structure of the crystal gradually changes from cubic symmetry to tetragonal symmetry, the very mathematical expressions describing the propagation of the deformation wave gradually change as well. A physical model which adequately describes such behavior would be interesting indeed.

c. Semiconductors—Germanium and Silicon. The TOE constants of semiconductors germanium and silicon exhibit the temperature behavior shown in Figs. 10 and 11. The TOE constant C_{111} is almost temperature independent. The combination $C_{123} + 6C_{144} + 8C_{456}$ goes to zero at low temperatures.

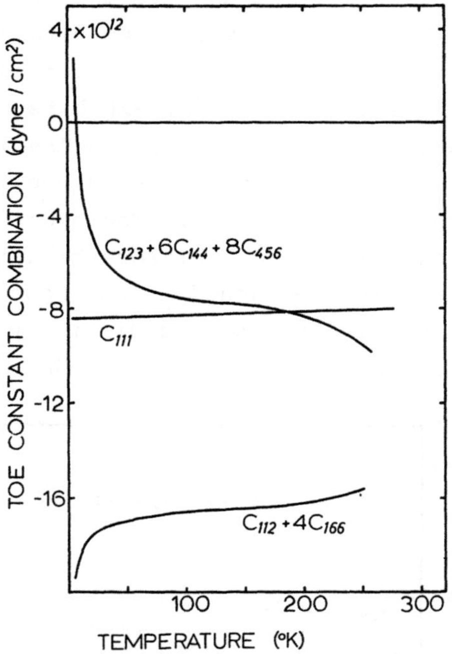

Fig. 10. Temperature dependence of combinations of TOE constants of germanium.

For a point of reference, one can observe that in a solid in which the interatomic forces are central forces and nearest-neighbor interactions predominate these quantities would be identically zero. Further, in such a solid the combination $C_{112} + 4C_{166}$ would have the value $\frac{5}{2} C_{111}$.

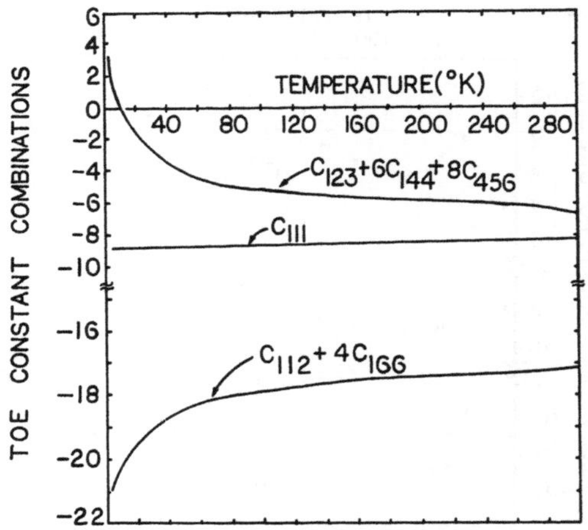

Fig. 11. Temperature dependence of combinations of TOE constants of silicon.

Examination of Figs. 10 and 11 reveals that both of those conditions are nominally met as the temperature approaches 0°K. Thus, in a general way, one might say that the central forces, nearest-neighbor model seems to be a reasonable description of the TOE constants of germanium and silicon as the temperature approaches 0°K. At higher temperatures a more complicated model is required.

d. Metal—Copper. For copper the requirements of the central forces, nearest-neighbor model appear to be met reasonably well throughout the temperature range 0 to 300°K, as seen in Fig. 12. At two temperatures, 40°K and 180°K, the combination $C_{123} + 6C_{144} + 8C_{456}$ appears to be identically zero. At these temperatures, therefore, the model is a very good description of copper.

VI. Conclusion

Measurement of the amplitudes of the fundamental and the harmonic of an initially sinusoidal nonlinear deformation wave can lead to some interesting observations about the way the waves propagate. Further, accurate determination of the amplitudes of the fundamental and the second harmonic components leads to values of combinations of TOE constants whose temperature dependence allows some fundamental observations about the nonlinear behavior of solids.

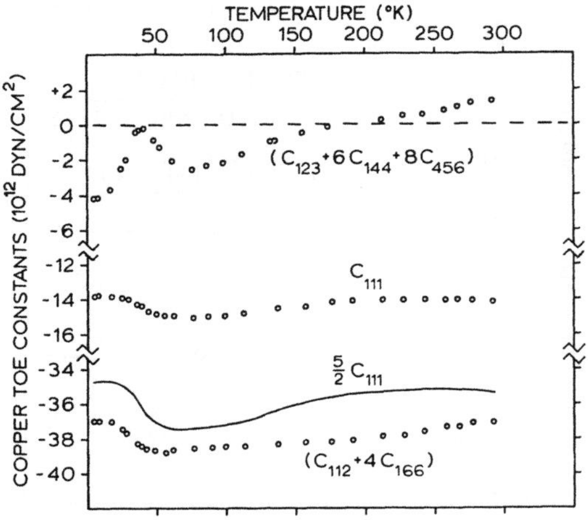

Fig. 12. Combinations of TOE constants of copper plotted as a function of temperature.

References

1. Murnaghan, F. D.: Finite deformation of an elastic solid. New York: John Wiley and Sons 1951.

2. Holt, Albert C. and Ford, Joseph: Theory of ultrasonic pulse measurements of third-order elastic constants for cubic crystals. J. Appl. Phys. 38 (1967) 42-50.

3. Breazeale, M. A. and Ford, Joseph: Ultrasonic studies of the nonlinear behavior of solids. J. Appl. Phys. 36 (1965) 3486-3490.

4. McSkimin, H.J.: Ultrasonic measurement techniques applicable to small solid specimens. J. Acoust. Soc. Am. 22 (1950) 413-418.

5. Gauster, W. B. and Breazeale, M. A.: Interference of finite amplitude ultrasonic waves in solids. J. Acoust. Soc. Am. 41 (1967) 860-861.

6. Bains, James A., Jr. and Breazeale, M. A.: Nonlinear distortion of ultrasonic waves in solids: Approach of a backward sawtooth. J. Acoust. Soc. Am. 57 (1975) 745-746.

7. Cantrell, John H. Jr. and Breazeale, M. A.: Ultrasonic investigation of the nonlinearity of fused silica for different hydroxyl-ion contents and homogeneities between 300 and 3°K. Phys. Rev. B 17 (1978) 4864-4870.

8. Mackey, James E. and Arnold, R. T.: Some combinations of third-order elastic constants for strontium titanate. J. Appl. Phys. 40 (1969) 4806-4811.

Acoustic Emission and Damage Accumulation in Solids

V.V.BOLOTIN, E.V.LOBANOV

USSR Academy of Sciences, Mechanical Engineering Research Institute, Moscow

Summary

A stochastic model of dispersed damage accumulation in solids is presented assuming a specimen as a set of microstructural elements with random properties. Estimates of the intensity and the spectral density of the acoustic emission are proposed based on the stochastic model. The Kaiser effect and the **power** law for the acoustic emission intensity can be obtained on the framework of the model proposed.

The up-to-date instrumentation opens a possibility to registrat very weak acoustic signals. The acoustic emission (AE) testing (AET) became a convenient approach to study deformation, damage and fracture phenomena both in laboratories and in operational conditions. AET is prospective for NDT of structures and machines, and for prediction their residual strength and life-time parameters. General principles, instrumentation and applications are presented in [1].

In this paper the relation is discussed between AE and the stochastic models of dispersed damage accumulation in solids proposed in [2] and developed in [3,4]. Since the AE signals are very weak, a comment seems to be necessary to propose this paper for the symposium devoted to nonlinear waves in solids. Firstly, the sources of AE (slips, micro- and macro-cracks) are of essentially nonlinear nature. Secondly, the AET includes as a necessary element a threshold detection procedure which in fact is a nonlinear transform of an initial signal.

The following data are used in AET: summation number of counts $N(t)$, the count rate $\dot{N}(t)$, the number of counts per cycle $N(n)$, the total logarithmic level $L(t)$, as well as AE

Nonlinear Deformation Waves
IUTAM Symposium Tallinn 1982
Editors U. Nigul, J. Engelbrecht

signatures and their Fourier transforms.

Let the detection threshold be chosen corresponding to damage of certain elements of a solid's structure: grains, crystals and intergrain boundaries in polycrystaline solids, segments of fiberes with the adjacent matrix in fibers reinforced composites [2]. Assume that the properties of elements are random, but all elements belong to one general sample. The number of damaged elements $N(t)$ in a certain volume of solid V is

$$N(t)=\sum_{k=1}^{N_o} \eta(t-\tau_k). \qquad (1)$$

Here N_o is a total number of elements in V, $\eta(t)$ is the Heaviside function, τ_k is the life-time of an arbitrary taken element beginning from $t=0$. Let $N(0)=0$, $N_o \gg 1$, and it is considered that $N(t) \gg 1$, but $N(t)/N_o \ll 1$. Since the density of damaged elements is small and the interaction of elements is negligible, the certain limit theorem is applicable. The probability distribution $F_N(N;t)$ for the number of damaged element is

$$F_N(N;t) \sim \Phi\left\{ (N-E[N(t)])/\sqrt{D[N(t)]} \right\}, \qquad (2)$$

where $\Phi(z)$ is the normalized function of Gaussian distribution The mathematical expection $E[N(t)]=N_oQ(t)$ and the dispersion $D[N(t)]=N_oQ(t)[1-Q(t)]$ are connected with the function $Q(t)=$ $=F_\tau(t)$ where $F_\tau(\tau)$ is the function of distribution of the life-time τ. When N_o is large, the variance of $N(t)$ is small in comparison with unity. Thus the semideterministic approximation is valid

$$N(t) \sim N_oQ(t). \qquad (3)$$

The proposed model can be applied to a wide range of loading and damaging processes. If no time effects are included, $N(t) \sim N_oF_r[S(t)]$. Here $F_r(r)$ is a probability distribution function for the characteristic strength r, $S(t)=\sup s(\tau)$, $\tau \in [0,t]$, $s(t)$ are nominal stresses or the load level. This model describes the Kaiser effect [1]. For delayed damage (the static fatigue) we must assume the existence of kynetic damage accumulation equations $d\psi/dt=f(\psi,s,r)$ for each element. Here $\psi(t)$ is a damage rate for an element, $\psi=0$ if the element is nondamaged, and $\varphi=1$ if the element is fractured. The right-hand-side depends on the nominal stress s and the strength parameter r. The function $Q(t)$ from Eq.(3) is the probability $Q(t)=P\{\varphi(t)\geqslant 1\}$. To describe the classic fatigue we must assume a kinetic equation $d\varphi/dn=f(\varphi,s_n,r)$ for each element (n is the number of cycles, and s_n is a characteristic stress of a cycle). The analogue of $Q(t)$ for n-th cycle is $Q(n)=P\{\varphi(n)\geqslant 1\}$, and the number of elements fractured during this cycle is $\Delta N(n)\sim N_o[Q(n)-Q(n-1)]$.

Now, we generalize the model taking into consideration the stress/strength spacial nonhomogeneity and the detection threshold. Let the AET device record only those signals $u(\underline{x},t)$ coming from the point $\underline{x}$, which level is above a certain level u_{th}. Instead of Eq.(3) we obtain

$$N(t)\sim\nu_o\int_{V_E(t)}Q(\underline{x},t)dV \qquad (4)$$

Here ν_o is a number of elements per volume unit, the function $Q(\underline{x},t)$ is equal to the probability distribution function $F_\tau(\tau|\underline{x},t)$ of the life-time τ under the given loading history $s(\underline{x},t)$. The area of integration in Eq.(4) is $V_E(t)=\{\underline{x}:u(\underline{x},t)\geqslant u_{th}\}$. In the case when the signal carries more than one count, we obtain

$$N(t)\sim\nu_o\sum_k\int_{V_E(t)}kP_k(\underline{x},t)Q(\underline{x},t)dV \qquad (5)$$

where $P_k(\underline{x},t)$ is the probability of k crossings of the level u_{th}.

To use Eqs.(4),(5) it is necessary to study properties of waves emitted from newborn defects. There is a serious problem of waves propagating from a crack associated with AET. The final size of a crack is a characteristic scale. We assume that a crack stops at the boundary of the structural element. The emission is of multiple nature, and the signal is subjected to dispersion because of the diffraction on nonhomogeneties. Coming to the surface, the signal originates Rayleigh waves which are the main information source in AET. The corresponding problems of wave dynamics result in the computation of multiple integrals. Therefore, semiempirical attenuation laws are of interest as

$$u=u_0(\underline{a}/\underline{x})^{1/2}\exp(-\psi\underline{x}). \qquad (6)$$

Here u_0 is a characteristic value of the signal, $\underline{x}$ is the epicentral distance, $\underline{a}$ is a characteristic length, and ψ is an efficient damping factor.

We consider an example of the sustained fracture of a specimen from a unidirectional fiber composite [3]. Let the fibers be elastic until fracture, and the matrix be elastic. Half-length of the structural element is

$$\lambda_e=\rho(E_f/2G_m)^{1/2}g(v_f), \qquad (7)$$

where ρ is the radius of fibers, E_f is the Young's modulus of fibers, G_m is the shear modulus of the matrix, $g(v_f)\sim 1$ is a function of the fibers volume ratio v_f. Let s be the nominal stress in the composite. The fibers strength r is a random value depending on the acting length $2\lambda_e$. Assume the three-parameter Weibull distribution is

$$F_r(r;\lambda)=1-\exp\left\{-\lambda\rho^{-1}\left[(r-r_0)/(r_c-r_0)\right]^{\alpha}\right\}, \qquad (8)$$

where r_0, r_c and α are constants. If $r<r_0$ then $F_r(r;\lambda)\equiv 0$. The kinetic equation for damage accumulation in a structural element under sustained loading is taken in the form related to the distribution (8):

$$d\varphi/dt=t_c^{-1}\left[(s-r_0)/(r-r_0)\right]^m. \qquad (9)$$

We assume $d\varphi/dt=0$ if $s<r_0$, and $\varphi=1$ if $s\geqslant r$. The constant t_c is a characteristic time of short tests.

Calculated summation number of AE is given in Fig.1. Here $r_0=0$, $m=4$, $\lambda_e=10^2$. Solid lines are drawn for $s=10^{-1}r_0$, and thin lines for $\dot{s}=10^{-3}r_c/t_c$. Values of α are plotted near the curves. The latter are drawn until $N(t)\approx N_0$, although the initial assumptions are valid only for sufficiently small $N(t)/N_0\ll 1$. Broken lines correspond to the moments of the loss of integrity of the composite [3]. The results are in the agreement with semi-empirical equations suggested by other authors [1]. For example, under constant loadings Eqs.(4), (7), (8) and (10) yield $N(t) \sim$ $\sim \text{const} \cdot s^{\alpha} t^{\beta}$. Here $\beta = \alpha/m$. Under loading with the constant velocity $N(t) \sim \text{const} \cdot t^{\alpha+\beta}$ etc. The exponents α and β can be found from the independent fatigue tests. The logarithmic

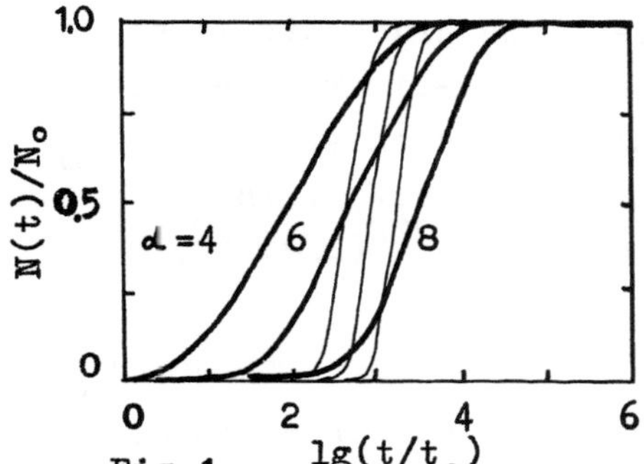

Fig.1

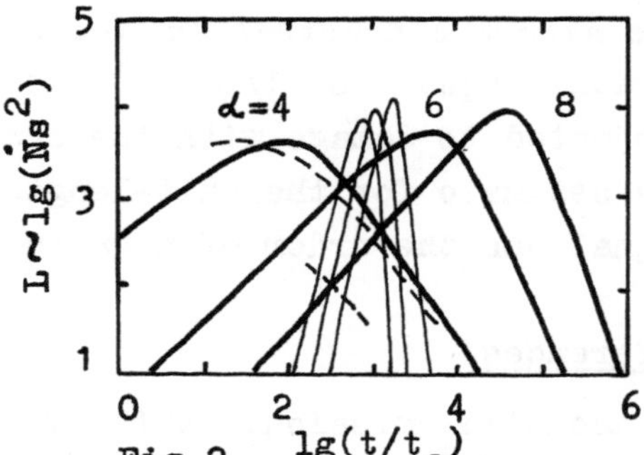

Fig.2

level $L = \text{const} \cdot \lg(\dot{N}s^2)$ is plotted in Fig.2 for the same numerical data. Fig.3 shows effect of the detection threshold. A long thin specimen is considered, and the attenuation law is taken as Eq.6. The nominal stresses velocity is $\dot{s} = 10^{-3} r_c/t_c$.

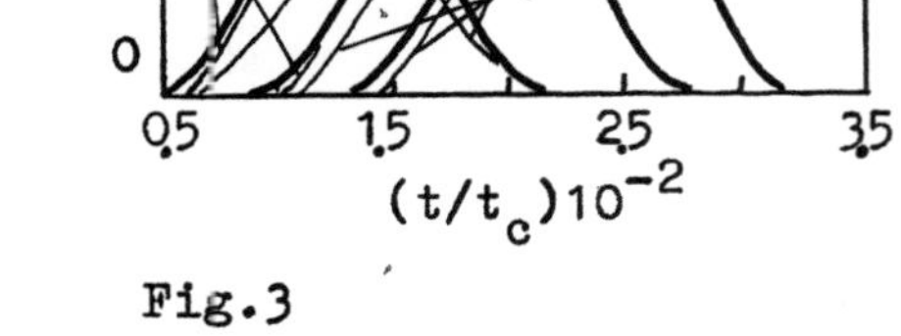

Fig.3

The solid lines correspond to the case when all signals are registered. The thin lines correspond to the threshold stresses equal to $0{,}1 r_c$ and $0{,}2 r_c$. When $\alpha = 4$, the detection effect is considerable, but at $\alpha = 8$ became negligible.

One of the applications area of AET is the detection of nucleas of macroscopic cracks and the control of their stable development. Both triangulation and spectral methods are used to locate a crack and to estimate its size. Let us show, how the detection threshold principles can be applied for this aim. As an example we take a "brush-like" transverse crack in a unidirectional fiber composite which model was suggested in [3]. If the matrix is elastic, so similar to (7) we obtain the following estimate for the half-length of a circular crack : $\lambda \sim l \cdot (E_f/2G_m)^{1/2} g(v_f)$. Here l is the radius of the crack. The amount of the elastic energy of a fiber entering the crack is $U \sim \rho^2 \lambda s^2/E_f$. The magnitude of AE signal from the rupture of a fiber is $u \sim const \times s(1/\rho)^{1/2}$, i.e. in $(1/\rho)^{1/2}$ times greater than one of the signal coming from a single dispersed rupture. The same effect takes place in polycrystalline solids taking into account [4] the stress concentration factor near the crack tip is approximately equal to $(1/\rho)^{1/2}$. If the matrix-fiber interface is subjected to damage with the characteristic shear stress τ_m^*, the estimate for the half-length is $\lambda \sim ls/(v_f \tau_m^*)$. The AE signal has the order of magnitude $u \sim const\ s^{3/2}(1/\rho)^{1/2}$.

References

1. Acoustic emission. -ASTM STP 505. -Philadelphia: ASTM, 1972. 337 p.

2. Bolotin V.V. Stochastic models of cumulative damage in composite materials.- In: Progress in fatigue and fracture. Oxford: Pergamon Press, 1976, pp. 103-111.

3. Bolotin V.V. United model of fracture of composites under sustained loading.- Mechanics of composite materials, 1981,N3, pp.405-420 (in Russian).

4. Bolotin V.V. Stochastic models of fracture with application to the reliability theory. - In: Structural Safety and Reliability. - Amsterdam-Oxford-New York: Elsevier, 1981, pp. 31-56.

Higher-Order Acoustic Nonlinearity Parameters of Fluids

L. BJØRNØ and K. BLACK *)

The Industrial Acoustics Research Group, Technical University of Denmark, Building 352, DK-2800 Lyngby, Denmark

*) Department of Fluid Mechanics

Summary

General thermodynamical expressions suitable for calculation on a computer are developed for the higher-order dimensionless nonlinearity ratios: B/A, C/A, D/A, E/A, F/A and G/A for fluids. Calculation of B/A, C/A and D/A for destilled water is performed based on available p-a-T relations and errors in earlier procedures are emphasized. Relations between Rao's exponent and B/2A are discussed.

Introduction

The nonlinear distortion of finite-amplitude signals during their propagation through single-phase fluids is caused by two main sources, a convective nonlinearity represented by the convective terms in the equations of motion and a material nonlinearity represented by the nonlinear equation of state of the fluid. For gases the convective contribution will be predominant, while for most liquids the material contribution will prevail [1].

Certain combinations of the temperature and pressure derivatives of the velocity of sound in liquids have been found to obey simple empirical relationships. Of primary importance among these is Rao's rule [2] for unassociated organic liquids from which Rao's exponent, as shown by Swamy [3], can be related directly to the pseudo-Grüneisen parameter for a liquid. Other thermodynamical relationships involving the liquid bulk modulus, such as Wada's rule, expressing the molecular compressibility, and the Tait equation, can also be related through simple expressions to Rao's constant, the so-called molecular velocity of sound [4-6]. Very recently Hartmann [7] suggested that another thermodynamical parameter, the second-order dimensionless nonlinearity ratio B/A for liquids [8] could be simply related to

Nonlinear Deformation Waves
IUTAM Symposium Tallinn 1982
Editors U Nigul, J Engelbrecht

Rao's exponent and to the pseudo-Grüneisen parameter. This relationship was derived under the assumption that the intermolecular potential energy is the dominant factor in the determination of the speed of sound and its derivatives. The interest in simple thermodynamical relationships and their parameters is in particular due to the fact that they form a key to the study of the molecular structure of liquids by producing information about physico-chemical quantities like the thermal and internal pressure of liquids, the intermolecular spacing, the anharmonic nature of the intermolecular potential etc.

The close connection excisting between the higher-order acoustic nonlinearity parameters of liquids and their molecular structure has been the motive power behind the desire to develop a thermodynamical procedure for a calculation of the dimensionless ratios of higher-order acoustic nonlinearity parameters based on knowledge about the pressure-temperature-sound velocity relations for liquids.

Theory

The material nonlinearity of a liquid may be expressed through the nonlinear relation between pressure and density for an adiabatic process [8-9]. An expansion in a Taylor series of the instantaneous pressure p about the equilibrium values $\rho = \rho_o$ and $S = S_o$ using the equation of state $p = p(\rho,S)$ for the liquid, where ρ and S are the thermodynamic state variables, density and specific entropy, respectively, leads to:

$$p = p_o + (\partial p/\partial \rho)_S(\rho-\rho_o) + (\partial^2 p/\partial \rho^2)_S(\rho-\rho_o)^2/2! + \\ + (\partial^3 p/\partial \rho^3)_S(\rho-\rho_o)^3/3! + \ldots\ldots\ldots + (\partial p/\partial S)_\rho(S-S_o) + \ldots \quad (1)$$

where p_o denotes the ambient pressure.

From (1) the isentropic equation of state ($S = S_o$) may be derived based on the fact that sound propagation is a substantially isentropic process. This leads to:

$$\Delta p = A(\Delta\rho/\rho) + B(\Delta\rho/\rho)^2/2! + C(\Delta\rho/\rho)^3/3! + D(\Delta\rho/\rho)^4/4! + \\ + E(\Delta\rho/\rho)^5/5! + F(\Delta\rho/\rho)^6/6! + G(\Delta\rho/\rho)^7/7! + \ldots \quad (2)$$

with $A = \rho_o(\partial p/\partial \rho)_S = \rho_o\, a_o^2$

$$B = \rho_o^2(\partial^2 p/\partial\rho^2)_S \qquad E = \rho_o^5(\partial^5 p/\partial\rho^5)_S$$
$$C = \rho_o^3(\partial^3 p/\partial\rho^3)_S \qquad F = \rho_o^6(\partial^6 p/\partial\rho^6)_S$$
$$D = \rho_o^4(\partial^4 p/\partial\rho^4)_S \qquad G = \rho_o^7(\partial^7 p/\partial\rho^7)_S$$

$\Delta p = p - p_o$ and $\Delta\rho = \rho - \rho_o$.

a_o denotes the infinitesimal amplitude velocity of sound at equilibrium pressure ($p = p_o$) and density ($\rho = \rho_o$).

The coefficients: A, B, ... G in (2) are acoustic nonlinearity parameters which divided by A form dimensionless nonlinearity ratios: B/A, C/A, ... G/A. Through introduction of two dimensionless, thermodynamical expressions ZN and B/2A defined by:

$$ZN = a_o^{N-1}(\partial^N a/\partial p^N)/(\partial a/\partial p)_S^N; \quad N = 1, 2, 3, \ldots \tag{3}$$

and

$$B/2A = \rho_o(\partial^2 p/\partial\rho^2)_S/2a_o^2 = \rho_o\, a_o(\partial a/\partial p)_S = $$
$$= \rho_o\, a_o(\partial a/\partial p)_T + a_o T\beta(\partial a/\partial T)_p/C_p \tag{4}$$

with the isobaric compressibility $\beta \equiv (\partial V/\partial T)_p/V$, $V = 1/\rho$ and C_p denoting the specific heat at constant pressure, the higher-order dimensionless nonlinearity ratios: C/A, D/A, E/A, F/A and G/A may, after some calculations, be written based on thermodynamical state variables and their derivatives as:

$$C/A = 3![1 + Z2/3][B/2A]^2 = (3/2)(B/A)^2 + 2a_o^3\,\rho_o^2(\partial^2 a/\partial p^2)_S$$
$$D/A = 4![1 + (11Z2 + Z3)/12][B/2A]^3 = 3(B/A)^3$$
$$+ 11(B/A)a_o^3\,\rho_o^2(\partial^2 a/\partial p^2)_S + 2a_o^5\,\rho_o^3(\partial^3 a/\partial p^3)_S$$
$$E/A = 5![1 + \{Z2(102 + 11Z2) + 18Z3 + Z4\}/60][B/2A]^4$$
$$F/A = 6![1 + \{Z2(954 + 292Z2 + 40Z3) + 246Z3$$
$$+ 27Z4 + Z5\}/360][B/2A]^5$$
$$G/A = 7![1 + \{Z2(18864 + Z2(10970 + 584Z2) + 2952Z3 + 134Z4)$$
$$+ Z3(6336 + 80Z3) + 1032Z4 + 76Z5 + 2Z6\}/5040][B/2A]^6 \tag{5}$$

A calculation of C/A .... G/A based on these expressions demands that the velocity of sound is known as a function of pressure for constant entropy, but due to the fact that most liquid data reported in literature are based on p-a-T relations, a transformation of ZN into a form suitable for p-a-T relations is necessary. This transformation has been performed and the ZN expressions for N = 1, 2, 3, 4 and 5 are:

$$Z1 = 1$$

$$Z2 = a_o[(\partial^2 a/\partial p^2)_T + 2(\partial^2 a/\partial p\partial T)(\partial T/\partial p)_S + (\partial^2 a/\partial T^2)_p(\partial T/\partial p)^2_S + (\partial a/\partial T)_p(\partial^2 T/\partial p^2)_S]/[(\partial a/\partial p)^2_T + 2(\partial a/\partial p)_T(\partial a/\partial T)_p(\partial T/\partial p)_S + (\partial a/\partial T)^2_p(\partial T/\partial p)^2_S]$$

$$Z3 = a_o^2[(\partial^3 a/\partial p^3)_T + (\partial T/\partial p)_S\{3(\partial^3 a/\partial p^2\partial T) + (\partial T/\partial p)_S \cdot \{3(\partial^3 a/\partial p\partial T^2) + (\partial^3 a/\partial T^3)_p(\partial T/\partial p)_S\}\} + 3(\partial^2 T/\partial p^2)_S \cdot \{(\partial^2 a/\partial T\partial p) + (\partial^2 a/\partial T^2)_p(\partial T/\partial p)_S\} + (\partial a/\partial T)_p \cdot (\partial^3 T/\partial p^3)_S]/[(\partial a/\partial p)_T + (\partial a/\partial T)_p(\partial T/\partial p)_S]^3$$

$$Z4 = a_o^3[(\partial^4 a/\partial p^4)_S + 4(\partial^4 a/\partial p^3\partial T)(\partial T/\partial p)_S + 6(\partial^4 a/\partial p^2\partial T^2) \cdot (\partial T/\partial p)^2_S + 4(\partial^4 a/\partial p\partial T^3)(\partial T/\partial p)^3_S + (\partial^4 a/\partial T^4)_p(\partial T/\partial p)^4_S + 6(\partial^2 T/\partial p^2)_S \cdot \{(\partial^3 a/\partial T\partial p^2) + 2(\partial^3 a/\partial T^2\partial p)(\partial T/\partial p)_S + (\partial^3 a/\partial T^3)_p(\partial T/\partial p)^2_S + (\partial^2 a/\partial T^2)_p(\partial^2 T/\partial p^2)_S\} + 4(\partial^3 T/\partial p^3)_S\{(\partial^2 a/\partial T\partial p) + (\partial^2 a/\partial T^2)_p(\partial T/\partial p)_S\} + (\partial a/\partial T)_p(\partial^4 T/\partial p^4)_s - 3(\partial^2 a/\partial T^2)_p(\partial^2 T/\partial p^2)^2_S]/ [(\partial a/\partial p)_T + (\partial a/\partial T)_p(\partial T/\partial p)_S]^4$$

$$Z5 = a_o^4[(\partial^5 a/\partial p^5)_T + 5(\partial^5 a/\partial p^4\partial T)(\partial T/\partial p)_S + 10(\partial^5 a/\partial p^3\partial T^2) \cdot (\partial T/\partial p)^2_S + 10(\partial^5 a/\partial p^2\partial T^3)(\partial T/\partial p)^3_S + 5(\partial^5 a/\partial p\partial T^4) \cdot (\partial T/\partial p)^4 + (\partial^5 a/\partial T^5)_p(\partial T/\partial p)^5_S + 10(\partial^2 T/\partial p^2)_S \cdot \{(\partial^4 a/\partial T\partial p^3) + 3(\partial^4 a/\partial T^2\partial p^2)(\partial T/\partial p)_S + 3(\partial^4 a/\partial T^3\partial p) \cdot (\partial T/\partial p)^2_S + (\partial^4 a/\partial T^4)_p(\partial T/\partial p)^3_S + 3(\partial^2 T/\partial p^2)_S \cdot \{(\partial^3 a/\partial T^2\partial p) + (\partial^3 a/\partial T^3)_p(\partial T/\partial p)_S\} + (\partial^2 a/\partial T^2)_p \cdot (\partial^3 T/\partial p^3)_S\} + 10(\partial^3 T/\partial p^3)_S\{(\partial^3 a/\partial T\partial p^2) + 2(\partial^3 a/\partial T^2\partial p) \cdot (\partial T/\partial p)_S + (\partial^3 a/\partial T^3)_p(\partial T/\partial p)^2_S + (\partial^2 a/\partial T^2)_p(\partial^2 T/\partial p^2)_S\} + 5(\partial^4 T/\partial p^4)_S\{(\partial^2 a/\partial T\partial p) + (\partial^2 a/\partial T^2)_p(\partial T/\partial p)_S\} + (\partial a/\partial T)_p(\partial^5 T/\partial p^5)_S - 5(\partial^2 T/\partial p^2)_S\{3(\partial^2 T/\partial p^2)_S$$

$$\cdot \{(\partial^3 a/\partial p \partial T^2) + (\partial^3 a/\partial T^3)_p (\partial T/\partial p)_S\} + 2(\partial^2 a/\partial T^2)_p$$

$$\cdot (\partial^3 T/\partial p^3)_S\}]/[(\partial a/\partial p)_T + (\partial a/\partial T)_p (\partial T/\partial p)_S]^5 \qquad (6).$$

Discussions and conclusions

The B/A .... G/A ratios may be calculated on a computer using available p-a-T relations for the computation of the ZN-values. For destilled water at 10^5Pa and at temperatures from 0°C to 30°C the values of B/A, C/A and D/A have been calculated using the expressions (5) with ZN-values computed from p-a-T relations found in the literature. These higher-order nonlinearity ratios are given in Table I and the calculated B/A-ratios are compared with experimental values found by Beyer [8]. Higher-order ratios than D/A could not be calculated based on destilled water data available as these data did not allow the formation of higher-order derivatives of the thermodynamical state variables. New p-a-T data, particularly produced aiming at the determination of the higher-order derivatives in ZN, are necessary for a computation of the ratios E/A, F/A and G/A.

As claimed by several authors [9], [10], sufficiently accurate values of C/A, D/A and higher-order ratios can be calculated using the first term of the expressions (5) involving a dependence on B/A only. A calculation of D/A at 10°C and 10^5Pa for destilled water gives for the individual terms in (5):

$$D/A = 3(B/A)^3 + 11(B/A)a_o^3\rho_o^2(\partial^2 a/\partial p^2)_S + 2a_o^5\rho_o^3(\partial^3 a/\partial p^3)_S$$

$$= 309 \quad + \quad 631 \quad - \quad 520 \quad = 420$$

which deviates strongly from the values calulated using only the first term as suggested in [9] and [10]. For comparison values of C/A and D/A calculated using the procedure suggested in [9] and [10] are given in Table I, which shows that the influences of the second term in C/A and the second and third term in D/A are not negligible and will lead to serious errors in the C/A and D/A values when not taken into consideration. The magnitude and the sign of these terms may at certain temperatures, see Table I, lead to an agreement between the C/A and D/A values calculated using the full expressions (5) and using the first term only, but in general all terms in (5) must be used for the determination of the higher-order nonlinearity ratios.

TABLE I.

	BJØRNØ & BLACK			BEYER [8]	THAKUR [10] COPPENS...[9]	
T °C	B/A	C/A	D/A	B/A (meas)	C/A	D/A
0	4.24	67.8	622	4.16	26.9	228.7
10	4.69	57.5	420	4.63	33.0	309.4
20	5.02	48.2	292	4.96	37.8	380.0
30	5.26	40.4	225	5.22	41.5	436.6

Destilled water at 10^5 Pa and at temperatures between 0 °C and 30 °C.
Comparison between measured and calculated, using (5), B/A-ratios and between calculated C/A and D/A values using all terms in the expression (5) and using the first term only.

In a recent paper [11] Madigosky et al. discussed Hartmann's [7] finding of a simple relationship between the Rao's exponent K and the B/2A ratio. Madigosky et al. calculated the B/2A ratio and Rao's exponent $K = (V/a)(\partial a/\partial T)_p/(\partial V/\partial T)_p$ for various flouro-carbon fluids at 10^5Pa and 0°C and 20°C and found that in general K does not equal 3 and B/A does not equal 2K which discripancies should be partly due to insufficiencies in the basic assumptions made during Hartmann's derivation of the relationship between K and B/A. Based on the thermodynamical relations used for the derivations of Z(N) the following relationship between K and B/A may be derived:

$$(\partial a/\partial p)_V = (\partial a/\partial p)_S - (\partial a/\partial T)_p(\partial V/\partial p)_S/(\partial V/\partial T)_p$$
$$= (a/V)(\partial V/\partial p)_S[K - B/2A] = -(V/A)[K - B/2A]$$

from which:

$$K - B/2A = -(a/V)(\partial a/\partial p)_V = -(a/V)[(\partial a/\partial p)_T + (\partial a/\partial T)_p \cdot (\partial T/\partial p)_V] \quad (7)$$

All previous experimental and theoretical results for a long series of fluids have shown that $(\partial a/\partial p)_T$ and $(\partial a/\partial T)_p$ are different from zero for almost all pressures and temperatures and calculations of the magnitudes and signs of the derivations in the bracket in (7) show that this bracket will only become equal to zero and thus lead to Hartmann's relation: K = B/2A, for a few p-a-T data, while in general B/2A ≠ K for all liquids studied.

References

1. Bjørnø, L.: Nonlinear Acoustics. In R.W.B.Stephens & H.G. Leventhall (Eds.), Acoustics and Vibration Progress Vol.2, Chapmann & Hall 1976.
2. Rao, M.R.: J.Chem.Phys. 9, [1941], 682.
3. Swamy, K.M.: Acustica, 29, [1973], 179.
4. Wada, Y.: J.Phys.Soc.Japan, 4, [1949], 280.
5. Jain, R.P. and Pandey, J.D.: Acustica, 32 [1975], 336.
6. Moelwyn-Hughes, E.A.: Physical Chemistry, 2 ed, Pergamon Press, 1961.
7. Hartmann, B.: J.Acoust.Soc.Amer., 65, (6), [1979], 1392.
8. Beyer, R.T.: J.Acoust.Soc.Amer., 32, (6), [1960], 719.
9. Coppens, A.B., Beyer, R.T., Seiden, M.B., Donohue, J., Guepin, F., Hodson, R.H. and Townsend, C.: J.Acoust.Soc.Amer., 38, [1965], 797.
10. Thakur, K.P.: Acustica, 39, [1978], 270.
11. Madigosky, W.M., Rosenbaum, I. and Lucas, R.: J.Acoust.Soc.Amer., 69, (6), [1981], 1639.

PART 5
COMBINED FIELDS AND APPLICATIONS

On Nonlinear Interaction of Solid Particles With a Sound Wave in Viscous Liquid

A.N.GUZ, A.P.ZHUK

Institute of Mechanics of the Academy of Sciences
of the Ukrainian SSR, Kiev

Summary

In the paper, a method is presented for determination of average hydrodynamic forces, acting on the solid particle in a sound field in viscous liquid. The method is based on a simplification of the basic system of hydrodynamic equations, in which both the non-linear and dissipative terms are retained. As a result, a non-linear relation is obtained for determination of the pressure in a liquid, vector potentials for the field of velocities of the viscous liquid being determined from the solution of the linear problem of diffraction. For the case of a spherical particle of small radius and for the case of low viscosity- the average hydrodynamic force is calculated, generated by sound field of the plane wave.

Determination of pressure in the sound wave

Let us consider the viscous compressible liquid, filling the entire space. The termal effects being neglected, the basic hydrodynamic relations in Euler variables have the following form [1,2]

$$\frac{d\vec{v}}{dt}+\frac{1}{\rho'}\vec{\nabla}p'-\frac{1}{3}\frac{\mu'}{\rho'}\vec{\nabla}(\vec{\nabla}\cdot\vec{v})-\frac{\mu'}{\rho'}\Delta\vec{v}=0, \qquad (1)$$

$$\dot{\rho}'+\vec{\nabla}\cdot(\rho'\vec{v})=0, \qquad (2)$$

$$p'=f(\rho'), \qquad (3)$$

where $\vec{v}$ - vector of the velocities field in the liquid, p' and ρ' - pressure and density at the given moment of time t , μ' - dynamic coefficient of viscosity. In the equation of motion (1) we account for the fact, that liquid satisfies the

Nonlinear Deformation Waves
IUTAM Symposium Tallinn 1982
Editors U Nigul, J Engelbrecht

Stokes condition. The relation

$$\vec{v} = \vec{\nabla}\Phi \tag{4}$$

that is valid for sound waves in liquids [3] , allows to represent the equation (1) in the following form

$$\frac{d\vec{v}}{dt} + \frac{1}{\varrho'}\vec{\nabla}p' - \frac{4}{3}\frac{\mu'}{\varrho'}\Delta\vec{v} = 0. \tag{5}$$

In acoustics [4,5] it is established that the relative disturbances $\varrho/\varrho_0 (\varrho = \varrho' - \varrho_0)$ and $p/p_0 (p = p' - p_0)$ of the density ϱ_0 and of the pressure p_0 in the initial state, together with the dynamic coefficient of the viscosity μ' , are values of the first order. Then, taking into account the relation $\mu'/\varrho' \approx \approx \mu'/\varrho_0 - (\mu'/\varrho_0)(\varrho/\varrho_0)$, we may simplify the equation (5), neglecting in it the terms of the higher order

$$\frac{d\vec{v}}{dt} + \frac{1}{\varrho'}\vec{\nabla}p' - \frac{4}{3}\frac{\mu'}{\varrho_0}\Delta\vec{v} = 0. \tag{6}$$

As the liquid is considered a barotropic one (3), the function may be introduced in the form [6]

$$\mathcal{P} = \int \frac{dp'}{\varrho'}, \tag{7}$$

which, using the relation (4), may be determined from the equation (6)

$$\mathcal{P} = -\dot{\Phi} - \frac{1}{2}(\vec{\nabla}\Phi)^2 + \frac{4}{3}\frac{\mu'}{\varrho_0}\Delta\Phi. \tag{8}$$

Restricting ourselves to the real case, we have

$$\frac{dp'}{d\varrho'} = a_0^2 \tag{9}$$

where a_0 - velocity of sound in a liquid.
We introduce the parameter [7] $s = \varrho/\varrho_0$ and use the method proposed in [7] . Expanding the relation (3) in the neighbourhood of the basic state into Taylor series, we obtain

$$p' = f(\varrho_0) + s f'(\varrho_0)\varrho_0 + \frac{1}{2} s^2 \varrho_0^2 f''(\varrho_0) + \cdots . \tag{10}$$

Taking into account the relation (10), the expression (7) is resolved relative to the parameter s

$$s \approx \frac{\mathcal{P}}{a_0^2} - \frac{1}{2} \frac{\varrho_0 f''(\varrho_0) - a_0^2}{a_0^2} \frac{\mathcal{P}^2}{a_0^4} + \cdots . \tag{11}$$

Now, taking into account the relation (11), the pressure in the sound field (10) is represented by

$$p' = p_0 + \varrho_0 \mathcal{P} + \frac{1}{2} \frac{\varrho_0}{a_0^2} \mathcal{P}^2 + \cdots , \tag{12}$$

Substituting the relation (8) into the expression (12) and neglecting the terms of the order of Mach number squares and higher, the following relation is obtained

$$p = \varrho_0 \left(\frac{\lambda' + 2\mu'}{\varrho_0} \Delta - \frac{\partial}{\partial t} \right) \Phi + \frac{1}{2} \frac{\varrho_0}{a_0^2} \dot{\Phi}^2 - \\ - \frac{1}{2} \varrho_0 (\vec{\nabla} \Phi)^2 - \frac{\lambda' + 2\mu'}{a_0^2} \dot{\Phi} \cdot \Delta \Phi , \tag{13}$$

where λ' - is the second viscosity coefficient.
Potential Φ in the expression (13) can be obtained using the relations (2),(6) and (9). Taking into account (4) and (9),we represent the equation (6) in the form

$$\frac{1}{a_0^2}\vec{\nabla}\dot{\Phi} + \frac{1}{2a_0^2}\vec{\nabla}(\vec{\nabla}\Phi)^2 + \frac{1}{\varrho'}\vec{\nabla}\varrho' - \frac{\lambda'+2\mu'}{\varrho_0 a_0^2}\vec{\nabla}(\Delta\Phi) = 0. \quad (14)$$

Taking the partial time derivative and integrating the expression obtained we derive

$$\frac{1}{a_0^2}\frac{d\dot{\Phi}}{dt} + \frac{1}{\varrho'}\ddot{\varrho}' - \frac{\lambda'+2\mu'}{\varrho_0 a_0^2}\Delta\dot{\Phi} = 0. \quad (15)$$

Upon scalar multiplication of the equation (14) by $\vec{v}$ and summing with expression (15), taking into account (2), the equation for determination of potentials is obtained

$$\frac{1}{a_0^2}\frac{d^2\Phi}{dt^2} - \frac{1}{2a_0^2}\frac{d(\vec{\nabla}\Phi)^2}{dt} - \frac{\lambda'+2\mu'}{\varrho_0 a_0^2}\frac{d(\Delta\Phi)}{dt} = \Delta\Phi. \quad (16)$$

Neglecting in this equation the terms of the order of Mach number squares, we obtain

$$\left[\left(1+\frac{\lambda'+2\mu'}{\varrho_0 a_0^2}\frac{\partial}{\partial t}\right)\Delta - \frac{1}{a_0^2}\frac{\partial^2}{\partial t^2}\right]\Phi = 0. \quad (17)$$

Evidently, in the computation of the pressure p (13) in a sound wave to a degree of accuracy within the Mach number squares the potential Φ may be used, determined from the linear equation (17) with the same precision. In computing the pressure with a higher level of accuracy, it is necessary to retain more terms in the equation (12) and potential Φ is determined from the equation (16).

Computation of hydrodynamic forces

When a solid particle is introduced into a liquid, in which a sound wave propagates – the reflected waves generate a complementary wave field. This complementary wave field possesses potential and vortex components, as in the viscous liquid on the particle surface the condition of adhesivity is satisfied. The total wave field producing the hydrodynamic forces that are applied to the solid particle - can be computed from the diffraction problem solution. As the potential of the velocities field in the sound wave is computed from the linear equation (17), - in the determination of potentials of the total wave field we restrict ourselves to the solution of the linear problem of diffraction, using the linearized theory [8,9] . Suppose that the centre of inertia of the solid particle is located at the origin of the immovable system of coordinates. We write the field of the total velocities of a liquid in the form

$$\vec{v} = \vec{\nabla}\Phi + \vec{\nabla} \times \vec{\Psi}, \tag{18}$$

where the scalar potential Φ is equal to the sum of potentials of the incident Φ_i and of the reflected Φ_d waves, and the vector potential $\vec{\Psi}$ is the vector potential of the reflected wave.
Relations for Φ_i and Φ_d are the solutions of the equation (17), and relation for $\vec{\Psi}_d$ - of the equation [9]

$$\left(\frac{\mu'}{\varrho_0}\Delta - \frac{\partial}{\partial t}\right)\vec{\Psi}_d = 0. \tag{19}$$

Potentials Φ_d and $\vec{\Psi}_d$ should satisfy the conditions of radiation, consistent with the physical picture observed in the liquid [3] . The expressions for Φ and $\vec{\Psi}$ are obtained taking into account the conditions of adhesion. For a stationary particle this condition requires that the vector $\vec{v}$ (18) on its surface is equal to zero. A free particle under the action of hydrodynamic forces will move. For such a particle the condition of adhesion has the form

$$\vec{v} = \vec{v}_s, \quad \vec{v}_s = \vec{V} + \vec{\omega} \times \vec{r}_s, \tag{20}$$

where $\vec{V}$ and $\vec{v}_S$ are velocity vector, respectively, of the centre of inertia and of the arbitrary point of the particle surface, $\vec{\omega}$ - is a vector of momentary angular velocity of the particle, $\vec{r}_S$ - radius-vector of the arbitrary point of particle surface. In that case, $\vec{v}_S$ is determined from the equations of particle movement. The vector of velocities of the field (18) obtained from the solution of the linear diffraction problem, allows to compute the average hydrodynamic action of the viscous liquid on the particle. For that purpose, in the expression for the tensor of stresses $\hat{\sigma}$ [9]

$$\hat{\sigma} = (-p + \lambda' \vec{\nabla} \cdot \vec{v}) \hat{E} + 2\mu' \hat{e}, \quad 2\hat{e} = \vec{\nabla}\vec{v} + (\vec{\nabla}\vec{v})^* \tag{21}$$

the pressure p should be found according to the relation (13), while the main vector $\vec{F}$ and the main moment $\vec{M}$ of hydrodynamic forces relative to the centre of inertia should be averaged in time

$$\vec{F} = \int_S \hat{\sigma} \cdot \vec{N} dS, \quad \vec{M} = \int_S \vec{r}_S \times (\hat{\sigma} \cdot \vec{N}) dS. \tag{22}$$

Here $\vec{N}$-is the unit vector of the normal to the surface S of the solid particle. It should be pointed out that in that case, for a free particle the displacement should be taken into account, that could be computed with sufficient accuracy by solving a linear diffraction problem. When displacements are taken into account, as boundary conditions require, the potentials Φ and $\vec{\Psi}$ should be referred to the reference system, connected with the particle. However, the derivative $\dot{\Phi}$ in the expression (13) for p should be computed in the stationary reference system, relative to which the particle movement is considered. Consequently, for the solution of the problem formulated the linearized theory [8,9] is of particular importance. It allows to determine the wave field potentials and the particle displacement that are used in computation of average forces. In that context we consider below some examples of the application of that theory for the determination of the reaction of compressible viscous liquid on the prescribed movement of a solid particle.

L o n g i t u d i n a l   v i b r a t i o n s   o f   a   s p h e r i c a l   p a r t i c l e . Suppose that a spherical particle of the radius a, located at the origin of a coordinate system (x_1, x_2, x_3) vibrates along the unit vector $\vec{e}_3$

$$\vec{v}_S = \vec{e}_3 v_0 \exp(-i\Omega t). \tag{23}$$

We connect with the particle the spherical system of coordinates (r, θ, φ) where the angle θ is calculated from the unit vector $\vec{e}_3$. Then the condition of adhesion (20) on the particle surface $r = a$ has the form

$$\begin{aligned} v_r &= v_0 P_1(\cos\theta) \exp(-i\Omega t), \\ v_\theta &= v_0 \frac{d}{d\theta} P_1(\cos\theta) \exp(-i\Omega t). \end{aligned} \tag{24}$$

For the case considered

$$\Phi = \Phi(r, \theta, t), \quad \vec{\Psi} = \vec{\nabla} \times \left[\vec{e}_r r \Psi_2(r, \theta, t)\right]. \tag{25}$$

The expressions for Φ and Ψ_2 , satisfying the equations (17) and (19), respectively, and the conditions of attenuation "at infinity" we represent in the form

$$\begin{aligned} \Phi &= A_1 z_1\left(\frac{r_1}{a} r\right) P_1(\cos\theta) \exp(-i\Omega t), \\ \Psi_2 &= A_2 h_1^{(1)}\left(\frac{r_2}{a} r\right) P_1(\cos\theta) \exp(-i\Omega t), \end{aligned} \tag{26}$$

$$r_1 = \bar{c}\left(1 - i\frac{\bar{c}^2}{R}\frac{\lambda' + 2\mu'}{\mu'}\right)^{-\frac{1}{2}}, \quad \bar{c} = a\frac{\Omega}{a_0},$$

$$r_2=(iR)^{\frac{1}{2}} \quad , R=\frac{\Omega a^2}{\nu'} .$$

Here $h_1^{(j)}(x)$ is the spherical Hankel function of the j-th kind, R - Reynolds number, $P_1(\cos\theta)$ - Legendre function, $z_1(x)\equiv h_1^{(1)}(x)$ with $Im\, r_1>0$ and $z_1(x)\equiv h_1^{(2)}(x)$ with $Im\, r_1<0$.
Satisfying the condition (24), we obtain

$$A_1=av_0\left[r_2 h_1^{(1)\prime}(r_2)-h_1^{(1)}(r_2)\right]\left[r_1 r_2 z_1'(r_1)h_1^{(1)\prime}(r_2)-\right.$$
$$\left.-2z_1(r_1)h_1^{(1)}(r_2)+r_1 z_1'(r_1)h_1^{(1)}(r_2)\right]^{-1},$$
$$A_2=av_0\left[r_1 z_1'(r_1)-z_1(r_1)\right]\left[r_1 r_2 z_1'(r_1)h_1^{(1)\prime}(r_2)-\right. \tag{27}$$
$$\left.-2z_1(r_1)h_1^{(1)}(r_2)+r_1 z_1'(r_1)h_1^{(1)}(r_2)\right]^{-1}.$$

Computing the projection on the unit vector $\vec{e}_3$ of the main vector of hydrodynamic forces (22) we obtain for the reaction of liquid

$$F_3=(X_{11}+iX_{12})\exp(-i\Omega t),$$
$$X_{11}+iX_{12}=-\frac{4}{3}\pi a^2\Omega \varrho_0 i\left[A_1 z_1(r_1)+A_2 h_1^{(1)}(r_2)\right]. \tag{28}$$

Below the case $Im\, r_1>0$ is considered. We proceed to compute the first term of the asymptotic representation of the reaction (28) at small Reynolds numbers, i.e. at $R\to 0$. Using the expressions for the determination of the spherical Hankel functions and taking into account

$$r_1\approx r_2\sqrt{\frac{\mu'}{\lambda'+2\mu'}}\left(1-\frac{1}{2}\frac{r_2^2}{\bar{c}^2}\frac{\mu'}{\lambda'+2\mu'}\right), \tag{29}$$

we obtain

$$X_{11} \approx 6\pi a \mu' v_0 \left(\frac{4}{3} \, \frac{\lambda' + 2\mu'}{2\lambda' + 5\mu'} \right)^{-1}, \; X_{12} \approx 0. \qquad (30)$$

Here X_{11} - the reaction of liquid at the start of each period, X_{12} - the reaction of liquid after a quarter of a period. The transition to incompressible liquid in relation (30) for X_{11} is impossible to realize , as in this derivation the approximate relation (29) for r_1 was used.

Determination of average hydrodynamic forces action on a spherical particle

Suppose that on a spherical particle of a radius a considered above, falls a plane sound wave

$$\Phi_i = A \exp[i(\beta x_3 - \omega t)]. \qquad (31)$$

For a free particle, with $r = a$, the condition (20)takes the form

$$v_r = V_3 \cos\theta, \quad v_\theta = -V_3 \sin\theta. \qquad (32)$$

For the case considered, we have [9]

$$\Phi = \Phi(r,\theta,t), \; \vec{\Psi} = \vec{\nabla} \times \vec{e}_r r \Psi_2(r,\theta,t). \qquad (33)$$

Let us expand the potential Φ_i in Legendre polynoms and represent the expressions for Φ and Ψ_2, satisfying the equations (17) and (19), respectively, in the following form

$$\Phi = \Phi_i + \Phi_d = \sum_{n=0}^{\infty} \left[A(2n+1) i^n j_n(\beta r) + \right. \qquad (34)$$

$$\left. + A_n h_n^{(1)}(\beta r) \right] P_n(\cos\theta) \exp(-i\omega t),$$

$$\Psi_2=\Psi_d=\sum_{n=0}^{\infty}B_n h_n^{(1)}(\alpha r)P_n(\cos\theta)\exp(-i\omega t),$$
$$\beta=k+i\varkappa\,,\quad \alpha=\sqrt{i\omega/\nu'}\,, \tag{35}$$

where Φ_d and Ψ_d - scalar potentials of waves reflection, $j_n(x)$ - spherical Bessel functions, $k=\omega/a_0$ - wave number, $\varkappa=(\lambda'+\mu')\omega^2/(2\varrho_0 a_0^3)$ - coefficient of attenuation, $\nu'=\mu'/\varrho_0$ - kinematic coefficient of viscosity.
The constants A_n and B_n are determined from the system (32). The components v_r and v_θ of the vector $\vec{v}$ are obtained by means of projection of the expression (18) on the unit vectors $\vec{e}_r$ and $\vec{e}_\theta$ and taking into account relations (33)

$$v_r=r^{-1}\sum_{n=0}^{\infty}[A\beta_{1n}+A_n\beta_{2n}+B_n\beta_{3n}]P_n(\cos\theta)\exp(-i\omega t),$$
$$v_\theta=r^{-1}\sum_{n=0}^{\infty}[A\delta_{1n}+A_n\delta_{2n}+B_n\delta_{3n}]\frac{d}{d\theta}P_n(\cos\theta)\exp(-i\omega t). \tag{36}$$

Here

$$\beta_{1n}=(2n+1)i^n[nj_n(\beta r)-\beta r j_{n+1}(\beta r)],\ \beta_{2n}=$$
$$=nh_n^{(1)}(\beta r)-\beta r h_{n+1}^{(1)}(\beta r),\ \beta_{3n}=n(n+1)h_n^{(1)}(\alpha r),$$
$$\delta_{1n}=(2n+1)i^n j_n(\beta r),\ \delta_{2n}=h_n^{(1)}(\beta r)\,,\ \delta_{3n}=$$
$$=(n+1)h_n^{(1)}(\alpha r)-\alpha r h_{n+1}^{(1)}(\alpha r).$$

The velocity of the translational displacement of the particle V_3 can be determined from the equation of motion

$$m\dot{V}_3=2\pi a^2\int_0^{\pi}(\sigma_{rr}\cos\theta-\sigma_{r\theta}\sin\theta)\sin\theta\,d\theta, \tag{37}$$

where $m=\frac{4}{3}\pi a^3 \varrho_1$-the mass, ϱ_1 - the average density of the particle. The expression in the right-hand side (37) is the projection on the axis Ox_3 of the hydrodynamic force (22) computed from the linearized theory. Consequently, from (21), taking into account (18), we derive

$$\sigma_{rr}=\frac{2\mu'}{r^2}\sum_{n=0}^{\infty}[A\alpha_{1n}+A_n\alpha_{2n}+B_n\alpha_{3n}]P_n(\cos\theta)\exp(-i\omega t),$$
$$\sigma_{r\theta}=\frac{2\mu'}{r^2}\sum_{n=0}^{\infty}[A\varepsilon_{1n}+A_n\varepsilon_{2n}+B_n\varepsilon_{3n}]\frac{d}{d\theta}P_n(\cos\theta)\exp(-i\omega t). \tag{38}$$

Here
$$\alpha_{1n}=(2n+1)i^n\left[\left(n^2-n-\frac{\alpha^2r^2}{2}\right)j_n(\beta r)+2\beta r j_{n+1}(\beta r)\right],$$
$$\alpha_{2n}=\left(n^2-n-\frac{\alpha^2r^2}{2}\right)h_n^{(1)}(\beta r)+2\beta r h_{n+1}^{(1)}(\beta r),\ \alpha_{3n}=$$
$$=n(n+1)\left[(n-1)h_n^{(1)}(\alpha r)-\alpha r h_{n+1}^{(1)}(\alpha r)\right],\ \varepsilon_{1n}=(2n+1)\times$$
$$\times i^n\left[(n-1)j_n(\beta r)-\beta r j_{n+1}(\beta r)\right],\ \varepsilon_{2n}=$$
$$=(n-1)h_n^{(1)}(\beta r)-\beta r h_{n+1}^{(1)}(\beta r),\ \varepsilon_{3n}=\left(n^2-1-\right.$$
$$\left.-\frac{\alpha^2r^2}{2}\right)h_n^{(1)}(\alpha r)+\alpha r h_{n+1}^{(1)}(\alpha r).$$

Now integrating the expression (37) and taking into account the relation (38), we obtain

$$V_3=\frac{\eta}{a}\left[3ij_1(\beta a)A+h_1^{(1)}(\beta a)A_1+2h_1^{(1)}(\alpha a)B_1\right]\exp(-i\omega t), \tag{39}$$
$$\eta=\varrho_0/\varrho_1.$$

Lastly, resolving the equations (32) relative to constants A_n and B_n, the following representations is obtained

$$A_n = A\frac{\beta_{3n}\delta_{1n}-\beta_{1n}\delta_{3n}}{\beta_{2n}\delta_{3n}-\beta_{3n}\delta_{2n}},\quad B_n = A\frac{\beta_{1n}\delta_{2n}-\beta_{2n}\delta_{1n}}{\beta_{2n}\delta_{3n}-\beta_{3n}\delta_{2n}},\quad r=a,\ n\neq 1$$

$$A_1 = A\frac{b_{31}d_{11}-b_{11}d_{31}}{b_{21}d_{31}-b_{31}d_{21}},\quad B_1 = A\frac{b_{11}d_{21}-b_{21}d_{11}}{b_{21}d_{31}-b_{31}d_{21}}, \tag{40}$$

where

$$b_{11}=3i\left[(1-\eta)j_1(\beta a)-\beta a j_2(\beta a)\right],\ b_{21}=(1-\eta)h_1^{(1)}(\beta a)-$$

$$-\beta a h_2^{(1)}(\beta a),\ b_{31}=2(1-\eta)h_1^{(1)}(\alpha a),\ d_{11}=3i(1-\eta)j_1(\beta a),$$

$$d_{21}=(1-\eta)h_1^{(1)}(\beta a),\ d_{31}=2(1-\eta)h_1^{(1)}(\alpha a)-\alpha a h_2^{(1)}(\alpha a).$$

The expressions obtained (34),(35) and (40) allow to compute the average hydrodynamic force. For that purpose, the projection of the force of action of the liquid (22) on the axis Ox_3 should be averaged in time. In that procedure, in the expression for the tensor of stresses $\hat{\sigma}$ the pressure p should be computed using the relation (13), taking into account the fact that the parameter $\dot{\Phi}$ is represented by

$$\dot{\Phi}=\dot{\Phi}(r,\theta,t)-V_3\cos\theta\frac{\partial}{\partial r}\Phi(r,\theta,t)+\frac{V_3}{r}\sin\theta\frac{\partial}{\partial\theta}\Phi(r,\theta,t). \tag{41}$$

We introduce the dimensionless parameter $\varepsilon=[\nu/(a^2\omega)]^{\frac{1}{2}}$ and consider the case with low viscosity, when the condition $\varepsilon \ll 1$ is satisfied. As a result, we can write

$$\beta a \approx ka,\quad \alpha a \approx \varepsilon^{-1}\frac{1+i}{\sqrt{2}} \tag{42}$$

and in computing of the spherical Hankel function of the argument (αa) the asymptotic representation may be used

$$h_n^{(1)}(z) \approx \frac{1}{z} \exp\left\{i\left[z-(n+1)\frac{\pi}{2}\right]\right\}. \tag{43}$$

Supposing a small radius of the spherical particle and omitting the cumbersome calculations for computation and averaging in time the expression for the intensity of the action of liquid on particles (22) we write finally

$$\bar{F} = \bar{F}_1\left[1+\varepsilon\frac{1}{(ka)^3}F_2(\eta)\right]. \tag{44}$$

Here, $\bar{F}_1 = 2\pi\varrho_0 A^2 \frac{1+2/9(1-\eta)^2}{(2+\eta)^2}(ka)^6$ is the average force, caused by sound pressure in ideal liquid [7] , $F_2(\eta) = \frac{96\eta^2-88\eta+7}{10{,}5\sqrt{2}[1+2/9(1-\eta)^2]}$.
It should be pointed out, that the expression (44) is obtained with the condition $\varepsilon \ll ka$ ($ka<1$, $\varepsilon \approx (ka)^4$) .
The effect of viscosity on the value of the average hydrodynamical force depends on the ratio η of densities of the liquid ϱ_0 and of particle material ϱ_1 and is represented by the function $F_2(\eta)$. Analysis of this function demonstrates that the viscosity increases the value of the average hydrodynamic force when η increases - for particles lighter that liquid. For particles heavier that liquid, a critical value η^* exists, at which the viscosity reduces to the maximal extent the value of the average hydrodynamic force.

References

1. Кочин, Н.Е.; Кибель, И.А.; Розе Н.В. Теоретическая гидромеханика, т.2. Москва: Физматгиз 1963.
2. Седов, Л.И. Механика сплошной среды, т.I. Москва: Наука, 1973.
3. Скучик, Е. Основы акустики, т.I. Москва:Мир 1976.

4. Виноградова, М.Б.; Руденко, О.В.; Сухоруков, А.П. Теория волн. Москва: Наука 1979.
5. Руденко, О.В.; Солуян, С.И. Теоретические основы нелинейной акустики. Москва: Наука 1975.
6. Седов, Л.И. Механика сплошной среды, т.2. Москва: Наука 1976.
7. King, L.V.; On the acoustic radiation pressure on spheres. Proc.Roy.Soc. A 147 (1934) 212-240.
8. Гузь, А.Н.: О представлении решений линеаризированных уравнений Стокса-Навье. ДАН СССР. 253 (1980) 825-827.
9. Гузь, А.Н.: Динамика твердых тел в сжимаемой вязкой жидкости (покоящаяся жидкость). Прикл.мех. 17, № 3 (1981) 3-22.

Wave Propagation in Elastic Membranes

MANFRED BRAUN

Fachbereich Maschinenbau
Universität Duisburg
Bundesrepublik Deutschland

Summary

The propagation of waves in a previously deformed elastic membrane is investigated. The membrane is considered as a simple 2-dimensional continuum, thus neglecting the dispersive effects of finite thickness. The strain-induced anisotropy of wave propagation is represented by slowness and ray curves. The three branches of each of these curves correspond to one flexural wave and two in-plane deformation waves. The intensity of an acceleration wave is governed by a transport equation, an ordinary differential equation valid along the rays of the propagating wave front. Its closed-form solution is presented for the case of a flat membrane subjected to an arbitrary homogeneous deformation.

Introduction

A pulse propagating in a thin sheet of elastic material disperses into a leading wave front with a rapid decay of disturbance behind it, and a diffusing front followed by an oscillatory disturbance. Only the latter is predicted by membrane theory, and it appears as a front of sharp discontinuity. In spite of this discrepancy, the non-dispersive wave propagation according to the membrane theory may serve as a useful first approximation to the more intricate situation in a real elastic sheet of finite thickness.

Let the material points of the membrane be identified by means of a system of convected surface parameters ξ^α, $\alpha = 1,2$. In its stress-free reference state the membrane is considered to form a smooth surface, given by its parametric representation

$$\underset{\sim}{x} = \underset{\sim}{x}(\xi^1, \xi^2) , \tag{1}$$

where $\underset{\sim}{x}$ denotes the position vector, from a fixed origin, of

Nonlinear Deformation Waves
IUTAM Symposium Tallinn 1982
Editors U. Nigul, J. Engelbrecht

a membrane point. If the membrane undergoes a time-dependent deformation, its shape will change to

$$\bar{\underset{\sim}{x}} = \bar{\underset{\sim}{x}}(\xi^1, \xi^2, t) \,, \tag{2}$$

and the velocity of a membrane point is $\underset{\sim}{v} = \partial \bar{\underset{\sim}{x}}/\partial t$. The partial derivatives of (1) and (2) with respect to the convected surface parameters ξ^α yield the associated base vectors

$$\underset{\sim\alpha}{a} = \frac{\partial \underset{\sim}{x}}{\partial \xi^\alpha} \,, \qquad \bar{\underset{\sim\alpha}{a}} = \frac{\partial \bar{\underset{\sim}{x}}}{\partial \xi^\alpha} \tag{3}$$

in reference and actual configurations, respectively. The dyadic

$$\underset{\sim}{F} = \bar{\underset{\sim\alpha}{a}} \otimes \underset{\sim}{a}^\alpha \tag{4}$$

represents the deformation gradient, a tensor of rank 2.

The elastic behaviour of the membrane is characterized by its strain-energy function. In case of material isotropy the strain energy, per unit area of the undeformed membrane, is a function

$$W = W(I_1, I_2) \tag{5}$$

of the two deformation invariants I_1, I_2 . A sheet of MOONEY-RIVLIN material under plane-stress conditions, for instance, has a strain-energy function of the form

$$W = \frac{1}{2}\mu h \left[(1-\alpha)(I_1 + \frac{1}{I_2}) + \alpha(I_2 + \frac{I_1}{I_2})\right] \,, \tag{6}$$

where μ and α are material parameters, and h denotes the thickness of the undeformed sheet. In the following, the strain-energy function will appear only through its derivatives

$$A_{i\ j}^{\alpha\ \beta} = \frac{\partial^2 W}{\partial F^i_{\ \alpha} \, \partial F^j_{\ \beta}} \,, \qquad A_{i\ j\ k}^{\alpha\ \beta\ \gamma} = \frac{\partial^3 W}{\partial F^i_{\ \alpha} \, \partial F^j_{\ \beta} \, \partial F^k_{\ \gamma}} \,, \tag{7}$$

known as the first- and second-order elasticities.

<u>Propagation Condition</u>

The propagation of a wave front can be described by an equation

$$t = \tau(\xi^{\alpha}) , \tag{8}$$

which assigns to any membrane point ξ^{α} the time of arrival of the wave front at this point. The partial derivatives

$$s_{\alpha} = \frac{\partial\tau}{\partial\xi^{\alpha}} \tag{9}$$

are components of the slowness vector $\underset{\sim}{s} = s_{\alpha}\underset{\sim}{a}^{\alpha}$ in the reference configuration. - In the sense of Lagrangian description it is preferable to consider the image of the wave front on the undeformed membrane, rather than the actual wave front itself.

At an acceleration wave the velocity field is continuous, while the acceleration vector undergoes a discontinuous change $[\dot{\underset{\sim}{v}}]$. It is convenient to introduce a scalar intensity a and a polarization vector $\underset{\sim}{d}$ of the wave by stipulating

$$[\dot{\underset{\sim}{v}}] = a\underset{\sim}{d} , \qquad \rho\underset{\sim}{d}\cdot\underset{\sim}{d} = 1 , \tag{10}$$

where ρ denotes the local mass density, per unit area of the undeformed membrane.

Results of the theory of acceleration waves can be adopted with the only reservation that the material points, represented by their coordinates ξ^{α}, form a 2-dimensional Riemannian, rather than a 3-dimensional Euclidean space. The propagation condition of an acceleration wave is, cf. [1],

$$(A_{i\ j}^{\ \alpha\ \beta} s_{\alpha} s_{\beta} - \rho g_{ij})d^{j} = 0 \tag{11}$$

with g_{ij} denoting the metric of 3-dimensional physical space.

Those slowness vectors $\underset{\sim}{s}$, which enforce a non-trivial solution of the homogeneous system (11), define a slowness curve in the local tangent plane of the undeformed membrane. An example is shown in Fig. 1. In any direction there are three possible slowness vectors. The corresponding polarization vectors $\underset{\sim}{d}$ are mutually orthogonal. Two of them are confined to the local tangent plane of the deformed membrane. They deviate from true longitudinal and transversal directions, unless the slowness vector points into a principal direction of stress and strain. One can, however, always distinguish between a faster quasi-longitudinal and a slower quasi-transversal wave.

The third polarization vector is orthogonal to the deformed membrane, and it corresponds to a flexural wave. Let $\underset{\sim}{\bar{n}}$ denote the unit normal vector and $\bar{b}_{\alpha\beta}$ the covariant components of the curvature tensor in any point of the deformed membrane. A simple calculation yields the compatibility condition

$$\left[\bar{b}_{\alpha\beta}\right] = \underset{\sim}{\bar{n}} \cdot \left[\underset{\sim}{\dot{v}}\right] s_\alpha \, s_\beta \, . \tag{12}$$

Thus an acceleration jump normal to the membrane is always accompanied by a discontinuity in curvature. The corresponding branch of the slowness curve is the ellipse

$$P^{\alpha\beta} s_\alpha \, s_\beta = \rho \, , \tag{13}$$

generated by the second PIOLA-KIRCHHOFF stress tensor

$$P^{\alpha\beta} = \bar{n}^i \, \bar{n}^j \, A_i{}^\alpha{}_j{}^\beta \, . \tag{14}$$

The two other branches of the slowness curve constitute an algebraic curve of 4th order.

Transport Equation

If the membrane ahead of the wave front is at rest, the intensity of an acceleration wave satisfies an equation of the form

$$(c^\alpha a^2)|_\alpha = q \, a^3 \, , \tag{15}$$

where $|_\alpha$ denotes the covariant derivative corresponding to the metric of the undeformed membrane; the components

$$c^{\alpha} = d^{i} d^{j} A_{i}{}^{\alpha}{}_{j}{}^{\beta} s_{\beta} \tag{16}$$

constitute the ray velocity $\underset{\sim}{c} = c^{\alpha} \underset{\sim}{a}_{\alpha}$ in the reference configuration; the coefficient, on the right-hand side of (15),

$$q = d^{i} d^{j} d^{k} A_{i}{}^{\alpha}{}_{j}{}^{\beta}{}_{k}{}^{\gamma} s_{\alpha} s_{\beta} s_{\gamma} \tag{17}$$

vanishes in the cases of flexural and transversal waves.

The ray vectors $\underset{\sim}{c}$ defined by (16), with $\underset{\sim}{d}$ and $\underset{\sim}{s}$ inserted from solutions of the propagation condition (11), generate a ray curve, which is polar reciprocal to the slowness curve. A pair of corresponding slowness and ray curves is depicted in Fig. 1.

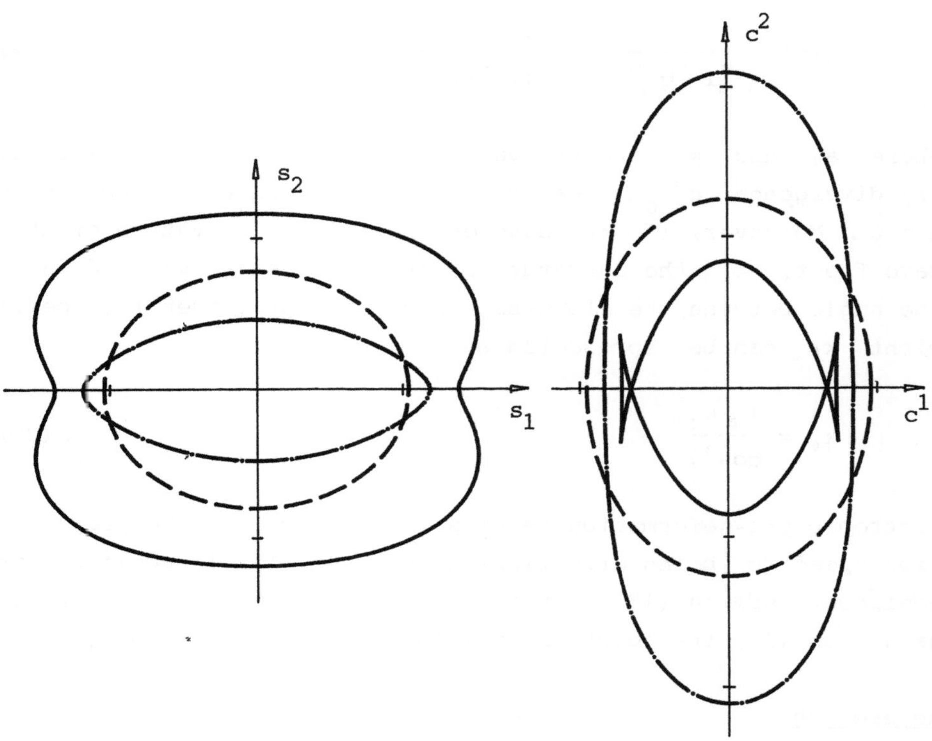

Fig.1. Slowness curve (left) and ray curve (right) of an elastic membrane composed of Mooney-Rivlin material ($\alpha = 3/8$). Principal stretches: $\lambda_1 = 2$, $\lambda_2 = 1$. Wave types: —·—·—·— quasi-longitudinal, ———— quasi-transversal, — — — — flexural.

Rays are vector lines of the velocity field $\underset{\sim}{c}$ on the membrane, they are generated by the differential equation $d\xi^\alpha/dt = c^\alpha$. If the intensity of an acceleration wave is traced along an individual ray, it satisfies the ordinary differential equation

$$2\frac{da}{dt} + c^\alpha|_\alpha \, a = q\, a^2 , \qquad (18)$$

which follows immediately from (15).

In the specific case of a flat membrane subjected to an arbitrary homogeneous deformation, the rays are straight lines, and the evolution of the ray divergence $c^\alpha|_\alpha$ along an individual ray admits an explicit representation. The transport equation has the solution

$$a(t) = \frac{a_0}{\left[1 + q\frac{a_0}{k_0}\left(1 - \sqrt{1+k_0 t}\right)\right]\sqrt{1+k_0 t}} , \qquad (19)$$

where a_0 and k_0 are the values of the intensity a and the ray divergence $c^\alpha|_\alpha$, respectively, at the initial time instant $t = 0$. Moreover, if κ_0 denotes the initial curvature of the wave front, κ_s the curvature of the slowness curve, and Δ the angle between the slowness and ray vectors, then the coefficient k_0 can be represented as

$$k_0 = \frac{\kappa_0 \kappa_s}{\cos^3 \Delta} . \qquad (20)$$

Isotropic pre-deformation results in $k_0 = c\kappa_0$, the propagation speed c being direction-independent. If, in addition, the nonlinear term in (18) is neglected, or vanishes ($q = 0$), we regain from (19) the solution derived by COHEN and BERKAL [2].

<u>References</u>

1. Braun, M.: Beschleunigungswellen in anisotropen hyperelastischen Stoffen. Acta Mechanica 19 (1974) 237-249.

2. Cohen, H.; Berkal, A.B.: Wave propagation in elastic membranes. Journal of Elasticity 2 (1972) 45-57.

Nonlinear Behaviour of Soils Under Dynamic Loading

D. MAKOVIČKA

Building Research Institute, Czech Technical University,
Prague, Czechoslovakia

Summary

The paper deals with the study of mechanical characteristics of sand in dependence on the magnitude and frequency of dynamic loading and on the ratio of the dynamic and static components of loading under repeated stress. A dynamic triaxial press was used to determine the mechanical quantities experimentally.

1. Method of Solution

To determine the mechanical quantities of the soils as functions of loading, a pseudo-triaxial press was used (Fig.1.), which enabled the soil samples to be loaded with a pulsating force of variable magnitude and frequency, acting along the longitudinal axis of the sample at a particular lateral prestressing of the sample.

The sand sample was inserted into the inner cylinder of the fixture. Along the longitudinal axis, one side of the sample made contact with the bottom of the fixture and the other side with the piston. The pulsating load along the longitudinal axis of the sample was generated in an Instron pulsation device by compressing the sample between the piston and the bottom of the triaxial. The axial and lateral loads of the sample were measured by fourth membrane box gauges an effective diameter of 50 mm, placed in the axis of the bottom and piston, and in the wall of the inner cylinder.

The space between the inner and outer cylinders was filled with pressure oil which actedated the lateral prestress of the sample via a network of orifices.

Nonlinear Deformation Waves
IUTAM Symposium Tallinn 1982
Editors U Nigul J Engelbrecht

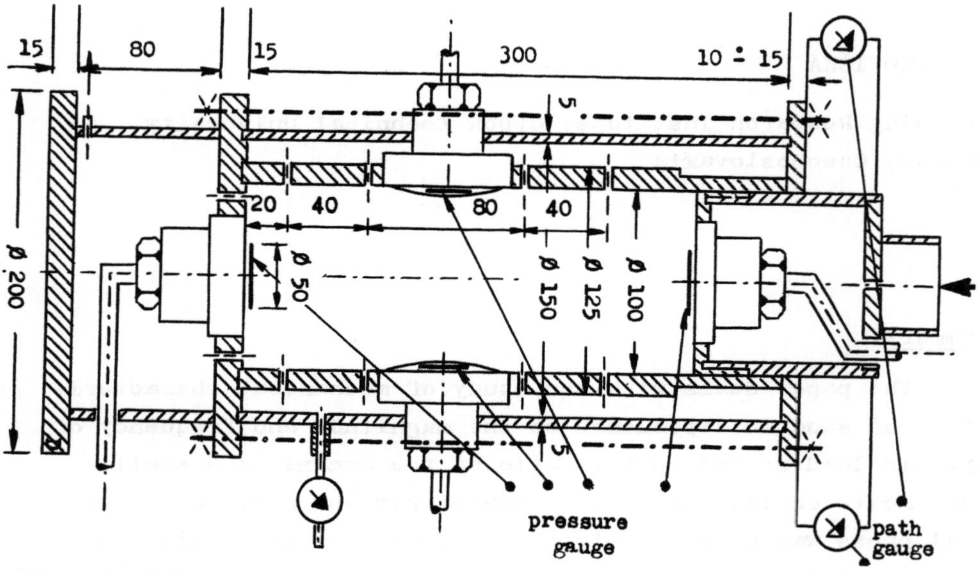

Fig. 1. Scheme of triaxial device

Figure 2 shows a schematic diagram of the pressure distribution underneath the piston. Under dynamic load, acting on the press piston

$$F(t) = F_{st} + F_d \sin 2\pi ft \qquad (1)$$

where F_{st} is the static prestress, F_d - the pulsating component of the force of frequency f; the pressure σ or the reaction of the sand sample to a given mean load $\sigma_m(t)$,

$$\sigma_m(t) = F(t)/S \qquad (2)$$

where measured, S - being the area of the piston. Since the reaction of the sand was measured with a box-gauge, diameter 50 mm, in fact the pressure was measured in the central part of the piston, above which the pressure $\sigma(t,n)$ is lower than $\sigma_m(t)$ as a result of the dome of breaking. The measured pressure can be described by the equation

$$\sigma(t,n) = \sigma_{st}(n) + \sigma_d(n) \sin 2\pi ft \qquad (3)$$

which corresponds to the mean pressure in a column 50 mm in diameter. In Eq. (3), σ_{st} and σ_d are the static and dynamic components of pressure, corresponding to components F_{st} and F_d, respectively, in dependence on the number of cycles n.

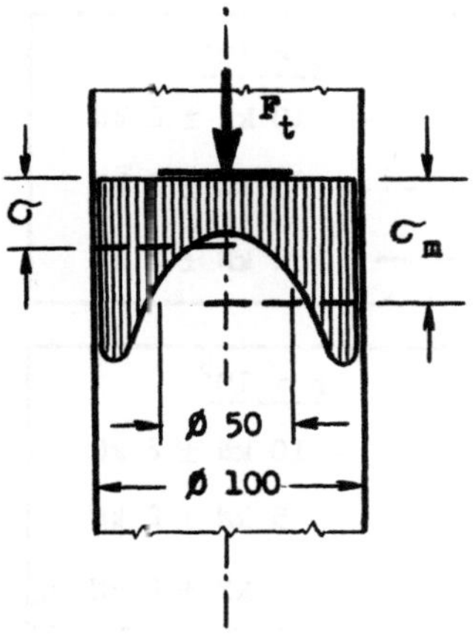

Fig.2. Pressure distribution underneath the piston

2. Results of Measurements

Fig.3 shows the measured pressure as a function of the number of cycles and of the loading frequency. These graphs document the creation of the dome of breaking which is a function of the structure of the medium. It may be claimed that the loading frequency has a decisive effect under compatible static and dynamic load components. The bonds within the structure as a result of the dynamic load acting on the medium and this is reflected in the resonance frequency of the piston column (of the particular sample after $5x10^5$ loading cycles), see Fig.4.

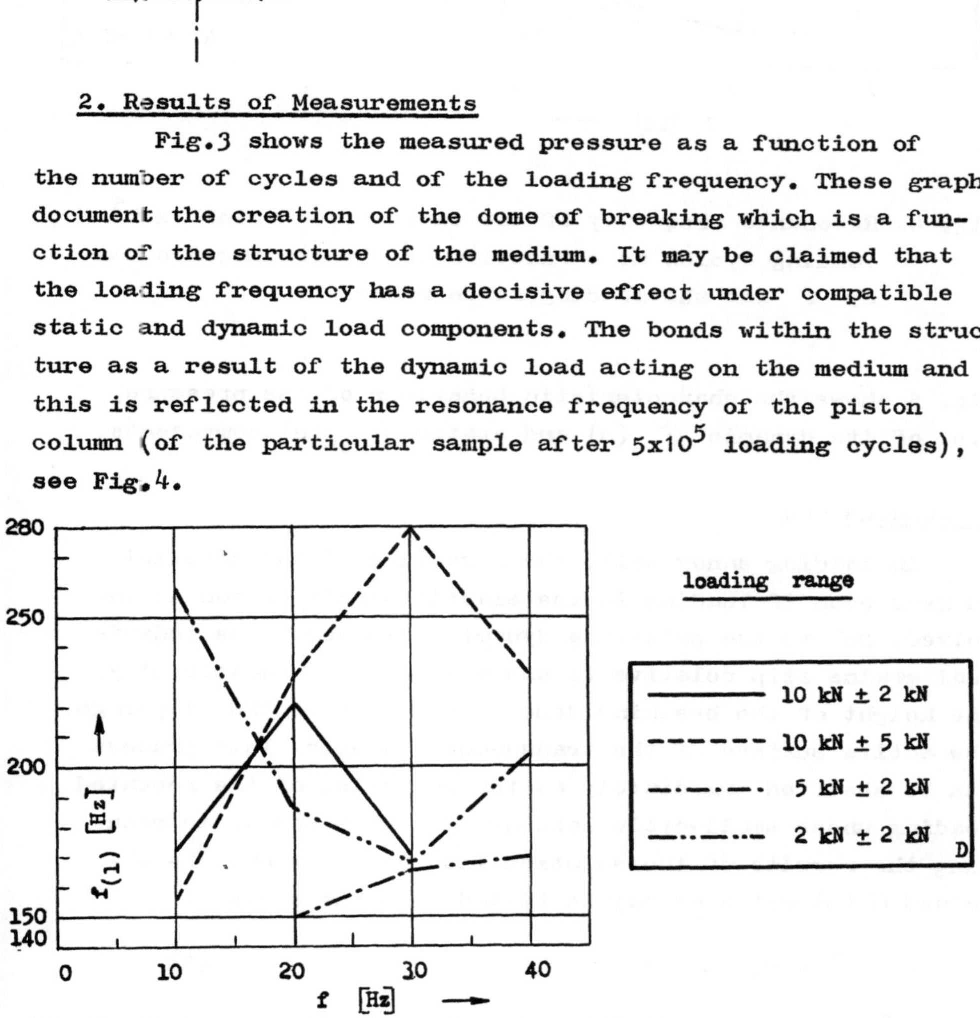

Fig.3. The pressure measured underneath the piston as a function of loading frequency after n cycles of repeated loading

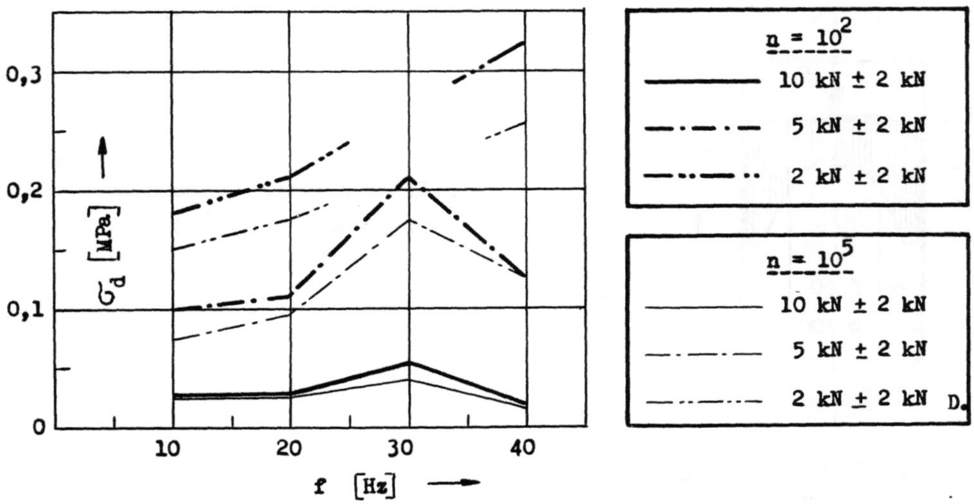

Fig. 4. Resonance frequency of the sample $f_{(1)}$ after 5×10^5 loading cycles as a function of the loading frequency f and the loading force range.

Fig. 5 shows the characteristic behaviour of the pressure, i.e. of its dynamic $\sigma_d(n)$ and static $\sigma_{st}(n)$ components.

3. Conclusions

In loading sandy soil, the structure of the material changes even if loading in the elastic strain region is involved. Due to the pulsating dynamic component, the individual grains slip relative to one another and, consequently, the height of the breaking dome, created above the diaphragm, the active surface of the transducer, changes. This process can be observed immediately at the beginning of the repeated loading under small-cycle loading. For purposes of introducing the results of the solution into calculation, the obtained $\sigma(n)$-patterns may be fitted with the curve

$$S = \beta \exp(\alpha n), \tag{4}$$

where β is a function of the manner of loading or of its force range, the ratio of the static and dynamic loading

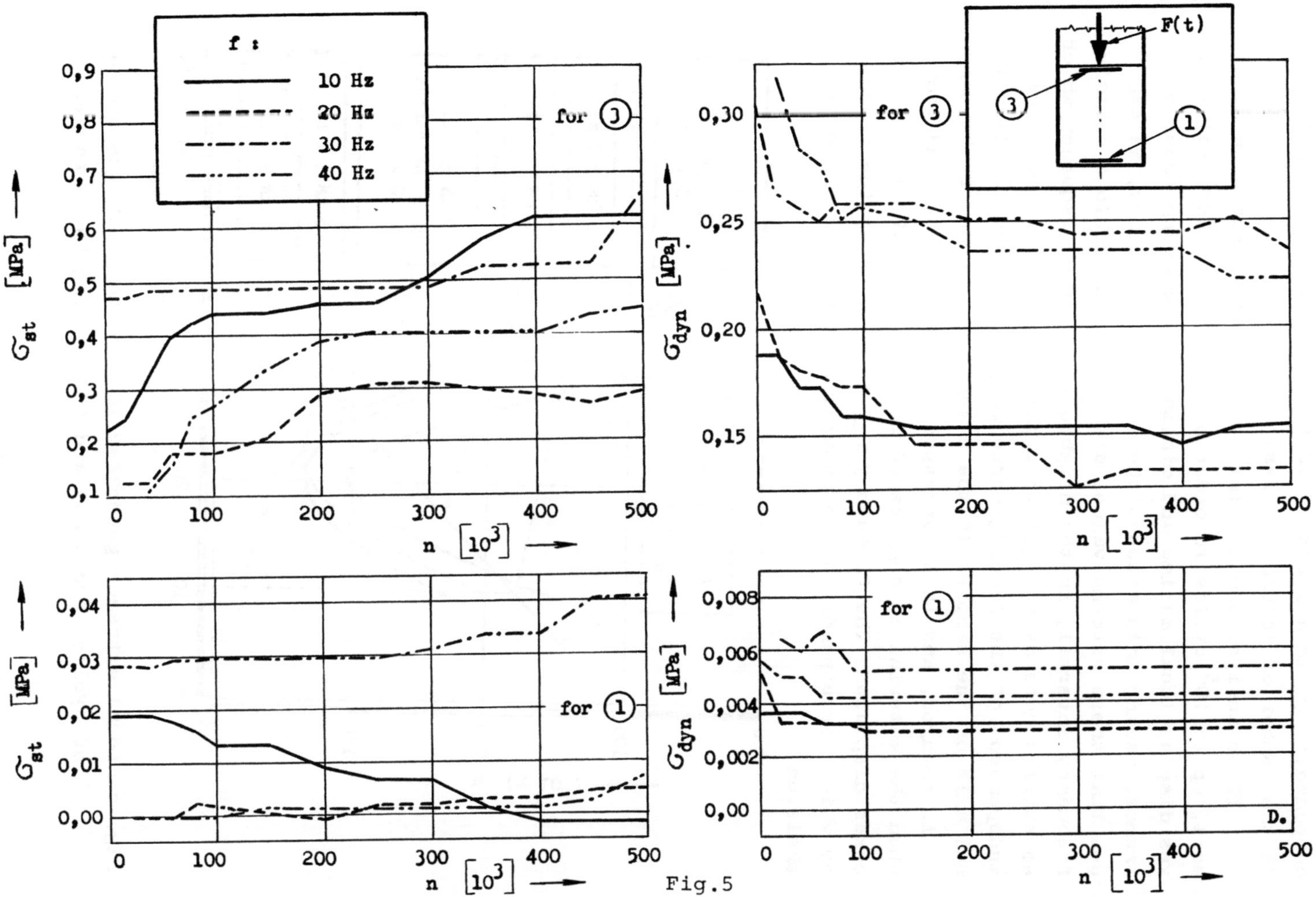
f :
10 Hz
20 Hz
30 Hz
40 Hz
for 3
for 1
F(t)
3
1
σst [MPa]
σdyn [MPa]
n [10^3]
0,9
0,8
0,7
0,6
0,5
0,4
0,3
0,2
0,1
0,30
0,25
0,20
0,15
0,04
0,03
0,02
0,01
0,00
0,008
0,006
0,004
0,002
0,00
0
100
200
300
400
500
D.

Fig.5

components, α is a function of the loading frequency and n is the number of loading cycles.

If we analyse the initial stage of behaviour of a sandy soil, up to 10^5 cycles in order of magnitude, the σ(n) - curves display local maxima and minima. By comparing these extremes, taking into account the exciting frequency, we can see that these extreme values are generated within a certain frequency interval, in the neighbourhood of a frequency which we shall denote by f_{cr} (fig.6). This frequency in practically independent of the loading force range along the longitudinal axis of the sample. If the lateral prestress of the sample is changed, and this corresponds to the actual conditions when the material is compressed in a rock massif, it also becomes displaced. The frequency f_{cr} is apparently the frequency which corresponds to the natural oscillations of the soil skeleton.

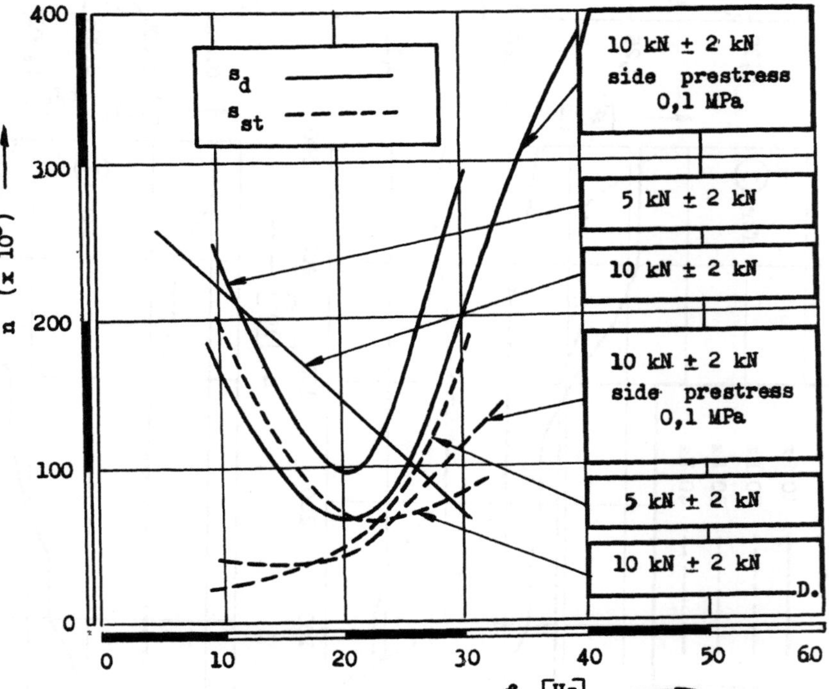

Fig. 6. Local extremes generated in dependence on the number of cycles, frequency and the loading force range.

Nonlinear Effects, Accompanying Seismic Waves Propagation

A.A.GVOZDEV, M.A.GRINFELD', N.V.ZVOLINSKI

Institute of Physics of the Earth, Institute of Problems of Mechanics, Moscow, USSR Academy of Sciences

The linear theory of isotropic elasticity is a theoretical foundation of the various branches of the Earth sciences: it is used in seismology, in the theories of the solid Earth tides and its free oscillations and so on.

On the other hand it is clear that fundamental analysis of the processes occured in an earthquake epicenter and the description of substance properties at the conditions of the Earth interiors are impossible without nonlinear theory. Rather poor contemporary development of these ideas in geophysics is a consequence of great deficiency in our knowledge of substance properties and extreme mathematical difficulties in solving nonlinear problems. We want to note only some reasons of up-to-date increase of geophysical interest to the nonlinear continuum mechanics :1)the higher accuracy of up-to-date instrumentation, 2)geophysical interest to the effects connected with changes in seismic velocities, 3)now, due to the rapid progress in nonlinear theory the simplest way of deriving the linear relations is the linearization of the exact nonlinear ones. In connection with the notes 1,2 we may refer to paper [1] devoted to experimental observations of changes in seismic wave velocities due to solid Earth tides. Due to the superposition principle it is impossible to use the linear theory of elasticity for the description of these effects. However, there is a possibility to use the nonlinear theory of acoustoelasticity for this purpose.

A characteristic feature of interrelation between the linear theory of elasticity and nonlinear one is that the linearization is usually made in dimensional form. So it is impossible to delineate explicitly the bounds of applicability of the

Nonlinear Deformation Waves
IUTAM Symposium Tallinn 1982
Editors U Nigul, J Engelbrecht

linear theory. In accordance with current viewpoint the static states and oscillations of a threedimensional elastic body are well described by the linear theory in the case of small deformations but this does not hold for the nonstationary processes of small disturbance propagation. The best examples of this statement are given by the theory of acceleration waves. The acceleration wave is understood as a sufficiently smooth surface. Displacement and its first derivatives are continuous across this surface, but the second derivatives and, in particular, accelerations have finite jumps. Acceleration waves may be theoretically studied both in linear and nonlinear cases. It seems to be natural to use these waves as a theoretical model for investigation of a nonlinear mechanism of growth and decay of seismic wave intensity, at least, at great distance from an earthquake.

There are three sources of nonlinearity in the exact equations of elastic motion. The first one is the physical nonlinearity, that is the nonlinearity of connection between the stress tensor and the tensor of finite deformations. The second one is the nonlinearity of the equations of momentum conservation. The third one is the geometrical nonlinearity, which in the case of threedimensional body is understood as the nonlinear connection between displacements U^i and the tensor of finite deformations U_{ij}. The isotropic elastic body is characterised by an isotropic tensor connection between the Cauchy stress tensor P^{ji} and the tensor of finite deformations. The full system of the equations of nonlinear elastic motion in Lagrangian coordinates x^i is written below:

$$mU^i_{tt}(x,t) = \left[\frac{m}{M} P^{jk}(\delta^i_k + U^i{}_{,k})\right]_{,j} + mp^i$$

$$P^{jq} = P^{jq}(U_{kl}), \quad U_{kl} = \frac{1}{2}(U_{k,l}+U_{l,k}+U^n{}_{,k}U_{n,l})$$

$$P^{ji} = f_o(x,I_M)X^{ji}+f_1(x,I_M)x^{ji}+f_2(x,I_M)X_{kl}x^{kj}x^{li},$$

$$I_1 = \frac{1}{2}x^{ijk}x^{pqr}x_{ip}x_{jq}X_{kr}, \quad I_2 = \frac{1}{2}x^{ijk}x^{pqr}x_{ip}X_{jq}X_{kr},$$

$$I_3 = \frac{1}{6}x^{ijk}x^{pqr}X_{ip}X_{j1}X_{kr}, \quad X_{kr} = x_{kr} + 2U_{kr}$$

Here m and x_{ij} the density and metric in initial configuration, ,i - symbol of covariant differentiation with respect to $x_{i,j}$,M and X_{ij} the density and metric-in actual configuration; I_T - the main invariants of tensor X_{ij}, mp^i-an external force. In the case of linear approximation the different well-known variants of the ray-method may be applied for the investigation of acceleration waves. As all variants of the ray-method founded on the use of the generalized functions or on the expansion in inverse powers of frequency series fail in the case of nonlinear problems then another variant of the ray-method is used in our study. This method [2-4] is based on the compatibility relations and it enables to investigate the evolution of acceleration wave as well as a weak shock wave up to the moment when caustic forms or the "gradient catastrophe"arises. Successively differentiating the equations of motion and using the compatibility relations one can derive an infinite system of equations in partial derivatives for vectors $H^i{}_M$

$$H^i{}_M = U^i{}_{,j_1 j_2 \dots j_M} \quad N^{j_1} n^{j_2} \dots n^{j_M}$$

We named this system the "determinative" one because it determines evolution of vectors $H^i{}_M$ which in its turn give the comprehensive information about jumps of derivatives of all orders. In the linear case the equations of "determinative" system lead to well-known "transfer equations". For acceleration waves in nonlinear hyperbolic problems the infinite "determinative" system has recurrent character, that is, M first equations completely determine M - 1 first nonzero vectors. In particular, the evolution of jump of acceleration is described by the following ordinary differential equation [2,3,5-8] (media in front of the wave is assumed to be at rest)

$$2\frac{dh}{dt} + h\frac{d \ln Jmc^5}{dt} - \frac{1}{mc} \; p^{ijklmn} e_i e_k e_m n_j n_l n_k h^2 = 0$$

Here h -modulus of the vector $H^i{}_2$, J - ray- divergence, e_i- unit vector parallel $H^i{}_2$, p^{ijklmn}- third derivative of elastic potential with respect to displacement gradient, n_i and c are the unit normal and the velocity of the wave-front. From this equation one can see the influence of the ray geometry on the change of wave intensity. This influence is just the same as in linear case. The difference of the linear case from the nonlinear one consists in the additional nonlinear term,which significantly changes the character of evolution and leads to occurence of "gradient catastrophe". In contrast to formation of a caustic the "gradient catastrophe" is an essentially nonlinear effect, which consists in growth of acceleration jump to infinity at a point where ray-divergence is not equal to zero. The simplest example of this kind is given by propagation of plane longitudinal wave into undisturbed region of an isotropic uniform medium. In accordance with linearized theory the intensity of such a discontinuity must remain constant but according to the strict nonlinear theory there are two opportunities. Either the intensity of jump decays to zero in an infinite time or it grows to infinity in a finite time t_{crit}. The deformations in described process might be small (at the wave front and ahead of it the deformations are exactly equal to zero). However, it is evident that the use of the linearized theory when the considered time intervals are greater than t_{crit} leads to erroneous conclusions. From experimental values of elastic moduli we obtained the next estimations for iron and glass:

iron: $h(t_o)t_{crit}=2.5\times10^4$ cm/sec, glass: $h(t_o)t_{crit}= 4\times10^4$ cm/sec

We studied the propagation of transverse waves too. Our main results are following: 1)kinematics of acceleration waves is well-described by the linear theory, 2)nonlinearity is an additional mechanism for change of seismic wave intensity, 3)in contrast to the linear theory of elasticity the nonlinear one predicts essentially different character of decay of P- and S-waves, 4)the law of the intensity change of transverse wave as well as the law of polarization plane rotation both in the linear and nonlinear cases are the same ones; nonlinearity leads to qualitative changes in the character of P-wave evolution, 5)long-term processes of disturbance propagation must be investigated with the help of nonlinear theory even if deformations are very small. In connection with 3) it is of interest to note that as a rule the amplitudes of P- and S- seismic waves differ considerably [9].

In the case of shock wave propagating in nonlinear media a recurrent character of "determinative" system is violated. But if we deal with weak shock wave and seek the solution of "determinative" system in form of small parameter series the recurrent character holds for each order of approximation. The relations corresponding to the first approximation essentially deviate from those of acoustic theory of shock waves but agree with the results for weak shock waves in fluid obtained by Landau and Whitham. Our approach allows to study weak shock waves in solids too and gives the opportunity to make the theory more definite with the help of next term of approximation.

There are also other branches where the adequate description is possible only with the use of the nonlinear theory even for any small deformations. For example, it is the theory of head waves. Linear theory of head waves was fulfilled in fifties-sixties; see, for example [10-12] . Nonlinear analysis [13] shows that intensity of head waves generated by acceleration ones are determined by the elastic moduli of third order. The approach developed in [13] is similar to linear one of Keller and Friedrichs [14].

REFERENCES

1. De Fazio, T.L., Aki, K., Alba, J. Solid earth tide and observed change in the in situ seismic velocity. Journ. Geoph. Res., 78 (1973), No 8.
2. Grinfel'd M.A. Ray method of calculating the wave front intensity in nonlinear elastic materials. J. Appl. Math. Mech., 42, 958-977 (1979).
3. Grinfel'd M.A. Amplification and attenuation of intensity of seismic wave fronts in isotropic nonlinear elastic media. Izvestiya Akad. Sciences, USSR. Physics of the Solid Earth, v. 5, N 7, (1979).
4. Grinfel'd M.A. Reflection-refraction of wave fronts on the boundary of nonlinear elastic media. Mechanics of solids 13(1978), N 5, 98-104.
5. Wright, T.W. Acceleration waves in simple elastic materials. Arch. Rat. Mech. Anal., 50 (1973), No 4.
6. Brown, M. Beschleunigungswellen in Anisotropen Hyperelastischen Stoffen. Acta Mech. 19 (1974), No 3-4.
7. Chen P.J. Growth and decay of waves in solids. Handbuch der Phys., a/3, v. 6, Springer-Verlag, 1973.

8. Wesolowski Z. Dynamical problems of nonlinear elasticity. Kiev, 1981 (in Russian).
9. Jeffreys H. The Earth. Cambridge Univ. Press, 1970.
10. Zaitsev L.P., Zvolinski N.V. Investigation of axisymmetrical head wave on flat boundary of two elastic fluids. Izv. Acad. Sci. USSR, Ser. Geoph., N 5, 1951 (in Russian).
11. Alexeev A.S., Babitch V.M., Gelchinsky B.Ya. Ray method calculation of intensity of wavefronts. In: Problems of dynamical theory of seismic wave propagation N 5, Leningrad, 1961 (in Russian).
12. Gvozdev A.A. On orientation of displacement vector in P-wave, propagating along a free boundary of elastic medium. Izv. Acad. Sci. USSR, Physics of the Solid Earth, N 2, 1965.
13. Grinfel'd M.A. On nonlinear theory of head wave. Dokl. Acad. Sc. USSR, N 3, 1980.
14. Friedrichs K.O., Keller J.B. J. Appl. Phys., v. 26, N 8, 961, 1955.

Nonlinear Elastic Surface Waves

D. F. PARKER and F. M. TALBOT

Department of Theoretical Mechanics, University of Nottingham,
University Park, Nottingham, NG7 2RD, England

Summary

Linear elastic surface waves having arbitrary profile of surface elevation propagate without change of form. They exhibit no frequency dispersion. We describe methods which determine profiles of nonlinear elastic surface waves which travel without change of form at speeds close to the Rayleigh speed c_R. We find that wavetrains are far from sinusoidal, even at small amplitude. A number of elevation profiles are presented. In each, the deviation of phase speed from c_R is proportional to amplitude.

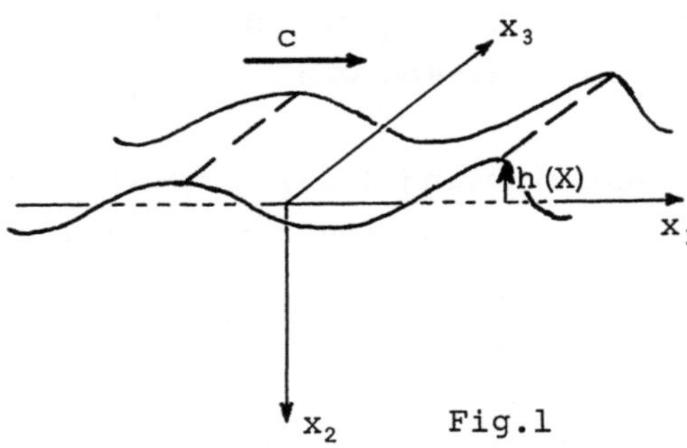

Fig.1

Formulation

In the isotropic elastic half-space $X_2 > 0$, with rectangular Eulerian coordinates (x_1, x_2, x_3) and Lagrangian coordinates (X_1, X_2, X_3), we consider plane-strain disturbances

$$x_i = X_i + \varepsilon u_i(X, X_2), \qquad u_3 \equiv 0; \qquad X := X_1 - ct$$

travelling without distortion at speed c in the x_1-direction (Fig.1). The velocity $\underset{\sim}{v}$, deformation gradient $\underset{\sim}{F}$ and Piola-Kirchhoff (engineering) stress $\underset{\sim}{\tau}$ have components

$$v_i = \frac{\partial x_i}{\partial t} = -\varepsilon c \frac{\partial u_i}{\partial X}, \quad F_{ij} = \delta_{ij} + \varepsilon \frac{\partial u_i}{\partial X_j}, \quad \tau_{ij} = \tau_{ij}(\underset{\sim}{F}). \qquad (1.1)$$

The momentum equations and traction-free boundary condition are

$$\frac{\partial \tau_{ij}}{\partial X_j} = \rho \frac{\partial v_i}{\partial t} = \varepsilon \rho c^2 \frac{\partial^2 u_i}{\partial X^2} \qquad \text{in} \quad X_2 > 0 \qquad (1.2)$$

$$\tau_{i2} = 0 \qquad \text{on} \quad X_2 = 0, \qquad (1.3)$$

Nonlinear Deformation Waves
IUTAM Symposium Tallinn 1982
Editors U Nigul, J Engelbrecht

for $i,j = 1,2$, where ρ is the reference density, assumed uniform.

We introduce stress functions $\alpha_i(X,X_2)$ $(i = 1,2)$ such that

$$\tau_{11} - \varepsilon\rho c^2 u_{1,1} = \varepsilon\alpha_{1,2}\,, \qquad \tau_{12} = -\varepsilon\alpha_{1,1}\,,$$
$$\tau_{21} - \varepsilon\rho c^2 u_{2,1} = \varepsilon\alpha_{2,2}\,, \qquad \tau_{22} = -\varepsilon\alpha_{2,1}\,, \tag{1.4}$$

where $_{;1} := \partial/\partial X$, $_{,2} := \partial/\partial X_2$. Without loss of generality, boundary conditions (1.3) may be taken as

$$\alpha_1(X,0) = 0 = \alpha_2(X,0). \tag{1.5}$$

Assuming that $\underset{\sim}{\tau}(\underset{\sim}{F})$ is suitably smooth, we split the stress $\underset{\sim}{\tau}$ as

$$\tau_{ij} = \varepsilon\{\lambda\delta_{ij}u_{k,k}+\mu(u_{i,j}+u_{j,i})\} + \varepsilon^2\mu N_{ij}(u_{k,\ell};\varepsilon). \tag{1.6}$$

In (1.6), the first bracket describes the linearly elastic stress while $\varepsilon^2\mu\underset{\sim}{N}$ includes *all* nonlinear corrections - but is small in comparison when $\varepsilon|u_{i,j}| \ll 1$.

After substituting (1.6) into (1.4) we separate the linear terms from the nonlinear terms and rearrange the system as

$$\frac{\partial\phi_1}{\partial X} - \frac{1}{A}\frac{\partial\psi_1}{\partial X_2} = \varepsilon\,\frac{N_{22}-N_{11}}{\rho c^2}\,, \qquad \frac{\partial\psi_1}{\partial X} + \frac{1}{A}\frac{\partial\phi_1}{\partial X_2} = -\varepsilon\,\frac{A^2N_{12}+N_{21}}{A\rho c^2}\,,$$
$$\frac{\partial\phi_2}{\partial X} - \frac{1}{B}\frac{\partial\psi_2}{\partial X_2} = \varepsilon\,\frac{B^2N_{22}-N_{11}}{B\rho c^2}\,, \qquad \frac{\partial\psi_2}{\partial X} + \frac{1}{B}\frac{\partial\phi_2}{\partial X_2} = -\varepsilon\,\frac{N_{12}+N_{21}}{\rho c^2}\,, \tag{1.7}$$

where

$$A^2 := 1 - \rho c^2/\mu, \qquad B^2 := 1 - \rho c^2/(\lambda+2\mu) > A^2, \tag{1.8}$$

$$u_1 := B^{-1}\phi_2 - \phi_1, \qquad \alpha_1 := -(2\mu-\rho c^2)A^{-1}\psi_1 + 2\mu\psi_2,$$
$$u_2 := A^{-1}\psi_1 - \psi_2, \qquad \alpha_2 := (2\mu-\rho c^2)B^{-1}\phi_2 - 2\mu\phi_1. \tag{1.9}$$

In view of (1.9), the boundary conditions (1.5) become

$$\psi_2 = \frac{1+A^2}{2A}\psi_1\,, \qquad \phi_2 = \frac{2B}{1+A^2}\phi_1 \qquad \text{on} \quad X_2 = 0\,. \tag{1.10}$$

The displacement $\underset{\sim}{u}(X,X_2)$ vanishes either (i) as $X_2 \to \infty$ (when $\underset{\sim}{u}$ is periodic in X) or (ii) as $X^2+X_2^2 \to \infty$, if there exist constants μ_1, μ_2 such that

$$\phi_1 \to 2\mu_1, \qquad \phi_2 \to 2B\mu_1, \qquad \psi_1 \to 2AB\mu_2, \qquad \psi_2 \to 2B\mu_2. \tag{1.11}$$

The surface elevation is $h(X) = \frac{1}{2}\varepsilon(1-A^2)A^{-1}\psi_1(X,0)$.

Linear Surface Waves (Rayleigh Waves)

In linear theory ($\varepsilon^2 \mu \underset{\sim}{N} \equiv \underset{\sim}{O}$), $\underset{\sim}{\tau}$ is indistinguishable from the Cauchy stress. Equations (1.7) assume Cauchy-Riemann form so that

$$\phi_1(X,X_2) = \beta(X,y), \quad \psi_1(X,X_2) = \gamma(X,y), \quad y := AX_2,$$

where $\beta(X,\eta)$, $\gamma(X,\eta)$ are *any* harmonic conjugate pair satisfying

$$\frac{\partial\beta}{\partial X} = \frac{\partial\gamma}{\partial\eta}, \quad \frac{\partial\beta}{\partial\eta} + \frac{\partial\gamma}{\partial X} = 0, \quad \eta > 0, \tag{2.1}$$

with $\beta, \gamma \to 0$ as (i) $\eta \to \infty$, or (ii) $X^2+\eta^2 \to \infty$. Equations $(1.7)_2$ and (1.10) can then be satisfied *only if*

$$\phi_2(X,X_2) = \frac{2B}{1+A^2}\beta(X,Y), \quad \psi_2(X,X_2) = \frac{2B}{1+A^2}\gamma(X,Y), \quad Y := BX_2,$$

$$(1+A^2)^2 = 4AB, \tag{2.2}$$

with $\mu_1 = 0 = \mu_2$. Equation (2.2) is the standard secular equation for the square of the propagation speed c. It is cubic in c^2 and gives just one positive root $c = c_R$ = Rayleigh speed $< c_S < c_L$. From (1.9), the corresponding displacements $u_i(X,X_2)$ are

$$u_1 = \frac{2\beta(X,Y)}{1+A_R^2} - \beta(X,y), \quad u_2 = \frac{\gamma(X,y)}{A_R} - \frac{2B_R\gamma(X,Y)}{1+A_R^2}, \tag{2.3}$$

($A_R^2 := 1-\rho c_R^2/\mu = 1-c_R^2/c_S^2$, etc.) with surface elevation profile

$$u_2(X,0) = \tfrac{1}{2}(1-A_R^2)A_R^{-1}\sigma(X), \quad \sigma(X) := \gamma(X,0). \tag{2.4}$$

A similar representation of waves having arbitrary $\sigma(X) \in \mathcal{C}^1$ is given in [1,2]. Sinusoidal waves follow from the choice $\sigma(X) = C\cos kX$, which gives

$$\beta(X,\eta) = -Ce^{-k\eta}\sin kX, \quad \gamma(X,\eta) = Ce^{-k\eta}\cos kX. \tag{2.5}$$

Nonlinear Surface Waves

The sinusoidal expressions (2.5) cannot be the leading terms in a solution to (1.4)-(1.6), expanded in powers of ε. Even when we write $c-c_R = O(\varepsilon)$, conditions (1.10) introduce terms such as

$$e^{+2kA_RX_2}\cos 2kX$$

which violate the decay requirement (1.11). This difficulty is resolved by treating (1.7), (1.10) and (1.11) as a system

$$u_{,X} - v_{,y} = \ell(X,y), \quad v_{,X} + u_{,y} = m(X,y), \quad y > 0, \tag{3.1}$$

$$U_{,X} - V_{,Y} = L(X,Y), \quad V_{,X} + U_{,Y} = M(X,Y), \quad Y > 0, \tag{3.2}$$

$$U(X,0) = u(X,0), \quad V(X,0) = Kv(X,0), \tag{3.3}$$

and deducing restrictions on ℓ, m, L, M, v(X,0) and K in order that solutions should be compatible with (1.11) for some μ_1, μ_2. In (3.1)-(3.3), the correspondence between variables is

$$\phi_1(X,X_2) = u(X,y), \quad \phi_2(X,X_2) = \frac{2B}{1+A^2}U(X,Y), \quad \varepsilon\frac{N_{22}-N_{11}}{\rho c^2} = \ell,$$
$$\psi_1(X,X_2) = v(X,y), \quad \psi_2(X,X_2) = \frac{2B}{1+A^2}V(X,Y), \quad \text{etc.}, \tag{3.4}$$

$$K := \frac{(1+A^2)^2}{4AB} = K(c^2), \quad y = AX_2, \quad Y = \Lambda y; \quad \Lambda := \frac{B}{A}. \tag{3.5}$$

By subtracting (3.1) from (3.2), we see that the differences

$$w(X,\eta) := U(X,\eta) - u(X,\eta), \quad \nu(X,\eta) := V(X,\eta) - v(X,\eta)$$

satisfy the boundary value problem

$$w_{,X} - \nu_{,\eta} = L(X,\eta) - \ell(X,\eta), \quad \nu_{,X} + w_{,\eta} = M(X,\eta) - m(X,\eta), \tag{3.6}$$

$$w(X,0) = 0, \quad \nu(X,0) = (K-1)v(X,0), \tag{3.7}$$

$$w \to (A^2-1)\mu_1, \quad \nu \to (1+A^2-2AB)\mu_2, \quad \begin{cases} \eta \to \infty & \text{(i)} \\ X^2+\eta^2 \to \infty & \text{(ii)}. \end{cases} \tag{3.8}$$

Since (3.6), $(3.7)_1$ and (3.8) relate w and ν *uniquely* to L-ℓ and M-m, we may express the boundary behaviour $\nu(X,0)$ as

$$\nu(X,0) = \mathcal{L}(L-\ell, M-m) + (1+A^2-2AB)\mu_2. \tag{3.9}$$

The linear functional $\mathcal{L}$ may be constructed by Fourier transformation, Green's functions, etc. as in [3,4]. When ℓ, m, L and M are expressed in terms of first partial derivatives of u, v, U and V through (1.6), (1.9) and (3.4), there results a consistency condition

$$(K-1)v(X,0) = (1+A^2-2AB)\mu_2 + \mathcal{L}(L-\ell, M-m). \tag{3.10}$$

This may be regarded as a nonlinear functional equation for v(X,0). It is *exact* and applies without restriction on ε. Additionally, $(3.8)_1$ yields an expression for $(A^2-1)\mu_1$.

Small Amplitude Solutions

For small amplitude waves, with $u_{i,j} = O(1)$, the quantities N_{ij} are $O(1)$. The leading approximation is (as for linear waves)

$$\phi_1 \simeq \beta(X,y), \quad \psi_1 \simeq \gamma(X,y), \quad \phi_2 \simeq \beta(X,Y), \quad \psi_2 \simeq \gamma(X,Y), \qquad (4.1)$$

so giving $w(X,\eta) = O(\varepsilon)$, $\nu(X,\eta) = O(\varepsilon)$, $K-1 = O(\varepsilon)$. Hence $c-c_R = O(\varepsilon)$, $A \simeq A_R$, $B \simeq B_R$ and (3.8) suggests that $\mu_1, \mu_2 = O(\varepsilon)$. To leading order, $\underset{\sim}{N}$ is quadratic in the first partial derivatives of $\gamma(X,A_R X_2)$ and $\gamma(X,B_R X_2)$, so that $(L-\ell, M-m)$ (and likewise μ_1) become *quadratic* in the boundary data $\gamma(X,0) = \sigma(X)$ for (2.1). We consider symmetric periodic profiles with

$$\gamma(X,\eta) = \sum_{n=1}^{\infty} C_n e^{-n\eta} \cos nX \qquad (4.2)$$

(period normalized as 2π). For a certain harmonic material [5] with Poisson ratio $\frac{1}{4}$, equation (3.10) yields eventually

$$\kappa C_n = \frac{Z}{n}\left\{\sum_{r=1}^{n-1} \alpha_{rn} r C_r (n-r) C_{n-r} + \sum_{r=1}^{\infty} \beta_{rn} r C_r (n+r) C_{n+r}\right\}, \qquad (4.3)$$

where $K-1 := \varepsilon\kappa$, $Z := \frac{1}{4}(\Lambda_R - 1)^{\frac{3}{2}} \Lambda_R^{-\frac{1}{2}}$ and

$$\alpha_{rn} = 1 - \frac{2n}{r+\Gamma(n-r)}, \quad \beta_{rn} = \frac{2n}{n+r} - \frac{2n}{r+\Gamma n} - \frac{2n}{n+\Gamma r}, \quad \Gamma := \frac{2}{1+\Lambda_R} < 1.$$

Solutions to this infinite quadratic system have either $\kappa = 0$ or $\kappa \propto$ amplitude. Truncated systems involving N unknowns have been solved numerically, giving convergence as N increases:

<u>$\kappa = 0$</u> $(c = c_R)$. We normalize, setting $C_1 = 1$. We find $C_{2p} = 0$ (the profiles are 'anti-periodic') as in Fig.2 (using N = 31).

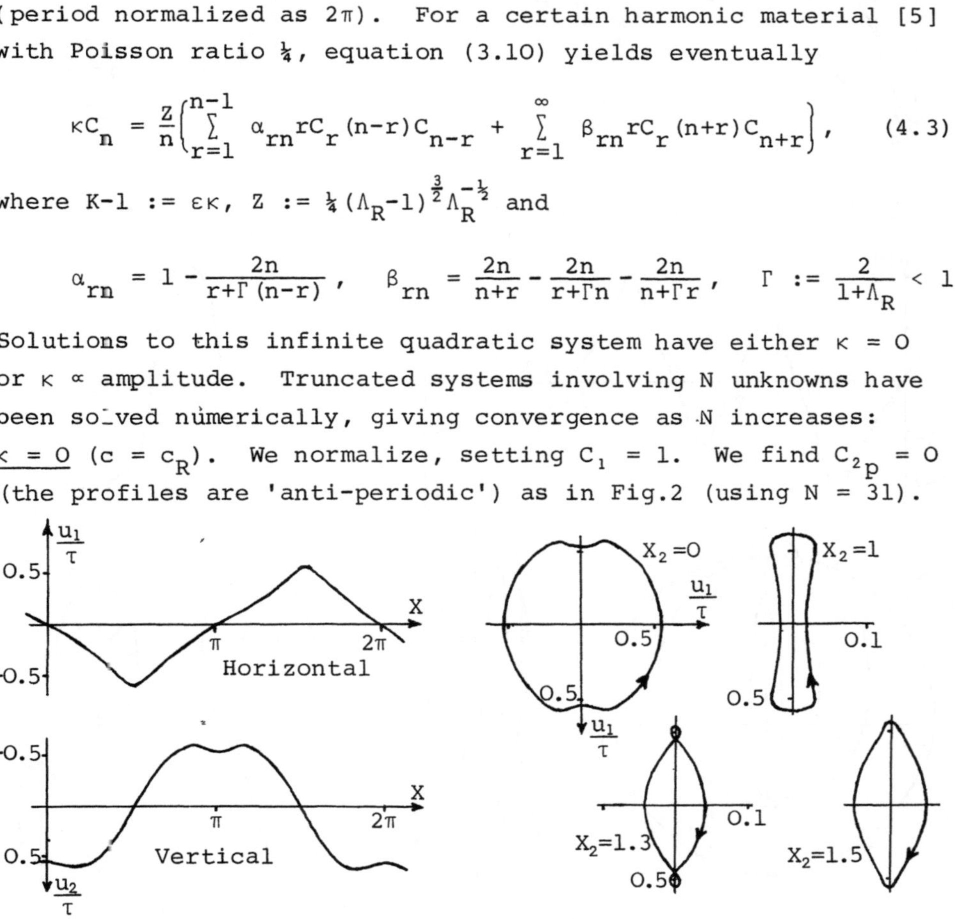

Fig.2 Surface displacements and particle orbits, $\kappa = 0$

$\kappa \neq 0$. For these waveforms we put $P_r = ZrC_r/\kappa$ in (4.3) and solve

$$P_n = \sum_{r=1}^{n-1} \alpha_{rn} P_r P_{n-r} + \sum_{r=1}^{\infty} \beta_{rn} P_r P_{n+r}. \tag{4.4}$$

Figures 3,4 show two solutions; in each, $c-c_R \simeq 0.3413c_R\varepsilon\kappa \propto$ amplitude/wavelength $\propto C_1$. When the profile is inverted, $c-c_R$

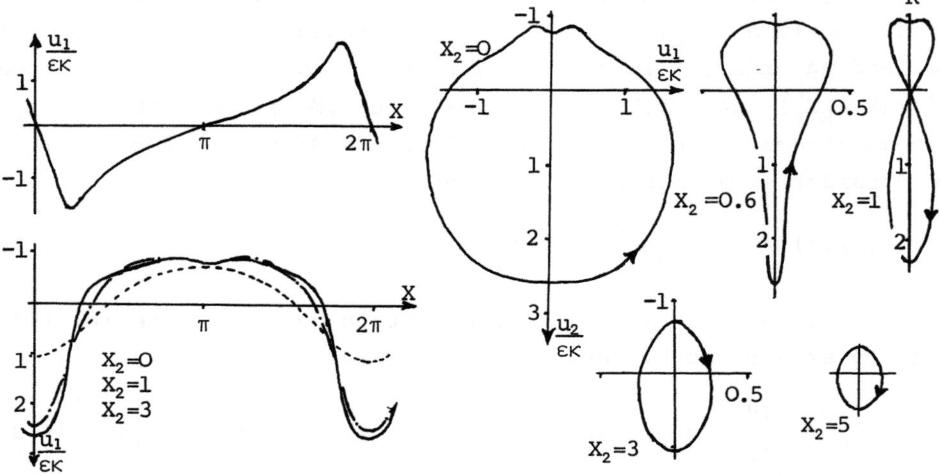

Fig.3 Displacements and particle orbits for solution I

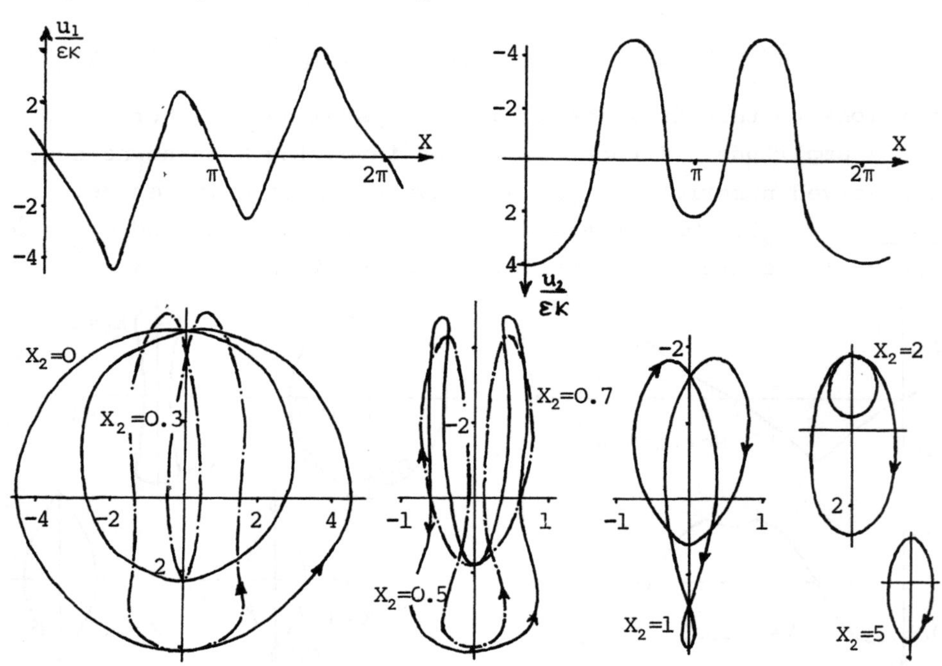

Fig.4 Surface displacements and particle orbits for solution II

changes sign. Motion at the surface is retrograde. At depths of approximately 0.1 to 0.2 wavelengths the particle orbits become contorted, before giving transition to elliptical paths at depths O(wavelength).

Fig.5

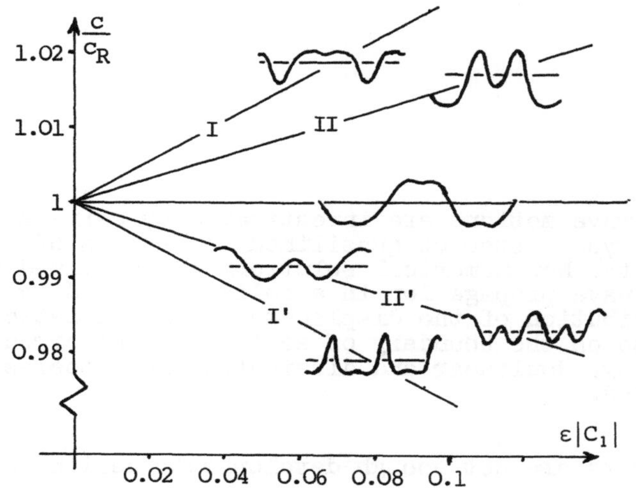

Figure 5 indicates how the perturbation $c-c_R$ in phase speed depends on the fundamental Fourier series coefficient $\varepsilon|C_1|$ for the *exceptional profiles* ($\kappa = 0$), for solutions I (Fig.3) and II (Fig.4) and for their inversions I', II'. Probably many other solutions to (4.3) exist - the corresponding profiles being small amplitude limits of profiles which vary gradually with amplitude. Non-periodic profiles may also exist. The Fourier sine transform of their surface elevation satisfies a nonlinear integral equation of convolution form, which is a natural generalization of (4.3).

References

1. Chadwick, P.: Surface and interfacial waves of arbitrary form in isotropic elastic media. J. Elast. 6 (1976) 73-80
2. Parker, D.F.: Investigations into nonlinear Rayleigh waves. Tech. Report 78-26, IAMS, University of British Columbia 1978
3. Talbot, F.M.: Ph.D. thesis, University of Nottingham 1982
4. Parker, D.F.: An analysis of nonlinear elastic waves of permanent form, to be published
5. John, F.: Plane strain problems for a perfectly elastic material of harmonic type. Comm. Pure Appl. Math. 19 (1960) 239-260

About Influence of Nonlinear Magnetoelastic Effects on Wave Propagation in an Elastic Electroconducting Medium

I.T.SELEZOV

Institute of Hydromechanics of the Academy of Sciences, Ukrainian SSR, Kiev

Summary

Transient wave motions are investigated by using a hyperbolic-parabolic type system of quasilinear equations of the magnetoelasticity. New numerical solutions are obtained which describe the wave propagation in a solid generated by a given initial distribution of the displacements and magnetic field, or the impulse on the boundary of semi-space and circular cylindrical cavity. Nonlinear and dissipative magnetoelastic effects are analysed.

1.Today there are not enough developments dealing with a nonlinear magnetoelastic wave propagation due to difficulties in solving nonlinear partial differential equations. In this communication numerical methods are applied to study transient nonlinear motions in a magnetoelastic solid. It is assumed that a process considered is isentropic, constitutional relations are linear, the medium is isotropic and nonmagnetizable. The effects of a displacement current $\dot{\vec{D}}$ and electrical charges - terms $\rho_e, \rho_e\vec{E}, \rho_e\dot{\vec{u}}$ ($\vec{E}$ is the electric field vector, $\vec{u}$ is the displacement vector) - are not taken into account. According to these assumptions the magnetoelastic motion is described by a coupled hyperbolic-parabolic system of equations [1] ,[2]

$$G\nabla^2\vec{u} + (G + \lambda)\vec{\nabla}(\vec{\nabla}\cdot\vec{u}) = \rho_c\ddot{\vec{u}} - P_H(\vec{\nabla}\times\vec{H})\times\vec{H}$$

$$R_m\dot{\vec{H}} = -\vec{\nabla}\times(\vec{\nabla}\times\vec{H}) + R_m\vec{\nabla}\times(\dot{\vec{u}}\times\vec{H}) \tag{1}$$

Nonlinear Deformation Waves
IUTAM Symposium Tallinn 1982
Editors U Nigul, J Engelbrecht

$$\vec{\nabla} \times \vec{H} = \vec{J}, \ \vec{\nabla} \times \vec{E} = -\dot{\vec{H}}, \quad \vec{\nabla}\cdot\vec{E} = 0, \quad \vec{\nabla}\cdot\vec{H} = 0 \tag{2}$$

$$\vec{J} = \gamma(\vec{E} + \dot{\vec{u}} \times \vec{H}), \quad \vec{D} = \varepsilon\vec{E}, \quad \vec{B} = \mu\vec{H} \tag{3}$$

where λ , G are Lame constants, ρ_c is the mass density, $\vec{H}$ is the magnetic field vector, $P_H = \mu H^2/\rho v^2$ is the magnetic pressure, $R_m = \mu\gamma l v$ is Reynolds magnetic number, l , v , ρ , H are the characteristic values. Boundary conditions in the interface of the media r and s are

$$\vec{n}\cdot(\vec{B}^r - \vec{B}^s) = 0, \quad \vec{n} \times (\vec{H}^r - \vec{H}^s) = 0$$

$$\vec{n}\cdot(\vec{D}^r - \vec{D}^s) = 0, \quad \vec{n} \times (\vec{E}^r - \vec{E}^s) = 0 \tag{4}$$

$$n_i \left[(\sigma^2_{ik} + T^2_{ik}) - (\sigma^s_{ik} + T^s_{ik})\right] = 0$$

where

$$T_{ik} = H_i B_k - 0.5\,\delta_{ik} H_j B_j$$

$$\sigma_{ik} = (\lambda\,\delta_{ij}\,\delta_{mn} + 2G\,\delta_{mi}\,\delta_{nj})\,\varepsilon_{mn}$$

$$\varepsilon_{mn} = 0.5\,(u_{m,n} + u_{n,m})$$

Initial conditions are of the form

$$\begin{aligned} \vec{u}(\vec{x},t)\Big|_{t=t_0} &= \vec{\varphi}_1(\vec{x}), \quad & \dot{\vec{u}}(\vec{x},t)\Big|_{t=t_0} &= \vec{\psi}_1(\vec{x}) \\ \vec{H}(\vec{x},t)\Big|_{t=t_0} &= \vec{\varphi}_2(\vec{x}), \quad & \dot{\vec{H}}(\vec{x},t)\Big|_{t=t_0} &= \vec{\psi}_2(\vec{x}) \\ &\ldots \end{aligned} \tag{5}$$

The system of equations (1) contains nonlinear terms coupling elastic and electromagnetic fields: $P_H\,(\vec{\nabla} \times \vec{H}) \times \vec{H}$ and $R_m \vec{\nabla} \times$ $\times(\dot{\vec{u}} \times \vec{H})$. Further the equations (1)-(5) are used to solve some problems.

2. Consider the infinite solid in a cartesian coordinate system $x_1=x$, x_2, x_3. We shall study one-dimensional dilatation motions in a direction of the axis Ox under action of the magnetic field $\vec{H}=(0,H,0)$. Then we have $\vec{u}=(u,0,0)$, $P_H=\mu H^2/(\lambda+2G)$, $R_m=\mu\gamma lc_e$, $c_e=\sqrt{(\lambda+2G)/\rho_c}$. In this case the equations (1) reduce to

$$
\begin{aligned}
&u_{tt} - c_e^2 u_{xx} = - P_H H H_x \\
&H_t - R_m^{-1} H_{xx} = - (Hu_t)_x
\end{aligned}
\tag{6}
$$

The Cauchy problem for a system (6) is formulated with assumptions that at the initial moment of time t=0 the magnetic field is H=const and the displacement distribution is given by

$$
u(x,0) = \begin{cases} a_0 \exp(-\alpha x^2) & \text{for } |x| \leq 0.5 \\ 0 & \text{for } |x| > 0.5 \end{cases}
\tag{7}
$$

Besides the cases of an initial velocity and magnetic field were considered. In new variables $s=u_t$, $w=u_x$ equations (6) are in the form

$$
\begin{aligned}
&w_t - s_x = 0, \quad s_t + (0.5\, P_H H^2 - w)_x = 0 \\
&H_t + (sH)_x - R_m^{-1} H_{xx} = 0
\end{aligned}
\tag{8}
$$

Numerical solutions of the problem (7), (8) were obtained by using a conservative finite-difference Godunov's scheme [3] . This scheme was explicit at s,w and stable if the Courant condition $\Delta x/\Delta t > 1$ is satisfied. Testing computations confirmed the scheme stability and showed that for a given value Δt the iteration process was unnecessary for defining more precisely the nonlinear term cofactors in (8) which were taken from low layer. Some results of numerical experiments are presented in fig.1. Due to a magnetic field action the extended tails appeared out of the head waves and amplitudes of these waves gradually decreased. A decrease of the conductivity to $R_m= 10^2$ slightly influences on a structure of the dilatation wave but strongly smoothes disturbances of the magnetic field. Further,

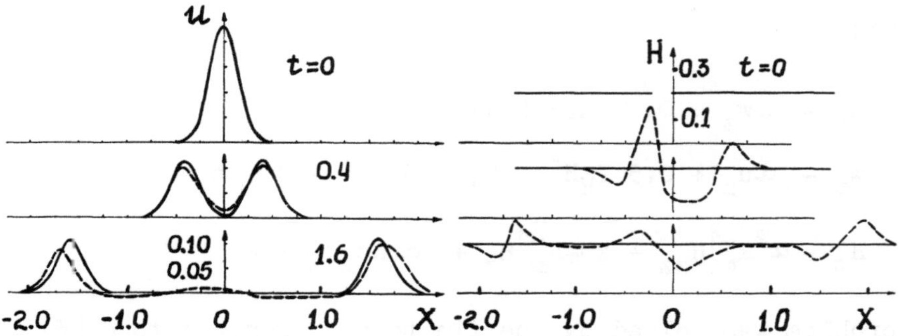

Fig.1. Propagation of displacement and magnetic field disturbances: $a = 0.234$, $\alpha = 25$, $P_H = 0.1$, $R_m = 10^2$.

if the nonlinear convective term is taken into account the symmetry of the wave propagation is broken up: the wave propagates in plus direction with a greater amplitude and velocity than in minus direction.

3.The semi-infinite magnetoelastic solid $x_1 = x \geq 0$ contacting with a vacuum $x < 0$ is considered. The magnetic field H_2 is switched on in the interface of the two mediums at the initial moment of time t=0. For $u_1 = u(x,t)$ and $H_2 = H(x,t)$ the initial boundary value problem is formulated. The problem is described by the equations (6) in a domain $0 \leq x < \infty$, $t > 0$ with conditions

$$u(x,0) = u_t(x,0) = H(x,0) = 0 \tag{9}$$

$$u_x(0,t) = 0, \quad u(\infty,t) = 0 \tag{10}$$

$$H(0,t) = 1, \quad H(\infty,t) = 0 \tag{11}$$

Similar problem but without taking into account of the nonlinear term $(Hu_t)_x$ was considered in [4] . The compressing reflection $z=1-\exp(-kx)$ which transforms a semi-infinite interval of change x $[0, \infty)$ into a finite interval of a new variable z $[0,1)$ is introduced. Also the new auxiliary variable $w(x,t) = \int_0^x u_t(\xi,t)d\xi$ is introduced. Then the problem is reduced to a solution of equations with the conditions

(9)-(11)

$$u_t - a w_z = 0, \quad a = k(1-z)$$
$$w_t - a u_z + 0.5\, P_H H^2 = 0, \quad 0 \leq z < 1, \quad t > 0 \tag{12}$$
$$H_t - a^2 R_m^{-1} H_{zz} + k a R_m^{-1} H_z + a(H u_z)_z = 0$$

This problem was solved by the finite-difference method [5] with a rectilinear net.

4.The infinite medium with a circular cylindrical cavity $r = a$ is considered in a cylindrical coordinate system r, θ, z connected with it. At instant t=0 on the boundary of cavity the longitudinal magnetic field changing in time as Heviside function is switched on. A corresponding initial-boundary value problem at the domain $a_1 \leq r < \infty$, $t > 0$ is

$$\frac{\partial^2 u_r}{\partial t^2} - \frac{\partial}{\partial r} \frac{1}{r} \frac{\partial}{\partial r}(r u_r) + \frac{P_H}{2} \frac{\partial H^2}{\partial r} = 0$$
$$\frac{\partial H}{\partial t} - \frac{1}{R_m} \frac{1}{r} \frac{\partial}{\partial r}\left(r \frac{\partial H}{\partial r}\right) + \frac{1}{r} \frac{\partial}{\partial r}\left(r H \frac{\partial H}{\partial r}\right) = 0 \tag{13}$$

$$u_r(r,0) = \frac{\partial u_r}{\partial t}(r,0) = H(r,0) = 0 \tag{14}$$

$$\frac{\partial u_r(a_1,t)}{\partial r} = u_r(\infty, t) = 0 \tag{15}$$

Numerical solutions of this problem were carried out on a base of the two-layered implicit finite-difference scheme [5]. As above the compressing reflection $\xi = b - \exp[Ek(r - a_1)]$ is introduced. Some results of the numerical computations are presented in fig.2. It shows that the boundary is vibrated with a decreasing amplitude due to a magnetoelastic dissipation of energy and waves going out from the cavity.

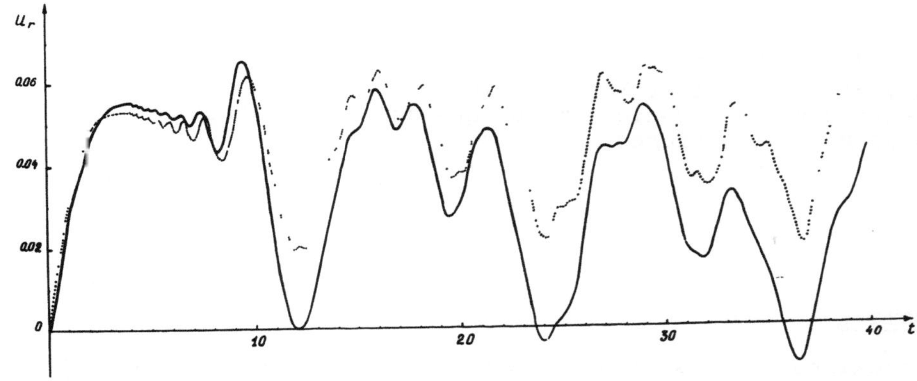

Fig.2. Radial oscillations of the cavity boundary: P_H=0.1, R_m=1.0 – ,10.0 ····

<u>References</u>

I. Селезов И.Т.: Некоторые приближенные формы уравнений движения магнитоупругих сред. Изв.АН СССР. Механ.тверд.тела. 5 (1975) 86-91.

2. Селезов И.Т., Селезова Л.В.: Волны в магнитогидроупругих средах. Киев: Наукова думка, 1975.

3. Численное решение многомерных задач газовой динамики. Под ред.Годунова С.К. Москва: Наука, 1976.

4. Moon F.C., Chattopadhyay S.: Magnetically induced stress waves in a conducting solid – theory and experiment. Trans. ASME E41 (1974) 641-646.

5. Самарский А.Г.: Введение в теорию разностных схем.Москва: Наука,1971.

Nonlinear Elastic Waves Generated by Electric Solitons in Ferroelectric Crystals

G.A.MAUGIN, J.POUGET

Université Pierre et Marie Curie,
Laboratoire de Mecanique Théorique associé au CNRS
75230 Paris 05, France

Summary

The existence of soliton-like solutions is exhibited on the basis of a lattice model equipped with permanent electric dipoles and a nonlinear continuum model for elastic ferroelectrics. Indeed, nonlinearities play a dominant role nearby the phase transition and an anharmonic analysis becomes necessary. Here the nonlinearities are in the order parameter associated with ferroelectricity (orientation of electric dipoles with respect to the atomic chain) while the elastic behaviour remains linear. Passing from the lattice description to a continuum allows one to obtain soliton solutions for the ferroelectric soft mode. In turn, these generate nonlinear elastic waves through piezoelectricity. The resulting wave corresponds to the propagation of a wall separating ferroelectric domains in the crystal.

1. Introduction

The essential property of ferroelectrics is to present a phase transition so that there exists an order parameter, the spontaneous polarization P_S, which strongly depends on temperature. We have a paraelectric phase, $P_S = 0$, for $T > T_C$ (Curie temperature) and a ferroelectric one, $P_S \neq 0$, for $T < T_C$ [1]. This phenomenon, in turn, affects the electromechanical properties of the materials. Indeed, these materials present a strong piezoelectric coupling for $T < T_C$ and this yields a coupling, through resonance, between acoustic phenomena and the so-called ferroelectric soft mode. The phase transition affects the soft mode to a large effect since its frequency ω_F [2] falls down to zero with $T \to T_C$ from below. In the process, the soft mode pulls down the acoustic mode of which the speed diminishes drastically[3] -[5]. Furthermore, as $\omega_F \to 0$, the corresponding amplitude (angle variation of P)

Nonlinear Deformation Waves
IUTAM Symposium Tallinn 1982
Editors U Nigul, J Engelbrecht

increases. Heretofore, the coupling between acoustic and soft modes was studied within the frame of a linearized theory, but nearby the phase transition it becomes necessary to envisage nonlinearities and an anharmonic analysis in which physicists are particularly interested from the experimental point of view.

Here we present the nonlinear equations which govern a mono-atomic chain equipped with electric dipoles. This elementary model, sufficient for our purpose, is based in the microphysics of crystals of the displacive type which has already allowed us to perform a complete study of coupled acoustic-soft modes [6]-[7]. The rotation of molecular groups equipped with *electric dipoles* is accounted for in addition to the atomic displacement. Such models provide a good approach to micropolar solids [8]-[9]. Then the continuum limit provides a set of nonlinear equations. Only the nonlinear ferroelectric behavior is retained but there is no difficulty, but analytical, to account for elastic nonlinearities as well. The *linearized* version of these equations compares favorably to a previous continuum model for elastic ferroelectrics [10-11] which has concluded with the existence of coupled acoustic-soft modes through piezoelectricity induced during the symmetry breaking by permanent dipoles in the basic ferroelectric state. Here, obviously, nonlinearities suggest, in certain conditions, the existence of *solitary waves* [12], more precisely, the interaction between an harmonic acoustic mode (phonon) and a ferroelectric soliton. The stability of this type of solutions shows that they correspond to the motion of a wall separating two ferroelectric domains, which in effect corresponds to a physical interpretation of the phase transition. In addition, the coupling exhibited allows one to generate a nonlinear acoustic wave by the intermediate of a ferroelectric soliton while no elastic nonlinearities are considered.

2. Model

In agreement with the usual picture of a prototype ferroelectric, barium titanate $BaTiO_3$ (Figure 1), we consider a monoatomic chain of which each point has a mass m and a microscopic dipole of strength $P = \delta Q$ (Q=electric charge of Ti; δ=relative displacement of positive and negative charges; $\delta \ll a$; a = lattice constant) so that at each lattice point n we have: (i) a longitudinal motion u_n, (ii) a transverse motion v_n and a rigid-body rotational motion θ_n associated with the orientation of electric dipoles [6]- cf. Figure 2. Only the case where all dipoles are initially aligned with the chain is considered. Let I be the inertia moment of the TiO_3 group. Nearest neighbor interactions are accounted for and represented by spring constants $k_\parallel$ and $k_\perp$ so that the usual deformation energy is given by

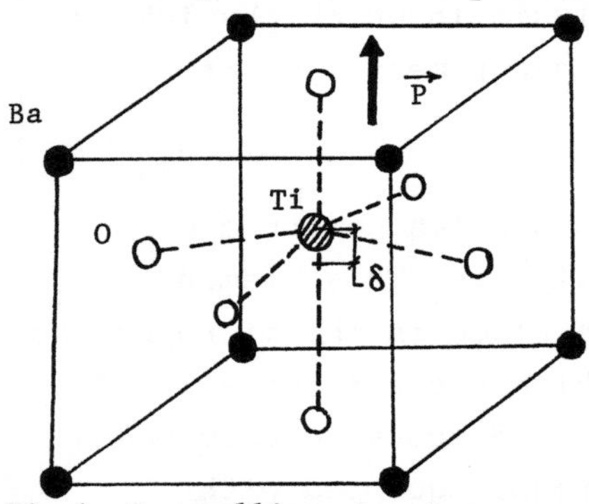

Fig.1.-Crystalline structure of ferroelectric $BaTiO_3$

$$W_{def} = \frac{1}{2} \sum_n k_\parallel [(u_{n-1}-u_n)^2 + (u_{n+1}-u_n)^2] + k_\perp [(v_{n-1}-v_n)^2 + (v_{n+1}-v_n)^2] . \quad (2.1)$$

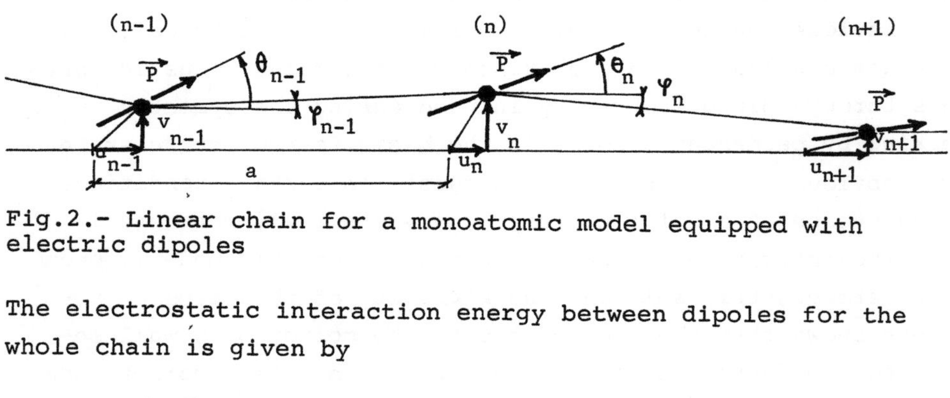

Fig.2.- Linear chain for a monoatomic model equipped with electric dipoles

The electrostatic interaction energy between dipoles for the whole chain is given by

$$W_{elec} = \frac{1}{2} \sum_n (W_n^{n-1} + W_n^{n+1}) \quad (2.2)$$

where W_n^{n-1} is the electrostatic energy of the couple $(n-1,n)$ of dipoles. On expanding up to terms of the fourth order in θ_n, we have (an expression of a similar form holds good for W_n^{n+1})

$$W_n^{n-1} = -\epsilon\Big[2 - 6\frac{u_n-u_{n-1}}{a} + 12\Big(\frac{u_n-u_{n-1}}{a}\Big)^2 - 3\Big(\frac{v_n-v_{n-1}}{a}\Big)^2 - \frac{1}{2}(\theta_{n-1}-\varphi_{n-1})^2$$

$$-\frac{1}{2}(\theta_n-\varphi_{n-1})^2 - \frac{1}{2}(\theta_n-\theta_{n-1})^2 + \frac{1}{24}\theta_n^4 + \frac{1}{24}(\theta_n-\theta_{n-1})^4 + \frac{1}{4}\theta_{n-1}^2\theta_n^2 + O(u_n^3, v_n^3, \theta_n^6)\Big] \quad (2.3)$$

where we have set

$$\epsilon = \frac{p^2}{4\pi\varepsilon a^2} \quad , \quad \varphi_{n-1} = \frac{v_n - v_{n-1}}{a} \tag{2.4}$$

and kept only nonlinearities with respect to θ_n since nonlinear couplings between u, v and θ would bring nothing essentially new in the forthcoming analysis. The kinetic energy and the Lagrangian of the system read

$$K = \frac{1}{2} \sum_n m[(\dot{u}_n^2 + \dot{v}_n^2) + I\dot{\theta}_n^2] \quad ,$$

$$L = K - (W_{def} + W_{elec}) \quad . \tag{2.5}$$

3. Motion equations

A. Lattice model: From eqs.(2.5) we obtain

$$m\,\ddot{u}_n = \hat{k}_{\parallel}\,(u_{n-1} + u_{n+1} - 2u_n) \quad ,$$

$$m\,\ddot{v}_n = \hat{k}_{\perp}\,(v_{n-1} + v_{n+1} - 2v_n) - (\epsilon/a)(\theta_{n+1} - \theta_{n-1}) \quad , \tag{3.1}$$

$$I\,\ddot{\theta}_n = \epsilon(\theta_{n-1} + \theta_{n+1} - 4\theta_n) + \frac{\epsilon}{2}\theta_n\left(\frac{2}{3}\theta_n^2 + \theta_n\theta_{n-1} + \theta_{n+1}^2\right) + \frac{\epsilon}{a}(v_{n+1} - v_{n-1})$$

where we have introduced stiffened spring constants

$$\hat{k}_{\parallel} = k_{\parallel} + (24\,\epsilon/a^2) , \qquad \hat{k}_{\perp} = k_{\perp} + (8\,\epsilon/a^2) \quad . \tag{3.2}$$

Equations $(3.1)_{2-3}$ couple the transverse displacement and the rotation while eq.$(3.1)_3$ is nonlinear. In a linear analysis, eqs.(3.1) exhibit the resonant coupling between the transverse acoustic mode and the ferroelectric-soft mode and the motion is stable insofar as we stay in the ferroelectric phase [6 - 7].

B. Continuum model: On taking the continuum limit of eqs.(3.1), we are led to

$$\rho\frac{\partial^2 u}{\partial t^2} = (\hat{k}_{\parallel}/a)\frac{\partial^2 u}{\partial x^2} \; , \; \rho\frac{\partial^2 v}{\partial t^2} = (\hat{k}_{\perp}/a)\frac{\partial^2 v}{\partial x^2} - \eta\frac{\partial\theta}{\partial x} \quad , \tag{3.3}$$

$$J\frac{\partial^2\theta}{\partial t^2} = \frac{\eta a^2}{2}\frac{\partial^2\theta}{\partial x^2} - \eta\theta + \frac{2}{3}\eta\theta^3 + \eta\frac{\partial v}{\partial x} \quad ,$$

where

$$\rho = m/a^3 \ , \quad J = I/a^3 \ , \quad \eta = 2\,\epsilon/a^3 \ .$$

We can define a polarization perturbation p_n of the dipole at site n so that $p_n = P\Theta_n$ and eqs. (3.3) then allow one to find an equation for the perturbation p in the volume polarization. Note that the coefficient of the Θ^3-term in $(3.3)_3$ is of the same order as the coefficient of the term linear in Θ. Only the amplitude variations of Θ justify introducing nonlinearities. Under weaker conditions for the expansion of the energy (2.3), hence a stronger nonlinearity in Θ, we could obtain for $(3.3)_3$ a sine-Gordon equation coupled with eq. $(3.3)_2$. In what follows we consider only the coupled equations $(3.3)_{2-3}$ since eq. $(3.3)_1$ admits usual harmonic solutions. On looking for propagative solutions functions of the variable $\xi = qx - \omega t$, eqs. $(3.3)_{2-3}$ yield

$$(\omega^2 - c_\perp^2 q^2)\frac{d^2 v}{d\xi^2} + \frac{\eta}{\rho} q \frac{d\Theta}{d\xi} = 0 \ ,$$

$$(\omega^2 - c_p^2 q^2)\frac{d^2\Theta}{d\xi^2} + \frac{\eta}{J}\Theta - \frac{2\eta}{3J}\Theta^3 - \frac{\eta q}{J}\frac{dv}{d\xi} = 0 \ , \qquad (3.4)$$

$$c_\perp^2 = \hat{k}_\perp / \rho a \ , \quad c_p^2 = \eta\, a^2/2J .$$

On eliminating v between eqs. (3.4), we obtain the following nonlinear equation for Θ:

$$\lambda(\omega, q)\frac{d^2\Theta}{d\xi^2} - \beta(\omega, q)\,\Theta^3 + \alpha(\omega, q)\,\Theta = 0 \ , \qquad (3.5)$$

where

$$\lambda(\omega,q) = (\omega^2 - \omega_p^2)(\omega^2 - \omega_\perp^2) \ , \ \omega_p^2 \equiv c_p^2 q^2 \ , \ \omega_\perp^2 = c_\perp^2 q^2 \ ,$$

$$\alpha(\omega,q) = (\eta/J)(\omega^2 - \hat{\omega}_\perp^2) \ , \ \hat{\omega}_\perp^2 = (c_\perp^2 - \frac{\eta}{\rho})q^2 \ , \qquad (3.6)$$

$$\beta(\omega,q) = (2\eta/3J)(\omega^2 - \omega_\perp^2) .$$

The coefficients λ, β and α depend on the pseudo-frequency a and the pseudo-wave number q.

4.Solitons

Equation (3.5) admits a first integral in the form

$$\lambda\left(\frac{d\theta}{d\xi}\right)^2 + V(\theta) = \kappa,\; V(\theta) = -\frac{1}{2}\beta\left(\theta - \frac{\alpha}{\beta}\right)^2, \kappa = 2C - \frac{\alpha^2}{2\beta} \quad , \qquad (4.1)$$

where C is an integration constant. Equation $(4.1)_1$ can be interpreted as the equation governing the motion of a particle of mass λ in a double-well potential $V(\theta)$ [14]-[15]. Note that the "mass" λ may be negative - cf. eq. (3.6). For $C \neq 0$ we have c n o i d a l w a v e s of which the study involves elliptic integrals. We consider the case $C = 0$ which provides s o l i t o n s o l u t i o n s. We have two different cases depending on the sign of λ , α and β .

(a) case $0 \leqslant \omega \leqslant \omega_P$: Then the solution of eq. (4.1) reads (sech=1/cosh)[13]

$$\theta(x,t) = \sqrt{\frac{2\alpha}{\beta}}\ \mathrm{sech}(qx - \omega t) \qquad (4.2)$$

and through eq. $(3.4)_1$,

$$v(x,t) = -\frac{4\eta^2}{3\rho J}\sqrt{\frac{2\alpha}{\beta}}\ \frac{q}{\beta}\ \tan^{-1}\left[\tanh \frac{1}{2}(qx - \omega t)\right] \quad , \qquad (4.3)$$

while ω and q must satisfy the "dispersion relation":

$$(\omega^2 - \omega_\perp^2)\left(\omega^2 - \omega_P^2 + \frac{\eta}{J}\right) + \frac{\eta^2}{\rho J} q^2 = 0 \ . \qquad (4.4)$$

This yields an approximate solution $\omega^2 \simeq \omega_P^2 - \frac{\eta}{J}$[graph (a) in Figure 3] . In this case there exists a $q = q_o$ such that $q_o = \sqrt{2}/a$ above which the solution (4.2)-(4.3) propagates.

(b) case $\omega_P < \omega < \hat{\omega}_\perp$ or $\omega > \omega_\perp$: Then the solution (4.2)-(4.3) is replaced by

$$\theta(x,t) = \sqrt{\frac{\alpha}{\beta}}\ \tanh(qx - \omega t),\; v(x,t) = -\frac{2\eta^3}{3\rho J}\sqrt{\frac{\alpha}{\beta}}\ \frac{q}{\beta}\ \log[\cosh(qx - \omega t)], \qquad (4.5)$$

with the dispersion relation

$$(\omega^2 - \omega_\perp^2)(\omega^2 - \omega_F^2) - \frac{\eta^2}{2J} q^2 = 0,\; \omega_F^2 = \omega_P^2 + \frac{\eta}{2J} \qquad .(4.6)$$

The solution (4.5) provides a coupling between a solitary transverse acoustic wave and a ferroelectric soliton. Note that the acoustic branch is the same one as the one found in a linear harmonic analysis while the soft-mode branch presents a cut-off frequency reduced by a factor $\sqrt{2}$ as compared to the harmonic case [cf. graphs (b) and (c) in Figure 3] .

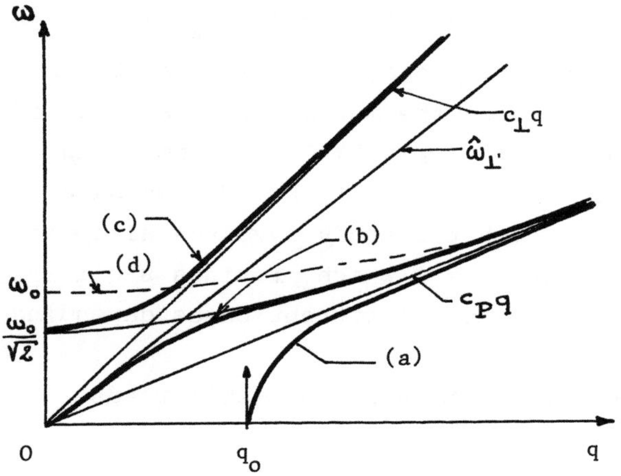

Fig.3. Dispersion curves (a) soliton solution (4.3)-(4.4); (b) and (c) soliton solution; (d) ferroelectric soft mode (harmonic analysis)

(c) stability of the solutions: It is of interest to know which one of the two above-obtained solutions is stable. To that purpose consider perturbations about a static solution in the form

$$\Theta(x,t) = \Theta_o(x) + \tilde{\Theta}(x,t) \quad , \quad v(x,t) = v_o(x) + \tilde{v}(x,t) \qquad (4.7)$$

with v_o and Θ_o solutions of eqs. $(3.3)_{2-3}$. At the first order in $\tilde{\Theta}$ and $\tilde{v}$, we obtain the system

$$\begin{bmatrix} -c_\perp^2 \frac{d^2}{dx^2} & \left(\frac{\eta}{\rho}\right)\frac{d}{dx} \\ -\frac{\eta d}{\rho dx} & -c_P^2 \frac{d^2}{dx^2} + U(x) \end{bmatrix} \begin{bmatrix} \hat{\tilde{v}} \\ \hat{\tilde{\Theta}} \end{bmatrix} = \omega^2 \begin{bmatrix} \hat{\tilde{v}} \\ \hat{\tilde{\Theta}} \end{bmatrix} , \qquad (4.8)$$

where $\hat{\tilde{v}}$ and $\hat{\tilde{\Theta}}$ are time-Fourier transforms of v and $\tilde{\Theta}$ and

$$U(x) = \frac{d^2\bar{V}'}{dx^2}(\Theta_o(x)) \qquad , \bar{V}(\Theta) = -\frac{\eta}{J}\left(1 - \frac{\Theta^2}{3}\right)^2 \qquad . \qquad (4.9)$$

The stability of the initial solution requires that ω be real, hence ω^2 positive. Note that v_o and $d\Theta_o/dx$ are eigenfunctions of the system (4.8) with eigenvalue $\omega = 0$ and constitute the fundamental state. It follows from this that the solution (4.2) is u n s t a b l e and has no physical meaning whereas the solution (4.5) is s t a b l e and can thus be interpreted as the motion of a wall which separates two domains in a ferroelectric sample[14]-[15].

5. Physical Interpretations and conclusion

The present lattice model, simple as it is, goes farther than previous ones [14]-[16] in that a variable associated with the ferroelectric behaviour (hence directly connected with the phase transition) and coupled with the displacement field intervenes. The continuum model allows one to treat the case where non-linearities are associated with the rotation of dipoles. The double-well potential in eqs. (4.1) - with $\alpha / \beta > 0$ - is responsible for the structural phase transition [14]-[15]. It is at the origin of strongly anharmonic effects observed experimentally by various means (neutron scattering, Raman scattering, infrared absorption) nearby the phase transition [14]. Another aspect of the present model is that it shows that a nonlinear elastic wave can be generated via a ferroelectric soliton while the mechanical behavior of the material remains practically linear. Other problems such as the interaction between solitons and the interaction between phonons and steady or moving ferroelectric walls could be studied on its bases.

References

1. Kittel,C.: Introduction to Solid State Physics,J.Wiley, New York (1971).
2. Cochran,W.: Advan.Phys.,9 (1970) 387.
3. Brody,E.M.; Cummins,H.Z.:Phys.Rev.Lett.,21 (1968) 1263.
4. Fleury,P.A.;Lazay,P.D.:Phys.Rev.Lett.,26 (1971) 1331.
5. Scott,J.F.:Rev.Modern.Phys.,46 (1974) 83.
6. Pouget,J.; Askar,A.;Maugin,G.A.: Phys.Rev.,B (submitted 1982) [monoatomic model].
7. Pouget,J.;Askar,A.; Maugin,G.A.:Phys.Rev.,B (submitted 1982) [diatomic model].
8. Askar,A.:Int.Engng.Sci., 10(1971) 293.
9. Askar,A.: Phys.Chem.Solids,34 (1973) 1901.
10. Maugin,G.A.; Pouget,J.:J.Acoust. Soc.Amer.,68 (1980) 575.
11. Pouget,J.; Maugin,G.A.: J.Acoust.Soc.Amer.,68(1980) 588.
12. Bishop,A.R.: in Solitons in Action,Ed.Lonngren,K.,Scott,A.; Academic Press, New York (1978).
13. Scott.A.C.;Chu,F.Y.F.;McLaughlin,D.W.,Proc.I.E.E.E.,61. (1973) 1443.
14. Aubry,S.: J.Chemical Phys., 62 (1975) 3217.
15. Aubry,S.: J.Chemical Phys., 64 (1976) 3392.
16. Krumhansl,J.A.; Schrieffer,J.R.:Phys.Rev.,B11(1975) 3535.

Deformation and Thermal Wave in a Long Rod Caused by Thermal Shock, Taking Into Account Large Temperature Differences and the Modified Law of Heat Conduction

A.SZEKERES

Department of Technical Mechanics
Technical University of Budapest

Summary

Based on our previous research we build up the general equations of thermoelasticity taking into account large temperature differencies and the modified law of the heat conduction. We consider further the thermal shock of a long rod and discuss a possible method for determining the relaxation time.

Introduction

In thermoelastic problems in general, and thus also in dynamic problems, it is customary to assume small changes in temperature in comparison to natural conditions. All the relationships used by previous authors relate to such an assumption. The idea of investigating what will be the difference if considerable changes in the temperature are assumed in dynamic problems has come up recently [1].

The one-dimensional dynamic problem of the heat conduction is described by the differential equation

$$aT_{xx} = T_t \tag{1}$$

on the basis of Fourier law of heat conduction.

The investigation of this equation shows that the temperature wave propagates at the infinite rate which, however, contradicts experience. (See the references of [2].)

To resolve this contradiction, the relationship

$$aT_{xx} = T_t + T_{tt} \cdot \tau \tag{2}$$

is suggested instead of Equation (1) of heat conduction. According to the Equation (2), the rate of propagation of temperature waves is a finite value,

$$v_T = \sqrt{a/\tau} \tag{3}$$

Nonlinear Deformation Waves
IUTAM Symposium Tallinn 1982
Editors U Nigul, J Engelbrecht

It is of certain interest (e.g. [3]) to couple the two problems mentioned above: how the equations of thermoelasticity describe the process if we take simultaneously into account large temperature differences and the modified law of heat conduction.

After this the program is following. First we introduce the modified law of heat conduction [2] and the general equation of heat conduction taking into account large temperature differences [1]. Then the general equations of thermoelasticity will be deduced and considering the problem of the thermal shock in a long rod, we turn to a possible experiment.

<u>Modified law of heat conduction</u>

In investigating an one-dimensional problem of dynamic heat conduction, the mechanical interaction may be neglected and the material may be assumed to be homogeneous and isotropic.

According to the 1st Law of Thermodynamics

$$\rho u_t = h_x . \tag{4}$$

In accordance with the experience, the constitutive equation will be written in the form

$$\varphi(u, T, T_x, T_t) = 0 . \tag{5}$$

The law of heat conduction will be also found empirically in the following form

$$\Psi(h, h_t, T, T_x, T_t) = 0 \tag{6}$$

A solution to the problem posed can be obtained by solving the system of the partial differential equations (4),(5),(6).

The following restrictions apply to the system (4)...(6): neither the constitutive equation nor the law of heat conduction can be time and space dependent and, therefore, x and t can not be explicitly included into (5) and (6). As far as the problem can be best described as a wave process, the systems (4), (5) and (6) must result in a second-order hyperbolic equation for the entire range concerning the body investigated and under any possible initial and boundary conditions.

Taking into consideration that (i) in compliance with the practice

$$\varphi_u = 1 \quad \text{and} \quad \Psi_h = 1 \tag{7}$$

and (ii) the resulting equation must be second-order hyperbolic, we get condition

$$(\varphi_{Tx} - \frac{\psi_{Tt}}{\rho})^2 + 4(\varphi_{Tt} + \psi_{ht}\varphi_t)\frac{\psi_{Tx}}{\rho} > 0. \tag{8}$$

Thus, it can be easily seen that, in their common form

$$u - cT = 0 \tag{9}$$

$$h - \varkappa T_x = 0 \tag{10}$$

the constitutive equation and the law of heat conduction can not apply simultaneously.

One of the opportunities is that the Equation (9) applies in its original form but the law of heat conduction is modified:

$$h + \tau h_t - \varkappa T_x = 0. \tag{11}$$

<u>The modified form of the "2nd Law" for large changes in temperature</u>

According to Nowacki and Biot $T_o\rho s$ is the amount of heat accumulated locally in the continuum which, due to the small change in temperature, i.e. $T_o = T_1$, can also be written as

$$q = T_1\rho s. \tag{12}$$

Consequently in accordance with the 2nd Law

$$dq = \rho T_1(s,\ldots)ds. \tag{13}$$

Further, the relationship dT/ds=0 is taken into consideration that permits to determine $T_o\rho s$.

For considerable changes in temperature, let us take the relationship

$$d(T_1 s) = dT_1 s + T_1 ds \tag{14}$$

as a basis for constructing the integrated formula for the 2nd Law. Assuming that T_1 is a function of entropy and other variables i.e.

$$T_1 = T_1(s,\ldots)$$

then, on the basis of (13) one might write:

$$\frac{1}{\rho}dq = d(T_1 s) - dT_1 s = d(T_1 s)\frac{\partial T}{\partial s}s ds. \tag{15}$$

After integration, when assuming that $q_o = 0$ and $s_o = 0$ we obtain

$$\frac{1}{\rho}q = T_1 s - \int_0^s \frac{\partial T}{\partial s} s ds \tag{16}$$

Note that on the basis of $dT_1/ds=0$ which corresponds to small changes in temperature, (12) follows from the expression (16) directly. To calculate the integral, let us now write the temperature as a function of entropy in the following form:

$$T_1=\sum_{i=0}^{\infty}\alpha_i s^i \tag{17}$$

Using this expression instead of (16) we obtain

$$\frac{1}{\rho}q=T_1 s-\sum_{i=0}^{\infty}\alpha_i\frac{i}{i+1}s^{i+1} \tag{18}$$

This is further used as the 2nd Law in the equations of thermoelasticity

The modified equations of thermoelasticity

The system of equations of thermoelasticity consists of the following equations: the geometric equation, the equation of motion, the 1st and the 2nd Law of Thermodynamics, and the heat conduction equation. The first three of these give the so-called general equation of motion, the last three - the general equation of heat conduction.

For the sake of convinience, the 1st Law of Thermodynamics can be written (with the intrinsic heat source neglected) as

$$s=c\ln\frac{T_1}{T_o}-\frac{1}{\rho}\beta^{kl}\varepsilon_{kl} \tag{19}$$

On the basis of Equation (11) and the definition of heat amount, the modified equation of heat conduction is

$$\tau\ddot{q}+\dot{q}=(\lambda^{kl}T_{,k})_{;l}. \tag{20}$$

After reducing the Equation(18), (19) and (20) yield

$$(\tau\ddot{T}+\dot{T})[c(1+\ln\frac{T_1}{T_o})-\frac{1}{\rho}\beta^{kl}\varepsilon_{kl}]+c\tau\frac{\dot{T}^2}{T_1}-\frac{\beta^{kl}}{\rho}[2\tau\dot{\varepsilon}_{kl}\dot{T}+T_1(\tau\ddot{\varepsilon}_{kl}+\dot{\varepsilon}_{kl})]-$$

$$-\sum_{i=o}^{\infty}i\alpha_i\{(c\ln\frac{T_1}{T_o}-\frac{1}{\rho}\beta^{kl}\varepsilon_{kl})^i[\frac{c}{T_1}(\tau\ddot{T}+\dot{T}-\tau\frac{\dot{T}^2}{T_1})-\frac{1}{\rho}\beta^{kl}(\tau\ddot{\varepsilon}_{kl}+\dot{\varepsilon}_{kl})]-$$

$$+i\tau[c\ln\frac{T_1}{T_o}-\frac{1}{\rho}\beta^{kl}\varepsilon_{kl}]^{i-1}[c\frac{\dot{T}}{T_1}-\frac{1}{\rho}\beta^{kl}\dot{\varepsilon}_{kl}]^2\}=\frac{1}{\rho}(\lambda^{kl}T_{,k})_{;l}, \tag{21}$$

This equation is the general equation of heat conduction for large temperature differences and for the modified law of heat conduction. The Equation (21) is in a good agreement with our earlier results. If $\tau=0$, we get the Equation (13) of (1), and if α_i 0, we get the Equation (31) of [2].

The general equation of motion, using the displacement instead of strain, is

$$\mu u_{i;jj}+(\mu+\lambda)u_{k;ki}-T,_{i}=\rho\ddot{u}_i \tag{22}$$

The equations (21) and (22), completed with the initial and boundary conditions, describe the problem of thermoelasticity mentioned above.

Thermal shock of a long rod

Let us apply the equations obtained above to the thermal shock in a long rod that is fixed on one end, thermally insulated, homogeneous and isotropic. If the heat is added at the free end of the rod in such a way that the temperature at the end of the rod changes in accordance with a definite time function, then the initial and boundary conditions are

$$T(o,t)=\nu t, \qquad u_x/o,t/=\alpha_t\, t, \tag{23}$$

$$T(l,t)=0, \qquad u(l,t)=0, \tag{24}$$

$$T(x,0)=0, \qquad u(x,o)=0 \quad \dot{u}(x,0)=0. \tag{25}$$

The system (21),(22) with the conditions (23)...(24) was solved by making use of the finite difference method.

The numerical results were obtained by using different approximations. First only the terms in the Equation (21) corresponding to i=0 were taken into consideration, then those corresponding to i=0,1 and so on up to 5. The calculations showed that, for given starting parameters, the displacement function u(x,t) and the temperature function T(x,t) were only slightly affected by the number of terms in the Equation (21).

Possible experiment

To determine unambiguously what an accuracy could be achieved by taking into consideration the relationships concerning considerable changes in temperature, it would be necessary to use the terms of higher order in equation (21). This, however, would require the knowledge of coefficients α_i. There is a considerably simple method for determination of these coefficients.

Let us take the formula

$$\rho\dot{s} > -\left(\frac{1}{T}h^k\right)_{;k} \tag{26}$$

of the 2nd Law of Thermodynamics as a basis, where the intrinsic heat source was neglected. Assuming that the process is reversible, we obtain for an one-dimensional problem

$$\rho\dot{s} = \frac{1}{T^2}\, T_x h - \frac{1}{T}\, h_x \tag{27}$$

However, according to the modified law of heat conduction (similarly on the basis of the assumptions made above),

$$\tau\dot{h} + h = -\varkappa T_x .$$

With temperature function $T(x,t)$ measured, $s(x,t)$ can be calculated on the basis of (27) and (28) after some mathematical difficulties overcome.

References

1. Szekeres, A.: The coupled dynamical problem of thermoelasticity in case of large temperature differencies. Lecture held on the 6th SMIRT conference, August 1981. Paris.
2. Szekeres, A.: Equation system of thermoelasticity using the modified law of thermal conductivity. Periodica Polytechnica, Vol. 24. No 3. 1980.
3. Parkus, H.: Oral communication, March 1981. Budapest.

Symbols

ρ - mass density

u - internal energy density

t^{kl} - Cauchy stress tensor

v_{kl} - strain rate tensor

h^k - intensity of heat flux

s - entropy density

α_{ij} - matrix of linear expansion coefficients

T_1 - temperature

T_o - temperature in normal state

$T = T_1 - T_o$

$\beta_{ij} = -(3\lambda+2\mu)\alpha_{ij}$

λ, μ - Lamé constants

α_t - coefficient of linear expansion

$\gamma = \alpha_t(3\lambda + 2\mu)$

ε_{kl} - strain tensor

q - heat

c - specific heat

u_k - displacement vector

λ^{kl} - matrix of heat conduction coefficients

$\varkappa$ - coefficient of heat conduction

α_i - constants of $T_1 = f(s)$

E - Young modulus

υ - temperature change rate

u - displacement

x - coordinate

t - time

$(\ )_x = (\ )/ x$

a - coefficient of temperature conduction

τ - relaxation time

Acceleration Waves in Nonlinear Generalized Thermoelasticity

TS. P. IVANOV

Faculty of Mathematics and Mechanics
University of Sofia

Summary

Propagation of acceleration waves and waves of discontinuities of higher order in nonlinear generalized thermoelasticity is considered. The possible velocities of the wave front and the transport equations which govern the growth and decay of the discontinuities are obtained. It is shown that shock waves could appear within a finite length of time when acceleration waves propagate in the medium. All waves of discontinuities of higher order decay to zero monotonically.

Basic Equations

In generalized thermoelasticity we deal with thermoelastic model that allow for a finite speed of heat propagation. Two different approaches have been developed in this field. Some authors [1-3] use a generalization based on the modified form of Fourier law in which they postulate the existance of relaxation time for the heat flux. On the other hand a generalization of the entropy pro duction inequality is used in [4-6]. One could see that these models correspond in a sense to the Maxwell and Kelvin-Voigt models in viscoelasticity and show different properties. Their comparison has been done in various aspects [7-8].

In the present paper we consider propagation of acceleration waves and waves of discontinuities of higher order in nonlinear generalized thermoelasticity. We assume that the constitutive equations are one-dimensional, generally nonlinear with respect to the deformation gradient and linear with respect to the temperature and its derivatives. Then

$$\sigma' = \rho_0 \dot{v}, \quad q' + \rho_0 \theta_0 \dot{\eta} = 0, \quad v' = \dot{\varepsilon}, \quad \tau_0 \dot{q} + q = -kT', \tag{1}$$

Nonlinear Deformation Waves
IUTAM Symposium Tallinn 1982
Editors U Nigul, J Engelbrecht

$$\rho_0\theta_0\eta=\gamma\theta_0\varepsilon+cT+h\dot{T},\quad \sigma=\psi(\varepsilon)-\gamma(T+\alpha\dot{T})$$

with $\varepsilon=u'$ and $v=\dot{u}$ where $u(X,t)$ is the displacement at time t in X-direction of a material point with a reference coordinate X. The prime denotes $\frac{\partial}{\partial X}$ and a superposed dot is used to mark $\frac{\partial}{\partial t}$, $\sigma(X,t)$ is the stress, $q(X,t)$-the heat flux, $\eta(X,t)$-the specific entropy change starting from the reference entropy, $T(X,t)$-the temperature change starting from the absolute reference temperature θ_0, ρ_0-the constant reference mass density. The differentiable function $\psi(\varepsilon)$ depends on both geometrical and physical nonlinearities. If we put $\tau_0=0$ in Eqs.(1) we shall obtain the model based on the generalized entropy production inequality where $k>0$, $\alpha>0$, $h>0$, $\alpha c\geq h$. The model based on the modified Fourier law is obtained by Eqs.(1) when $\alpha=h=0$ and $\tau_0>0$, $k>0$, $c>0$ [7].

When $\tau_0=0$ we can rewrite Eqs.(1) in the form

$$\sigma'-\rho_0\dot{v}=0,\quad \dot{\varepsilon}-v'=0,\quad h\dot{\sigma}-(h\frac{d\psi}{d\varepsilon}+\alpha\gamma^2\theta_0)\dot{\varepsilon}-\alpha\gamma q'=\frac{c\alpha-h}{\alpha}(\psi-\gamma T-\sigma),$$
$$\alpha\gamma\dot{q}+k\,\frac{d\psi}{d\varepsilon}\varepsilon'-k\sigma'=-\gamma q \tag{2}$$

where T satisfies the equations

$$kT'=-q,\quad \alpha\gamma\dot{T}=\psi-\gamma T-\sigma, \tag{3}$$

and

$$\rho_0\theta_0\eta=\gamma\theta_0\varepsilon+cT+\frac{h}{\alpha\gamma}(\psi-\gamma T-\sigma) \tag{4}$$

provided $\gamma\neq0$. Here the last equation in (2) is the necessary and sufficient condition for existence of a solution of Eqs.(3) with respect to T.

When $\alpha=h=0$ we use Eqs.(1) in the form

$$\frac{d\psi}{d\varepsilon}\varepsilon'-\gamma T'-\rho_0\dot{v}=0,\quad \gamma\theta_0\dot{\varepsilon}+c\dot{T}+q'=0,\quad \tau_0\dot{q}+kT'=-q,\ \dot{\varepsilon}-v'=0 \tag{5}$$

where

$$\sigma=\psi-\gamma T,\quad \rho_0\theta_0\eta=\gamma\theta_0\varepsilon+cT. \tag{6}$$

The systems of equations (2) and (5) are both hyperbolic provided $\frac{d\psi}{d\varepsilon}>0$ [9].

Waves of Discontinuities

At any instant t we consider the plane $X=Vt$ across which the functions are continuous but their derivatives may be discontinuous. The constant $V\neq 0$ is the velocity of the wave front. We expand all functions ahead of and behind the wave front. For instance the function $f(X,t)$ has the following expansions:

$$f(X,t)=f_0(t)+\sum_{k=1}^{\infty}\frac{1}{k!}f_k^{\pm}(t)\xi^k \tag{7}$$

for $\xi>0$ and $\xi<0$ respectively where $\xi=X-Vt$. Let $|f|=f_--f_+$ denote the jump of the function $f(X,t)$ across $\xi=0$ where $\pm$ correspond to limits taken from the sides $\xi>0$ and $\xi<0$. Then

$$\left|\frac{\partial^k f}{\partial X^{k-p}\partial t^p}\right|=(-V)^p f_k(t),\quad 0\leqq p\leqq k \tag{8}$$

where $f_k(t)=f_k^-(t)-f_k^+(t)$ for $k=1,2,3,\ldots.$

We assume that all functions in Eqs.(2) or Eqs.(5) and their derivatives of order less then n, where $n\geqq 1$, are continuous but at least one of the n-th order derivatives is discontinuous. Let us substitute all functions in Eqs.(2) and Eqs.(5) with the expressions written in the form (7) and substract the corresponding equations and also the derivatives of these equations with respect to ξ for $\xi=0$. We obtain for every $m\geqq n$ the sets of equations

$$A_{ij}f_j^{(m)}=P_i^{(m-1)},\quad B_{ij}g_j^{(m)}=Q_i^{(m-1)}\qquad (i,j=1,2,3,4) \tag{9}$$

corresponding to (2) and (5)where

$$||A_{ij}||=\left\|\begin{matrix}1 & 0 & 0 & \rho_0 V\\ 0 & V & 0 & 1\\ hV & -(h\psi_0'+\alpha\gamma^2\theta_0)V & \alpha\gamma & 0\\ k & -k\psi_0' & \alpha\gamma V & 0\end{matrix}\right\|,$$

$$||B_{ij}||=\left\|\begin{matrix}\Psi_0' & -\gamma & \rho_0 V & 0\\ -\gamma\theta_0 V & -cV & 0 & 1\\ 0 & k & 0 & -\tau_0 V\\ V & 0 & 1 & 0\end{matrix}\right\|, \tag{10}$$

$$||f_j^{(m)}||=||\sigma_m,\varepsilon_m,q_m,v_m||^T,\quad ||g_j^{(m)}||=||\varepsilon_m,T_m,v_m,q_m||^T,$$

$$||P_i^{(n-1)}||=||0,0,0,0||^T,\quad ||Q_i^{(n-1)}||=||0,0,0,0||^T,$$

$$||P_i^{(m)}||=||\rho_0\dot{v}_m,\dot{\varepsilon}_m,\frac{h-c\alpha}{\alpha}(\psi_0'\varepsilon_m+\frac{\gamma}{k}q_{m-1}-\sigma_m)+h\dot{\sigma}_m-(h\psi_0'+\alpha\gamma^2\theta_0)\dot{\varepsilon}_m,$$
$$\alpha\gamma\dot{q}_m+\gamma q_m||^T,\quad ||Q_i^{(m)}||=||\rho_0\dot{v}_m,-\gamma\theta_0\dot{\varepsilon}_m-c\dot{T}_m,-q_m-\tau_0\dot{q}_m,\dot{\varepsilon}_m||^T,$$
$$||P_i^{(2n-1)}||=||\rho_0\dot{v}_{2n-1},\dot{\varepsilon}_{2n-1},\frac{h-c\alpha}{\alpha}(\psi_0'\varepsilon_{2n-1}+\frac{\gamma}{k}q_{2n-2}-\sigma_{2n-1})$$
$$+h\dot{\sigma}_{2n-1}-(h\psi_0'+\alpha\gamma^2\theta_0)\dot{\varepsilon}_{2n-1}+\frac{(2n-1)!}{n!}hV\psi_0''\varepsilon_n^2,\alpha\gamma\dot{q}_{2n-1}+\gamma q_{2n-1}$$
$$+\frac{(2n-1)!}{n!}k\psi_0''\varepsilon_n^2||^T,\quad ||Q_i^{(2n-1)}||=||\rho_0\dot{v}_{2n-1}-\frac{(2n-1)!}{n!}\psi_0''\varepsilon_n^2,$$
$$-\gamma\theta_0\dot{\varepsilon}_{2n-1}-c\dot{T}_{2n-1},-q_{2n-1}-\tau_0\dot{q}_{2n-1},\dot{\varepsilon}_{2n-1}||^T,\quad n\leqq m<2n-1$$

where $\psi_0^{(k)}=\left.\frac{d^k\psi}{d\varepsilon^k}\right|_{\varepsilon=\varepsilon_0(t)}$ and a superscript T denotes the transposed matrix. In order to obtain the expressions for $P_i^{(m)}$ we use

$$-kT_m=q_{m-1}\qquad \text{for } m\geqq n. \tag{11}$$

This follows from the first equation in (3) when the procedure applied for the equations in (2) and (5) is used. The expressions for $P_i^{(m)}$ and $Q_i^{(m)}$ when $m\geqq 2n$ could be obtained in a similar way and they depend only on discontinuities of order m and less than m. The condition for the existence of a nontrivial solution of Eqs. (9) for m=n is the vanishing of the determinants

$$(\alpha\gamma)^{-1}|A_{ij}|\equiv(hV^2-k)(\rho_0V^2-\psi_0')-\alpha\gamma^2\theta_0V^2=0,$$
$$|B_{ij}|\equiv(\tau_0cV^2-k)(\rho_0V^2-\psi_0')-\tau_0\gamma^2\theta_0V^2=0. \tag{12}$$

Provided $\psi_0'>0$ it can be shown that each of the equations in (12) have four different, real and nonvanishing roots with respect to V. Then

$$f_j^{(m)}=R_jp_m+S_j^{(m-1)},\quad g_j^{(m)}=r_jq_m+s_j^{(m-1)}\qquad \text{for } m\geqq n \tag{13}$$

where $S_j^{(n-1)}=s_j^{(n-1)}=0$, $S_j^{(m-1)}$, $s_j^{(m-1)}$ (m>n) are particular solutions of Eqs.(9) and depend only on discontinuities of order less than m; p_m and q_m are functions of t; R_j and r_j are components of nonvanishing right eigenvectors of the matrices $||A_{ij}||$ and

$||B_{ij}||$ respectively

$$||R_j||=||\rho_0 V^2(k-hV^2),k-hV^2,k\gamma\theta_0 V,V(hV^2-k)||^T,$$
$$||r_j||=||k-\tau_0 cV^2,\tau_0\gamma\theta_0 V^2,V(\tau_0 cV^2-k),k\gamma\theta_0 V||^T. \qquad (14)$$

Here V is a root of the corresponding equation in (12). The components of nonvanishing left eigenvectors denoted by L_j and l_j respectively are

$$||L_j||=||k-hV^2,\rho_0 V(hV^2-k),V,-1||,$$
$$(15)$$
$$||l_j||=||k-\tau_0 cV^2,\tau_0\gamma V,\gamma,\rho_0 V(\tau_0 cV^2-k)||.$$

Let us multiply the sets of equations in (9) for m>n by the corresponding components of the eigenvectors in (15) and sum the equations obtained in each of the sets. With the help of (13) we deduce after some computations that the transport equations for $m\geq n$ have the form

$$A\dot{p}_m-B_m p_m^2+Lp_m=D_{m-1};\quad a\dot{q}_m-b_m q_m^2+lq_m=d_{m-1} \qquad (16)$$

where

$$A=2V\{\rho_0(hV^2-k)^2+\alpha k\gamma^2\theta_0\},\ L=\{(\alpha c-h)V^2+k\}\gamma^2\theta_0 V,$$
$$a=2V\{\rho_0(\tau_0 cV^2-k)^2+\tau_0 k\gamma^2\theta_0\},\ l=k\gamma^2\theta_0 V,$$
$$B_n=\begin{cases}(hV^2-k)^3\psi_0'' & \text{for } n=1\\ 0 & \text{for } n>1\end{cases},\quad b_n=\begin{cases}(\tau_0 cV^2-k)^3\psi_0'' & \text{for } n=1\\ 0 & \text{for } n>1\end{cases} \qquad (17)$$

$$B_m=b_m=0 \quad \text{for } m>n,\ D_{n-1}=d_{n-1}=0$$

and D_{m-1}, d_{m-1} when m>n, depend on discontinuities of order less than m.

Conclusions

Both models of generalized thermoelasticity allow for propagation of acceleration waves and waves of discontinuities of higher order. For the model based on the generalized entropy production inequality we have $\left|\frac{\partial^n T}{\partial X^p \partial t^{n-p}}\right|=0$ when $0\leq p\leq n$ for a singular surface of order n, as follows from (8) and (11). For the model using modified Fourier's law we obtain from (8), (13) and (14) that

$$\left|\frac{\partial^n T}{\partial X^p \partial t^{n-p}}\right|\equiv\frac{\tau_0 V}{k}\left|\frac{\partial^n q}{\partial X^p \partial t^{n-p}}\right|\neq 0 \quad \text{for } 0\leq p\leq n. \qquad (18)$$

The transport equations (16) for both models are linear and homogeneous for m=n but linear and nonhomogeneous for m>n provided n>1. Then the waves of discontinuities of an arbitrary order n>1 could propagate with two different positive velocities, determined from (12), and magnitudes of the jumps that decay monotonically to zero since constants A, L, a and l are all positive. These waves appear only when the corresponding derivatives of the functions in the boundary conditions are discontinuous. When n=1 acceleration waves could propagate with the same two positive velocities. The transport equations (16) are again linear and nonhomogeneous for m>1 but for m=1 they are nonlinear and homogeneous provided $\psi''_0 \neq 0$. That means that shock waves with discontinuities in the functions themselves could appear within a finite length of time nevertheless that the functions in the boundary conditions are continuous as it is shown in [7] for the case ψ''_0=const. In the classical thermoelasticity the above conclusions are also valid exept that now there exists only one velocity.

References

1. Cattaneo, C.: Sulla condizione del calore. Atti del Seminario Mat. e Fis. Univ. Modena 3 (1948) 83-101.
2. Lord, H.W.; Shulman, Y.: A generalized dynamical theory of thermoelasticity. J. Mech. Phys. Solids 15 (1967) 299-309.
3. Fox, N.: Generalized thermoelasticity. Int. J. Eng. Sci. 7 (1969) 437-445.
4. Muller, I.: The coldness, a universal function in thermoelastic bodies. Arch. Rat. Mech. Anal. 41 (1971) 319-332.
5. Green, A.E.; Laws, N.: On the entropy production inequality. Arch. Rat. Mech. Anal. 45 (1972) 47-53.
6. Green, A.E.; Lindsay, K.A.: Thermoelasticity. J. of Elasticity 2 (1972) 1-7.
7. Ivanov, Ts.P.: Acceleration waves in generalized thermoelasticity. IV National Congress on Theoretical and Applied Mechanics, vol.I, Sofia: Bulgarian Academy of Sciences 1981.
8. Ivanov, Ts.P.: Shock waves in generalized thermomechanics. Prikladnaja Mechanika (Kiev) 17 (1981) 27-33.
9. Ivanov, Ts.P.: Waves of discontinuities in temperature rate dependent thermoviscoelasticity and thermoelasticity. Proceedings of Symposium on Nonlinear waves of deformation , vol.II, Tallin: Academy of Sciences of the Estonian SSR 1978.

Solitary Waves in Density Stratified Fluids

R. GRIMSHAW

Department of Mathematics,
University of Melbourne,
Parkville,
Victoria, 3052, Australia.

Summary

Evolution equations for weakly nonlinear long waves propagating in stratified shear flows are presented. These equations are either of the Korteweg-de Vries type for shallow fluids, or the Benjamin-Davis-Ono type for deep fluids. The effects of dissipation and/or varying background are included. Solitary wave solutions are described and applications to the atmosphere and ocean discussed. Since observations of solitary waves often show amplitudes in excess of that allowed for by the weakly nonlinear theory, various extensions of this theory to higher order in amplitude are described.

§1. Introduction

The evolution of long nonlinear waves propagating in a stratified shear flow is a topic which has attracted much recent interest. To the lowest order in amplitude these waves are governed by a linear long wave equation, but the cumulative and competing effects of nonlinearity and dispersion may lead to the formation of one or more solitary waves. These waves are common occurrences on the thermocline of lakes, fjords or in coastal waters (see, for instance, Farmer [1] or Apel [2]); they have also been observed on the nocturnal inversion in the atmosphere (Christie, Muirhead and Hales [3]).

We shall consider waves propagating horizontally in an inviscid and incompressible fluid, with the x-axis in the direction of propagation. We shall then suppose that the horizontal waveguide is characterized by a basic density profile $\rho_0(z)$ and a basic shear flow $u_0(z)$ in the wave direction. The waveguide is bounded below by a rigid boundary $z = -h$, and above by a free surface whose equilibrium position is $z = \zeta_0$. Long waves of small amplitude are characterized by two small parameters; one, α, is

Nonlinear Deformation Waves
IUTAM Symposium Tallinn 1982
Editors U Nigul, J Engelbrecht

the ratio of the wave amplitude to the vertical scale of the waveguide and measures nonlinearity; the other ϵ, is the ratio of the vertical scale of the waveguide to the wavelength and measures dispersion. To the lowest order in amplitude, the vertical particle displacement at any point in the fluid is given by

$$\eta = \alpha A(\tau, \theta)\phi(z). \tag{1}$$

Here the amplitude A is a function of a slow time variable τ, and a phase variable θ which describes slow spatial modulations in a frame which moves with the linear long wave phase speed c_0. Thus we put

$$\tau = \epsilon\alpha t, \quad \theta = \epsilon(x - c_0 t). \tag{2}$$

The vertical modal function $\phi(z)$ is given by

$$\left.\begin{aligned} &\frac{\partial}{\partial z}\{\rho_0(c_0 - u_0)^2\frac{\partial\phi}{\partial z}\} + \rho_0 N^2\phi = 0, \quad \text{for} \quad -h < z < \zeta_0, \\ &\text{and} \quad \phi = 0 \quad \text{at} \quad z = -h, \; \phi = \sigma(c_0 - u_0)^2\frac{\partial\phi}{\partial z} \quad \text{at} \quad z = \zeta_0. \end{aligned}\right\} \tag{3}$$

It is normalized so that its maximum absolute value is one. Throughout all variables are non-dimensional based on a length scale h_1 (a typical vertical dimension) and a time scale N_1^{-1} (where N_1 is a typical value of the Brunt-Väisälä frequency); the parameter σ is $h_1 N_1^2 g^{-1}$ and is small in the Boussinesq approximation, while N(z) is the non-dimensional Brunt-Väisälä frequency $-(\sigma\rho_0)^{-1}\frac{d\rho_0}{dz}$. In general there are in an infinite number of vertical modes $\phi(z)$ each satisfying equation (3) with a distinct phase speed c_0. Hereafter we select just one of these modes and determine the evolution of its amplitude $A(\tau, \theta)$.

When the horizontal waveguide has limited vertical extent there is an analogy with shallow water theory, and the appropriate evolution equation is the Korteweg-de Vries (KdV) equation

$$\frac{\partial A}{\partial \tau} + \nu A\frac{\partial A}{\partial \theta} + \lambda \frac{\partial^3 A}{\partial \theta^3} = 0, \tag{4}$$

for which the balance between nonlinearity and dispersion requires that $\alpha = \epsilon^2$ (see, for instance, Benney [4]). The derivation is by a multi-scaling perturbation method. The details are described by Benney [4], Maslowe and Redekopp [5], or Grimshaw [6], where it is shown that the coefficients ν and λ are given by

$$\left.\begin{aligned} I\nu &= 3\int_{-h}^{\zeta_0} \rho_0(c_0 - u_0)^2\left(\frac{\partial\phi}{\partial z}\right)^3 dz, \\ I\lambda &= \int_{-h}^{\zeta_0} \rho_0(c_0 - u_0)^2\phi^2 dz, \\ I &= 2\int_{-h}^{\zeta_0} \rho_0(c_0 - u_0)\left(\frac{\partial\phi}{\partial z}\right)^2 dz. \end{aligned}\right\} \qquad (5)$$

In these expressions it has been assumed that there are no critical layers (i.e. c_0 lies outside the range of $u_0(z)$), a sufficient condition for which is that the local Richardson number is everywhere greater than $\frac{1}{4}$. Note that for right-going waves ($c_0 > \max u_0$), the coefficient λ is always positive, but the coefficient ν can in general take either sign. However, for internal waves in the Boussinesq approximation ($\sigma \to 0$)

$$I\nu = \int_{-h}^{\zeta_0} \rho_0\phi^3\{-2\frac{\partial N^2}{\partial z} - \frac{3N^2}{(c_0 - u_0)}\frac{\partial u_0}{\partial z} - (c_0 - u_0)^2\frac{\partial}{\partial z}\left[\frac{\partial^2 u_0}{\partial z^2}(c_0 - u_0)^{-1}\right]\}dz + O(\sigma). \qquad (6)$$

Thus in the absence of a shear flow, and for a near-surface thermccline ($\partial N^2/\partial z > 0$), ν is negative for first mode waves which are consequently waves of depression; that is, they face away from the free surface.

However, when the horizontal waveguide is abutted by a deep fluid region within which the density is constant and the shear flow zero, it was shown by Benjamin [7] and Davis and Acrovos [8] that a different theory is needed. To be specific suppose that thw waveguide retains a vertical scale h, determined by the structure of $\rho_0(z)$ and $u_0(z)$, but that the lower boundary is displaced to $z = -H/\epsilon$; other configurations are possible, but will not be considered here. The appropriate evolution equation is

$$\left.\begin{aligned} &\frac{\partial A}{\partial \tau} + \nu A\frac{\partial A}{\partial \theta} + \delta\, \mathcal{L}\left(\frac{\partial A}{\partial \theta}\right) = 0 \ , \\ \text{where} \quad &\mathcal{L}(A) = -\frac{1}{2\pi}\int_{-\infty}^{\infty} |\tau|\coth(|\tau|H)\exp(i\tau\theta)\,\mathcal{F}(A)\,d\tau, \\ \text{and} \quad &\mathcal{F}(A) = \int_{-\infty}^{\infty}\exp(-i\tau\theta)A\,d\theta \ , \end{aligned}\right\} \quad (7)$$

for which the balance between nonlinearity and dispersion requires that $\alpha = \epsilon$. The derivation is again by a multi-scaling perturbation method, now supplemented by an asymptotic matching procedure between the horizontal waveguide region and the deep fluid region. The details are described by Maslowe and Redekopp [5], Grimshaw [6], or Tung, Ko and Chang [9], where it is shown that the coefficient ν is again given by (5), with the proviso that $-h$ is replaced by $-\infty$, and the modal function $\phi(z)$ is again defined by (3), with the proviso that the boundary condition at $z = -h$ is replaced with the condition that $\partial\phi/\partial z \to 0$ as $z \to -\infty$. The coefficient δ is given by

$$I\delta = (\rho_0 c_0^2 \phi^2)_{z \to -\infty} \ . \qquad (8)$$

In the limit $H \to 0$, (7) can be shown to reduce to the KdV equation (4), and in the limit $H \to \infty$ (that is, $\coth|\tau|H \to 1$), equation (7) becomes the Benjamin-Davis-Ono (BDO) equation. Note that for right-going waves the coefficient δ is always positive. However, since ν is again given by equation (6), it follows that first mode waves travelling on a near-surface thermocline in the absence of a shear flow are waves of depression.

In addition to the solitary wave solution which we shall exhibit in the next section, it is well known that the KdV equation has N-soliton solutions, and is amenable to the inverse scattering transform technique. Similarly (7) possesses a solitary wave solution, N-soliton solutions, and is likewise amenable to the inverse scattering transform technique. However, we do not propose to pursue these topics here; rather our concern is the effect of a slowly varying environment and dissipation on the solitary wave solutions. Thus in §2(a) we shall consider the modifications to the KdV equation needed to incorporate these effects, and in §2(b) we shall undertake the same task for the

BDO equation. Since many observed internal solitary waves have large amplitudes we shall in §3(a) extend the KdV solitary wave to second order in amplitude, and in §3(b) we shall do likewise for the BDO equation. Other extensions to the theory can be considered; for instance, Grimshaw [10] has derived the analogous equations to (4) and (7) for a compressible fluid.

§2. Slowly varying solitary waves

(a) KdV: Let us now suppose that the environment consisting of the basic shear flow, the basic density profile and the fluid depth vary slowly on time and space scales of order β^{-1}. If the effect of this slow variation is to be comparable with the effects of weak nonlinearity and dispersion, then we must choose $\beta = \epsilon\alpha$ (i.e. ϵ^3 for the KdV case). The main consequence of this is that the linear long wave phase speed c_0 will now vary on these long length and time scales, thus defining a set of rays whose trajectories in time and space define the wave direction and phase. In this context the variable τ is replaced by a variable which defines evolution along a ray, and the phase variable θ is replaced by a variable which is transversal to the ray path. Suppose, for instance that the environment varies spatially in just one direction, for which the slow space variable is

$$X = \epsilon\alpha x. \tag{9}$$

Thus $c_0 = c_0(X)$ and assuming that the wave propagates in this same direction we may put (see Grimshaw [6])

$$\tau = \int_0^X \frac{dX'}{c_0(X')}, \quad \theta = \frac{1}{\alpha}(\tau - \epsilon\alpha t). \tag{10}$$

It has been shown by Grimshaw [6] that the generalization of the KdV equation (4) is

$$\left.\begin{aligned} &\frac{\partial A}{\partial \tau} + \frac{1}{2}\frac{A}{\gamma}\frac{\partial \gamma}{\partial \tau} + \frac{\nu}{c_0} A \frac{\partial A}{\partial \theta} + \frac{\lambda}{c_0^3}\frac{\partial^3 A}{\partial \theta^3} = 0, \\ \text{where} \qquad &\gamma = c_0^2 I, \end{aligned}\right\} \tag{11}$$

and the coefficients ν, λ and I are given defined by equation (5), but are now functions of τ. The significance of the coefficient γ

is that γA^2 is proportional to the wave action flux in the X-direction; if $\gamma^{\frac{1}{2}}A$ is introduced as a new variable in equation (11), it becomes a KdV equation of the same form as (4), but with coefficients which are functions of τ. When the environment varies in more than one spatial direction, or contains a temporal variation, the equation analogous to (11) is the two-dimensional KdV equation, or Kadomtsev-Petviashvili equation, with variable co-efficients (for a derivation, see Grimshaw [6]).

Unlike the constant-coefficient KdV equation (4), the variable-coefficient equation (11) does not in general possess exact solutions analogous to the N-soliton solutions of equation (4). Here we propose to discuss one class of asymptotic solutions to equation (11), the slowly varying solitary wave. This arises when the coefficients in (11), λ, ν, c_0 and γ are slowly varying functions of τ; that is, they vary slowly with respect to a scale defined by the wavelength of the solitary wave. The slowly varying solitary wave can be constructed either by a multi-scale perturbation method applied directly to equation (11) (Grimshaw [11]), or by an adaptation of the inverse scattering transform technique (Karpmen and Maslov [12], or Kaup and Newell [13]). Both methods produce the same result, so here we shall only describe that result. The slowly varying solitary wave is to leading order,

$$\left.\begin{aligned} & A = a\,\mathrm{sech}^2 k\psi, \\ \text{where}\quad & \psi = \theta - \int_0^\tau \frac{1}{3}\frac{\nu a}{c_0}\,d\tau, \\ \text{and}\quad & \nu a = 12\lambda\,\frac{k^2}{c_0^2}\,. \end{aligned}\right\} \tag{12}$$

Here the amplitude a, and the parameter k (c_0/k is a measure of the local wavelength), are slowly varying functions of τ; also note that $\frac{1}{3}\nu a$ is the increment in phase speed above the linear long wave phase speed c_0. The slow variation in the amplitude is determined from the above-mentioned perturbation methods, and is given by

$$\left.\begin{aligned} &\gamma \frac{a^2}{k} = \text{constant}, \\ \text{or} \qquad &\frac{I^2 c_0^2 \lambda a^3}{\nu} = \text{constant}. \end{aligned}\right\} \tag{13}$$

This last relation is obtained from the last equation in (12), and the definition of γ in (11). Equation (13) is most readily interpreted by observing that equation (11) possesses the conservation law

$$\frac{\partial}{\partial\tau} \int_{-\infty}^{\infty} \gamma A^2 d\theta = 0 . \tag{14}$$

Substitution of (12) into (14) produces (13), which thus describes the conservation of wave action flux. However, equation (11) also possesses the conservation law

$$\frac{\partial}{\partial\tau} \int_{-\infty}^{\infty} \gamma^{\frac{1}{2}} A d\theta = 0. \tag{15}$$

It is readily seen that with a given by (13), equation (15) cannot be satisfied by the solitary wave (12) alone. The resolution of this is that the slowly varying solitary wave loses mass to a shelf, or tail, region which has the form of a linear long wave trailing behind the solitary wave, whose amplitude at the rear of the solitary wave is given by (see Grimshaw [11])

$$A_1(\tau) = \frac{3c_0}{\nu k^2} \frac{\partial k}{\partial\tau} . \tag{16}$$

Note that when k is increasing with τ, the shelf has the same sign as the solitary wave. The shelf occupies the region $\psi < 0$, and is given by

$$\left.\begin{aligned} &\left\{\frac{\gamma(\tau_0)}{\gamma(\tau)}\right\}^{\frac{1}{2}} A_1(\tau_0) , \\ \text{where} \qquad &\theta = \int_0^{\tau_0} \frac{1}{3} \frac{\nu a}{c_0} d\tau . \end{aligned}\right\} \tag{17}$$

Throughout this discussion it has been implicitly assumed that the coefficients ν, λ, γ and c_0 are constants for $\tau < 0$ (i.e. the environment varies only in the region $\tau > 0$). It then follows from (17) that the shelf region extends from the solitary wave, located at $\psi = 0$ (see (12)), to the location $\theta = 0$, at which point it has zero amplitude. It can now be readily verified that the slowly varying solitary wave (12), together with the shelf (17), will satisfy the conservation law (15).

It is somewhat curious that the conversation law (15) is not the conservation law for mass flux derived from the exact equations for an inviscid incompressible fluid. Assuming that only the vertical mode $\phi(z)$ is invoked, it may be shown that this latter law takes the form

$$\frac{\partial}{\partial\tau}\int_{-\infty}^{\infty} c_0\phi(\zeta_0)A\ d\theta = 0\ , \tag{18}$$

in which the integrand differs from that in (15) by the factor $I^{-\frac{1}{2}}\phi(\zeta_0)$. It has been shown by Miles [14] in the case of a surface solitary wave that this necessitates the introduction of a reflected linear long wave. The same conclusion applies here although for internal waves $\phi(\zeta_0)$ is $0(\sigma)$ and hence usually quite small (it is zero if the upper boundary is rigid), and consequently the reflected wave is likewise quite small.

Next we shall consider some applications of (13) which describes the variation of the amplitude with X. First suppose that the basic shear flow is zero ($u_0 = 0$), and the density stratification is that for a two-layer fluid; which has a constant density ρ_1 to $z = -d$, and then a constant density ρ_2. We define the Boussinesq parameter σ so that $\sigma\ \Delta\ \rho = 2(\rho_2 - \rho_1)/\rho_2 + \rho_1$ where $\Delta\rho$ is a measure of the density jump at the interface. Then, in the Boussinesq approximation $\sigma \to 0$, we find that

$$\left.\begin{aligned} c_0^2 &= \Delta\rho\frac{d(h-d)}{h}\ , \quad I = \frac{2c_0h}{d(h-d)}\ , \\ I\lambda &= \tfrac{1}{3}c_0^2h, \quad I\nu = \frac{3c_0^2h(2d-h)}{d^2(h-d)^2}\ . \end{aligned}\right\} \tag{19}$$

It follows from (13) that

$$a^3 = \frac{(\text{constant})(2d-h)}{(\Delta\rho)^2d^2(h-d)^2}\ . \tag{20}$$

Supposing that the solitary wave is initially in deeper water ($d \ll h$), (20) describes the change in amplitude as the solitary wave progresses into shallow water, that is, as h decreases; the amplitude increases to a maximum at h = 3d, and then decreases to zero at h = 2d. At this latter point, the nonlinear coefficient ν (19) is zero, and it is necessary to include a cubic nonlinear term in (11) to determine the structure of the solitary wave. In §3(a) we show that the coefficient of this cubic term is negative, from whence it follows that the solitary wave cannot exist beyond the point when ν is zero. (20) also shows that the amplitude increases when the intensity of the stratification, here represented by $\Delta\rho$, decreases.

Next suppose that the basic shear flow is zero and the density stratification is uniform so that the Brunt-Väisälä frequency is a constant. Again using the Boussinesq approximation we find that

$$\left.\begin{aligned} c_0^2 &= \left(\frac{Nh}{s\pi}\right)^2, \quad I = c_0 h\left(\frac{s\pi}{h}\right)^2 , \\ \lambda &= \frac{1}{2}\, c_0 \left(\frac{h}{s\pi}\right)^2, \quad \nu = -\frac{1}{3}\,\frac{\sigma c_0 N^2}{s\pi}\,\{8 + (-1)^s\} , \end{aligned}\right\} \tag{21}$$

where s = 1, 2, 3, ... is the mode number. Note that the coefficient ν is here $O(\sigma)$ as $\sigma \to 0$, which implies that the amplitude should be rescaled by σ^{-1}; indeed Gear and Grimshaw [15] have shown that in this instance it is permissible to choose the amplitude parameter $\alpha = \sigma$, and consequently $\epsilon = \sigma^{\frac{1}{2}}$. It now follows from (13) that

$$a^3 = \frac{(\text{constant})}{N^2 h^4} . \tag{22}$$

When dissipative effects are included, (11) is replaced by

$$\frac{\partial A}{\partial \tau} + \frac{1}{2}\,\frac{A}{\gamma}\,\frac{\partial \gamma}{\partial \tau} + \frac{\nu}{c_0}\, A\, \frac{\partial A}{\partial \theta} + \frac{\lambda}{c_0^3}\,\frac{\partial^3 A}{\partial \theta^3} + \mu \mathcal{V}(A) = 0, \tag{23}$$

where $\mathcal{V}(A)$ is a pseudo-differential operator with respect to θ, and μ is a coefficient (a slowly varying function of τ). For instance, when the dominant dissipative effect is the frictional drag at the rigid boundary z = -h, it may be shown that (Grimshaw [6])

$$\left.\begin{aligned}\mathcal{V}(A) &= \frac{1}{2\pi}\int_{-\infty}^{\infty}(-i\tau)^{\frac{1}{2}}\exp(i\tau\theta)\,\mathcal{F}(A)\,d\tau . \\ \text{and}\qquad I\mu &= \left(\frac{\hat{\nu}}{|c_0|\epsilon^5}\right)^{\frac{1}{2}}|c_0 - u_0|^{3/2}\left(\frac{\partial\phi}{\partial z}\right)^2 \quad \text{at } z = -h.\end{aligned}\right\} \tag{24}$$

Here we recall that $\mathcal{F}(A)$ is the Fourier transform of A (see (7)), and $\hat{\nu}$ is the coefficient of viscosity in the bottom boundary layer; note that we require $\hat{\nu}$ to be $O(\epsilon^5)$. For the slowly varying solitary wave, (13) is replaced by

$$\left.\begin{aligned}\frac{\partial}{\partial\tau}\left(\frac{\gamma a^2}{k}\right) + \mu f k^{\frac{1}{2}}\left(\gamma\frac{a^2}{k}\right) &= 0, \\ \text{where}\qquad f = \frac{3}{4\pi}\int_{-\infty}^{\infty}\left|\frac{\tau}{2}\right|^{\frac{1}{2}}|\mathcal{F}(\text{sech}^2\theta)|^2 d\tau &\approx 1{\cdot}08.\end{aligned}\right\} \tag{25}$$

Assuming for simplicity that the coefficients λ, ν, γ and μ are constant, it follows that the amplitude a decays as $12\lambda\nu^{-1}$ (constant $+ \frac{1}{6}\mu f\tau c_0^{\frac{1}{2}})^{-4}$. Even allowing for a turbulent boundary layer (i.e. $\hat{\nu}$ is an eddy viscosity), it is found that for most observed solitary waves, the calculated rate of decay is quite slow, and the solitary waves will persist over long distances. The conservation law (11) still holds in the presence of dissipation and so a shelf region is formed, whose magnitude at the rear of the solitary wave is

$$A_1(\tau) = \frac{3c_0}{\nu k^2}\left\{\frac{\partial k}{\partial\tau} + \mu f k^{3/2}\right\} . \tag{26}$$

The shelf region is then again given by (17). It follows from (18) then while the amplitude decays, the shelf grows as (constant $+ \frac{1}{6}\mu f\tau c_0^{\frac{1}{2}}$) and will ultimately dominate the solitary wave.

<u>(b)BDO</u>: The inclusion of a slow variation in the environment is similar to the KdV case described in (a). The slow space variable is again (9), but now $\alpha = \epsilon$, and the environment varies on the scale ϵ^{-2}. τ and θ are again defined by (10) and the variable-coefficient BDO equation is (see Grimshaw [4])

$$\frac{\partial A}{\partial\tau} + \frac{1}{2}\frac{A}{\gamma}\frac{\partial\gamma}{\partial\tau} + \frac{\nu}{c_0}A\frac{\partial A}{\partial\theta} + \frac{\delta}{c_0^2}\mathcal{L}\left(\frac{\partial A}{\partial\theta}\right) = 0 , \tag{27}$$

where the operator $\mathcal{L}$ is defined by (7) with $\coth(|\tau|H) \to 1$. The slowly varying solitary wave solution has been discussed by Grimshaw [16], and to leading order is given by

$$\left.\begin{aligned} A &= a/1 + k^2\psi^2 \;, \\ \text{where} \qquad \psi &= \theta - \int_0^\tau \frac{1}{4}\,\frac{\nu a}{c_0}\,d\tau \;, \end{aligned}\right\} \qquad (28)$$

and

$$\nu a = 4\delta\,\frac{k}{c_0}\;.$$

The slow variation of the amplitude is again determined from the conservation law for wave action flux (14), from which we again obtain the first equation in (13); thus the amplitude variation is

$$Ic_0\,\frac{\delta a}{\nu} = \text{constant}. \qquad (29)$$

The conservation law (15) also holds, and so a shelf region is formed, whose amplitude at the rear of the solitary wave is given by (see Grimshaw [16])

$$A_1^-(\tau) = -\,\frac{4\pi c_0^2}{\gamma^{\frac{1}{2}}\delta k}\left(\frac{\gamma^{\frac{1}{2}}\delta}{c_0\nu}\right)_\tau . \qquad (30)$$

The remaining part of the shelf is then given by (17) provided that in the definition of τ_0, $\frac{1}{3}$ is replaced by $\frac{1}{4}$. Also, as in the KdV case the conservation law (15) is not that for mass flux, which is again given by (18), provided that in the integrand $\phi(\zeta_0)$ is replaced by $\phi(\zeta_0) - \phi(-\infty)$. Consequently there will again be a reflected wave.

As an application of equation (29) first consider a two-layer fluid in the absence of a shear flow, which has a constant density ρ_1 to $z = -d$, and a constant density ρ_2 in the deep fluid region. With the Boussinesq parameter σ defined as in the KdV case, we find in the Boussinesq approximation that $c_0^2 = \Delta\rho d$, $Id = 2c_0$, $2\delta = c_0 d$ and $2\nu d = -3c_0$. Consequently the amplitude variation is given by $a\Delta\rho d^2$ = constant. Next suppose that there is again no shear flow, and that the Brunt-Väisälä frequency is constant to a depth $z = -d$, and zero in the deep fluid region. In the Boussinesq approximation we find that

$$c_0^2 = \frac{N^2 d^2}{(s+\frac{1}{2})^2\pi^2}, \quad I = c_0 d \;\frac{(s+\frac{1}{2})^2\pi^2}{d^2}, \qquad I\delta = c_0^2, \quad \nu = -\frac{2c_0}{d}. \tag{31}$$

Consequently the amplitude variation is given by aN^2d^3 = constant. Both these examples suggest that the amplitude will increase as the depth of the stratified layer decreases, and also as the stratification decreases.

Dissipative effects can be included in (27) in a similar manner to the KdV case. For instance, when the dissipative operator $\mathcal{Y}(A)$ has the form given by (24), we find that the amplitude is governed by (25) where now $f \approx 0{\cdot}87$. When δ, ν, γ and μ are constant the amplitude a decays as $4\delta\nu^{-1}(\text{constant} + \frac{1}{2}\,\mu f\tau c_0^{\frac{1}{2}})^{-2}$, which, as in the KdV case, is generally found to give a slow rate of decay. The shelf is again given by (30), and so, for constant coefficients, there is no shelf formation.

Finally let us consider the consequences of radiation damping. This arises when the deep fluid region, instead of having constant density, is weakly stratified with a Brunt-Väisälä frequency ϵN_0. The evolution equation again (27) but the operator $\mathcal{L}$ is now

$$\mathcal{L}(A) = -\frac{1}{2\pi}\int_{-\infty}^{\infty} (\tau^2 - N_0^2)^{\frac{1}{2}} \exp(i\tau\theta)\,\mathcal{F}(A)\,d\tau. \tag{32}$$

It is now possible for internal gravity waves to propagate, and the radiation condition requires that $(\tau^2 - N_0^2)^{\frac{1}{2}}$ be either real and positive, or have an imaginary part with the same sign as τ. For the slowly varying solitary wave (13) is replaced by

$$\frac{\partial}{\partial\tau}\left(\gamma\frac{a^2}{k}\right) + \frac{64}{3}\,\gamma\frac{\delta^3 N_0^3}{\nu^2 c_0^4}\, g\left(\frac{2N_0}{k}\right) = 0, \tag{33}$$

where $$g(x) = 1 + \frac{3\pi}{2x}\{L_2(x) - I_2(x)\}.$$

Here L_2 is a modified Struve function, and I_2 a modified Bessel function. For the derivation of (33) see Grimshaw [16], or Maslowe and Redekopp [5]. For small x, $g(x) \approx 1$ and assuming that γ, δ, ν and c_0 are constants, (33) gives a linear rate of decay for the amplitude. In contrast to boundary layer friction, we find that (33) gives quite rapid rates of decay, suggesting that internal solitary waves are unlikely to be observed in the deep ocean. For instance, in the example of the uniformly stratified layer where the coefficients are given by (31), the time scale estimated from (33) is $\frac{3}{16}\cdot\frac{a}{d}\cdot(\frac{N}{N_0})^3(s\pi + \frac{\pi}{2})^2 N^{-1}$, which is often comparable to the Brunt-Väisälä period $2\pi N^{-1}$.

§3. Second order solitary waves

(a) KdV: The solitary wave (12) is the first term in an amplitude expansion. Here we propose to continue the expansion to the second order in amplitude following the development of Gear and Grimshaw [15], to whom the reader is referred for further details. Considering only the solitary wave in a constant environment, (1) is replaced by

$$\left.\begin{aligned} \eta &= \alpha A(\psi)\phi(z) + \alpha^2\eta_2(\psi, z) + \ldots, \\ \text{where} \quad \psi &= \epsilon\kappa(x - ct), \end{aligned}\right\} \tag{34}$$

and we recall that for this KdV case $\alpha = \epsilon^2$. Simultaneously the phase speed c, and κ (the wavelength is given by $(k\kappa\epsilon)^{-1}$) are expanded in the form

$$\left.\begin{aligned} c &= c_0 + \alpha c_1 + \alpha^2 c_2 + \ldots \\ \kappa &= 1 + \alpha\kappa_1 + \ldots \quad . \end{aligned}\right\} \tag{35}$$

Here c_0 is the linear long wave speed, and at first order A is given by (12), where now

$$\nu a = 3c_1 = 12\lambda k^2. \tag{36}$$

Note that k differs here by a factor c_0 from the k used in §2. At the second order we find that

$$\eta_2 = A_2(\psi)\phi(z) + c_1 A\beta_1(z) + A^2\beta_2(z), \tag{37}$$

where β_1, β_2 are functions known completely once the modal function $\phi(z)$ has been determined, and A_2 satisfies the equation

$$-c_1A_2 + \nu AA_2 + \lambda\frac{\partial^2 A_2}{\partial\psi^2} = \sigma_1 A + \sigma_2 A^2 + \sigma_3 A^3 + c_2 A - 2\kappa_1 \frac{\partial^2 A_2}{\partial\psi^2}. \quad (38)$$

Here σ_1, σ_2, σ_3 are coefficients known completely in terms of integrals of $\phi(z)$. Note that the left-hand side of (38) is the linearized form of the steady KdV equation.

The coefficients c_2, and κ_1 are determined so that A_2 decays as $|\psi| \to \infty$, and satisfies the condition $A_2(0) = 0$, so that the amplitude of the modal function $\phi(z)$ is measured exactly by αa. It follows that

$$\left.\begin{aligned} A_2 &= \frac{3\sigma_3 a^2}{2\nu}\, \text{sech}^2 k\psi \tanh^2 k\psi \;, \\ c_2 &= -\sigma_1 - \frac{2}{3}\, a\sigma_2 - \frac{1}{2}\, a^2\sigma_3 \;, \\ \nu\kappa_1 &= -\sigma_2 - \frac{3}{4}\, a\sigma_3 \;. \end{aligned}\right\} \quad (39)$$

Further progress depends on evaluating the coefficients σ_1, σ_2 and σ_3 for specific cases (see Gear and Grimshaw [15]). However, in many cases of interest, it is found that $c_2 < 0$ (note that $c_1 > 0$), $\kappa_1 < 0$ and $a\sigma_3/\nu < 0$, These results imply that the phase speed is decreased relative to the first order theory, the wave-length is increased and the wave is narrower at the crest. These qualitative trends are consistent with observations of large amplitude solitary waves.

In some special cases it is found that the nonlinear coefficient ν (5) is small relative to λ; for instance in the special case of the two-layer fluid (19), this occurs for $h \approx 2d$, or in the special case of the uniformly stratified fluid, ν is $O(\sigma)$ as $\sigma \to 0$. A re-scaling is required, and it is shown by Gear and Grimshaw [15] that with α of $O(\nu)$, the equation for $A(\psi)$ is

$$-c_2 A + \frac{1}{2}\,\hat{\nu} A^2 - \sigma_3 A^3 + \lambda\frac{\partial^2 A}{\partial\psi^2} = 0. \quad (40)$$

Here $c_1 = 0$ (see (36) as $\nu \to 0$), and σ_3, λ are evaluated in the limit $\nu \to 0$, while $\hat{\nu}$ is given by the limit of $\nu\alpha^{-1}$ as $\alpha \to 0$. The solitary wave solution of (40) is

$$A = \frac{a\,\text{sech}^2 k\psi}{1-b\tanh^2 k\psi}\,,$$

$$\text{where}\quad c_2 = \frac{1}{3}\hat{\nu}a - \frac{1}{2}\sigma_3 a^2 = 4\lambda k^2,\quad b = \left(\frac{2\hat{\nu}}{3\sigma_3 a} - 1\right)^{-1}. \tag{41}$$

For small b this agrees with the result obtained above (39). For the special case of the two-layer fluid (coefficients given by (19)) we find that $\sigma_3 > 0$, and $\hat{\nu}$ is $\gtrless 0$ as $h \gtrless 2d$, giving solitary waves of elevation (depression); however, there is no solitary wave when $\hat{\nu} = 0$.

(b) BDO: A similar amplitude expansion has been carried out for this case by Grimshaw [17]; in this case $\alpha = \epsilon$. Equations (34), (35) again apply, the first order solitary wave is given by (28), where now

$$\nu a = 4c_1 = 4\delta k\,. \tag{42}$$

η_2 is again given by an expression of the form (37) and A_2 satisfi the equation

$$-c_1 A_2 + \nu A A_2 + \delta\mathcal{L}(A_2) = \sigma_1 A + \sigma_2 A^2 + \sigma_3 A^3 + c_2 A - \kappa_1 \delta\mathcal{L}(A)\,. \tag{43}$$

The solution for A_2 is found to be (Grimshaw [17])

$$A_2 = \frac{\sigma_3 a^2}{8\nu} K(\psi),$$

$$\text{where}\quad K(\psi) = \text{Re}\left\{\frac{(5-J_0)}{1-ik\psi} + \frac{(3-J_0)}{(1-ik\psi)^2} - \frac{8}{(1-ik\psi)^3} + 2\left(\frac{1+ik\psi}{1-ik\psi}\right)^2 J\right\}, \tag{44}$$

$$\text{and}\quad J(\psi) = \int_0^\infty \frac{\exp(-z)\,dz}{(1-ik\psi+z)^2}\,,\quad J_0 = J(0)\,.$$

The function $K(\psi)$ has been normalized so that $K(0) = 0$, is positive for all $\psi > 0$ and $\to 0$ as $(k\psi)^{-4}$ as $|k\psi| \to \infty$. Also,

$$c_2 = -\sigma_1 - \frac{1}{2}a\sigma_2 - \frac{1}{4}a^2\sigma_3 - \frac{1}{16}a^2\sigma_3 J_0\,,$$

$$\nu\kappa_1 = -2\sigma_2 - \frac{13}{8}a\sigma_3 - \frac{1}{8}a\sigma_3 J_0\,. \tag{45}$$

Further progress again depends on evaluating the coefficients σ_1, σ_2, σ_3 for specific cases (see Grimshaw [17]). However in most cases of interest it is found that $c_2 < 0$, $\kappa_1 < 0$, and

$a\sigma_3/\nu < 0$, which imply that the phase speed is decreased relative to the first order theory, the wavelength is increased and the wave is narrower at the crest.

References

1. Farmer D.M.; Observations of long nonlinear internal waves in a lake. J. Phys. Ocean. 8 (1978), 63-73.

2. Apel J.R.; Satellite sensing of ocean surface dynamics. Ann. Rev. Earth Planet. Sci., 8 (1980), 303-342.

3. Christie D.R., Muirhead K.J. and Hales A.L.; On solitary waves in the atmosphere. J. Atmos. Sci., 35 (1978), 805-825.

4. Benney D.J.; Long non-linear waves in fluid flows. J. Mathematical Phys., 45 (1966), 52-63.

5. Maslowe S.A. and Redekopp L.G.; Long nonlinear waves in stratified shear flows. J. Fluid Mech. 101 (1980), 321-348.

6. Grimshaw R.; Evolution equations for long nonlinear internal waves in stratified shear flows. Studies in Applied Math., 65 (1981), 159-188.

7. Benjamin T.B.; Internal waves of permanent form in fluids of great depth. J. Fluid Mech., 29 (1967), 559-592.

8. Davis R.E. and Acrivos A.; Solitary internal waves in deep water. J. Fluid Mech., 29, 593-607.

9. Tung K.-K., Ko D.R.S. and Chang J.J.; Weakly nonlinear internal waves in shear. Studies in Applied Math., 65 (1981), 189-221.

10. Grimshaw, R.; Solitary waves in a compressible fluid. Pageoph., 119 (1980/81), 780-797.

11. Grimshaw R.; Slowly varying solitary waves I. Korteweg-de Vries equation. Proc. R. Soc. Lond., A368 (1979), 359-375.

12. Karpman V.I. and Maslov E.M.. Perturbation theory for solitons. Zh. Exsp. Teor. Fiz., 73 (1977), 537-559.

13. Kaup D.J. and Newell A.C.; Solitons as particles, oscillators and in slowly changing media: a singular perturbation theory. Proc. Roy. Soc. Lond., A361 (1978), 413-446.

14. Miles J.W.; On the Korteweg-de Vries equation for a gradually varying channel. J. Fluid Mech., 91 (1979), 181-190.

15. Gear J.A. and Grimshaw R.; A second-order theory for solitary waves in shallow fluids. Maths. Res. Rep. No. 8 (1982).

16. Grimshaw R.; Slowly varying solitary waves in deep fluids, Proc. R. Soc. Lond., A376 (1981), 319-332.

17. Grimshaw R.; A second-order theory for solitary waves in deep fluids. Phys. Fluids, 24 (1981), 1611-1618.

On Propagation of Waves Containing Discontinuities

V.V.MARKOV

Steklov Mathematical Institute
Academy of Sciences
Moscow, USSR

Let a heat conductivity process be described by the equation with piece wise constant coefficient of temperature conductivity $\varkappa$:

$$\partial T/\partial t = \varkappa_1 \partial^2 T/\partial X^2 \qquad (\partial T/\partial t \leqslant 0)$$
$$\partial T/\partial t = \varkappa\, \partial^2 T/\partial X^2 \qquad (\partial T/\partial t \geqslant 0)$$

Let us consider for this equation the problem of heat propagation due to the supply of finite amount of heat Q in the point X=0 at t=0 under assumptions that the change of the sign of $\partial T/\partial t$ takes place only once and that at initial instant T=T(X,0)=0 (X≠0) and a heat supply or a heat loss is absent at t>0.

Evidently, this problem is self-similar one (the self-similar variable is $\xi = X/\sqrt{\varkappa t}$) and the amount of heat contained between any moving surfaces, on which ξ =const, is constant. In particular, the amounts of heat Q_1 and Q_2 containing respectively before and after the point of the jump of $\varkappa$ are constants.

If the non-dimensional coordinate of the jump is ξ_0, the solution, as it is easy to verify, will have a form:

$$T = \frac{1}{\sqrt{\varkappa t}} Q_1 \left\{ 2C_{V1} \int_0^{\xi_0} \exp(-\xi^2/4\varepsilon) d\xi \right\}^{-1} \exp(-\xi^2/4\varepsilon) \qquad (0 \leq \xi \leq \xi_0)$$
$$T = \frac{1}{\sqrt{\varkappa t}} Q_2 \left\{ 2C_V \int_{\xi_0}^{\infty} \exp(-\xi^2/4) d\xi \right\}^{-1} \exp(-\xi^2/4) \qquad (\xi_0 \leq \xi < \infty) \qquad (1)$$

where $\varepsilon = \varkappa_1/\varkappa$.

The value ξ_0 , in connection with the requirement about the only change of the sign of $\partial T/\partial t$, must satisfy to a condition:

Nonlinear Deformation Waves
IUTAM Symposium Tallinn 1982
Editors U. Nigul, J. Engelbrecht

$\sqrt{2} < \xi_c < \sqrt{2\varepsilon}$, and the solution exists when $\varepsilon \geq 1$.
One can see from (1) the complete flow of energy in point ξ_c is zero, but T and $\partial T/\partial X$, in the general case, have the jump being determined by values ξ_0, Q_1, Q_2, C_{V1}, C_V, $\varkappa$, $\varkappa_1$.
If one demands the continuity of temperature than this condition gives a relation between $k=(Q_1 C_V)/(Q_2 C_{V1})$ and ξ_0:

$$k=\sqrt{\varepsilon}\,\Phi(\xi_0/2\sqrt{\varepsilon})\Phi'(\xi_c/2)\{\Phi'(\xi_0/2\sqrt{\varepsilon})[\sqrt{\pi}-\Phi(\xi_c/2)]\}^{-1} \qquad (\Phi(z)=2\int_0^z \exp(-z^2)dz)$$

The graph of this relation is shown for $\varepsilon=4$ in the figure.

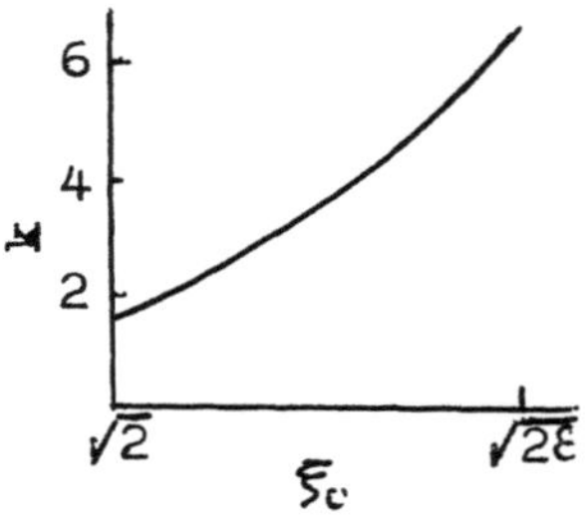

Let us notice that upon the condition of the continuity of temperature the ratio of temperature gradients on the right and on the left of the jump is constant and equals $1/\varepsilon$.

It is necessary to note in the conclusion that this problem has been considered in the book by G.I.Barenblatt "Similarity, Self-similarity and Intermediate Asymptotics" (Gidrometeoizdat, Leningrad, 1978). However the solution suggested in this book is erroneous. The error is connected with the erroneous condition of simultaneous continuity of temperature and its gradient in the point of the jump of $\varkappa$. Indeed, the condition of the continuity of energy flow must be satisfied on the jump propagating at a speed D :

$$[D\, C_V T + \lambda \partial T/\partial X] = 0$$

One can see from here that the supply or the loss of energy takesplace in the presence of the jump of C_V upon the condition of continuity of λ, T and $\partial T/\partial X$.

Workshop on Applications in Nonlinear Wave Propagation

Held at the Technical University Tallinn, 25 August, 1982

Following communications have been displayed on posters:

1. V.G.Andrejev, O.V.Rudenko (USSR)
   Nonlinear interactions in media with artificial attenuation and dispersion
2. K.I.Artamonov, A.T.Fedorchenko (USSR)
   Numerical investigation of nonlinear sound-vortex interactions in subsonic flows
3. V.L. Berdichevskii, V.G.Sutirin (USSR)
   Waves in microinhomogeneous media
4. Yu.A. Berezin (USSR)
   Nonlinear gasdynamical processes in neutron matter with non-monotonic equations of state
5. A.S.Kosmodamiansky, V.I.Storozhev (USSR)
   Nonlinear anharmonic effects of elastic wave propagation in crystal plates
6. E.A.Krasilshchikova (USSR)
   The method of Volterra integral equations at diffraction problems
7. S.V.Meleshko, V.P.Shapeev, N.N.Yanenko (USSR)
   The problem of the breakdown of an arbitrary discontinuity
8. I.A.Ogorodnikov (USSR)
   Equation of heterogeneous medium motion. Waves in gas-liquid medium
9. G.Turchetti (Italy)
   Perturbation series and Padé approximants for multisoliton solutions of KdV equation
10. V.V.Viktorov, Yu.V.Kulinich, A.S.Chursin (USSR)
    Electromagnetic wave generation by nonlinear deformation waves

Nonlinear Deformation Waves
IUTAM Symposium Tallinn 1982
Editors U Nigul, J Engelbrecht

11. V.M.Bobrenko, A.N.Koutsenko (USSR)
    Acoustic elasticity coefficients of ultrasonic waves and their application in non-destructive testing
12. A.A.Ravasoo (USSR)
    Deformation waves in a medium with an inhomogeneous initial deformation
13. A.M.Bragov, Yu.G.Korotkikh, A.K.Lomunov, Yu.B.Purtov (USSR)
    Experimental and theoretical analysis of elastic-plastic wave processes in split bars
14. V.I.Kondaurov, L.V.Nikitin (USSR)
    The propagation of stress waves in viscoelasticplastic finite strain medium
15. P.V.Makarov, T.M.Platova, V.A.Skripnjak (USSR)
    Investigation of the shock wave propagation in the media with relaxation
16. Kh.A.Rakhmatulin, N.Mamadaliev (USSR)
    Interaction of elastic-plastic bodies with targets during the impact

The List of Unpublished Reports

The following lectures have been held at the Symposium but the texts published elsewhere:

1. K.Harumi (Japan)
   Computer simulation of elastic waves
2. V.A.Krasilnikov, V.I.Pavlov (USSR)
   Hamilton approach to the nonlinear mathematical models of hydrodynamic media
3. V.N.Kukudjanov (USSR)
   On numerical solution of the wave propagation problems in elasto-plastic media with finite deformations

Nonlinear Deformation Waves
IUTAM Symposium Tallinn 1982
Editors U Nigul, J Engelbrecht

IUTAM Symposia

New Volumes

Three-Dimensional Turbulent Boundary Layers
Symposium, Berlin, Germany,
March 29 – April 1, 1982
Editors: H. Fernholz, E. Krause
1982. 288 figures. XV, 389 pages
ISBN 3-540-11772-5

Creep in Structures
3rd Symposium, Leicester, UK,
September 8–12,1980
Editors: A. R. S. Ponter, D. R. Hayhurst
1981. 243 figures. XVI, 615 pages
ISBN 3-540-10596-4

Physical Non-Linearities in Structural Analysis
Symposium Senlis, France
May 27–30, 1980
Editors: J. Hult, J. Lemaitre
1981. 109 figures. XI, 287 pages (30 pages in French). ISBN 3-540-10544-1

Unsteady Turbulent Shear Flows
Symposium Toulouse, France,
May 5–8, 1981
Editors: R. Michel, J. Cousteix, R. Houdeville
1981. 283 figures. XXI, 424 pages
ISBN 3-540-11099-2

Laminar-Turbulent Transition
Symposium Stuttgart, Germany,
September 16–22, 1979
Editors: R. Eppler, H. Fasel
1980. 289 figures. XVIII, 432 pages
ISBN 3-540-10142-X

Physics and Mechanics of Ice
Symposium Copenhagen, August 6–10, 1979
Technical University of Denmark
Editor: P. Tryde
1980. 159 figures, 13 tables. XIV, 378 pages
ISBN 3-540-09906-9

International Association for Hydraulic Research
Practical Experiences with Flow-Induced Vibrations
Symposium Karlsruhe, Germany,
September 3–6, 1979, University of Karlsruhe
Editors: E. Neudascher, D. Rockwell
1980. 429 figures, 217 charts, 29 tables.
XXII, 849 pages. ISBN 3-540-10314-7

International Commision on Acoustics
American Institute of Aeronautics and Astronautics
Mechanics of Sound Generation in Flows
Joint Symposium Göttingen, Germany,
August 28–31, 1979
Max-Planck-Institut für Strömungsforschung
Editor: E.-A. Müller
1979. 177 figures, 6 tables. XV, 300 pages
ISBN 3-540-09785-6

High Velocity Deformation of Solids
Symposium Tokyo, Japan,
August 24–27, 1977
Editors: K. Kawata, J. Shiori
1978. 230 figures, 20 tables. XVIII, 452 pages
ISBN 3-540-09208-0

Dynamics of Multibody Systems
Symposium Munich, Germany,
August 29–September 3, 1977
Editor: K. Magnus
1978. 107 figures. XVI, 376 pages
ISBN 3-540-08623-4

Buckling of Structures
Symposium Cambridge, USA, June 17–21, 1974
Editor: B. Budiansky
1976. 214 figures. VIII, 398 pages
ISBN 3-540-07274-8

Symposium Transsonicum II
Göttingen, Germany, Semptember 8–13, 1975
Editors: K. Oswatitsch, D. Rues
1976. 324 figures. XVI, 574 pages
ISBN 3-540-07526-7

Dynamics of Rotors
Symposium Lyngby, Denmark, August 12–16, 1974
Editor: F. I. Niordson
1975. 195 figures. XII, 564 pages
ISBN 3-540-07384-1

MIX
Papier aus verantwortungsvollen Quellen
Paper from responsible sources
FSC® C105338

If you have any concerns about our products,
you can contact us on
ProductSafety@springernature.com

In case Publisher is established outside the EU,
the EU authorized representative is:
Springer Nature Customer Service Center GmbH
Europaplatz 3, 69115 Heidelberg, Germany

Printed by Libri Plureos GmbH
in Hamburg, Germany